MOLECULAR BIOLOGY

MOLECULAR BIOLOGY

G.P. JEYANTHI

Professor
Department of Biochemistry,
Biotechnology and Bioinformatics
Avinashilingam University for Women
Coimbatore, TN

MJP PUBLISHERS

Cataloguing-in-Publication Data

Jeyanthi, G.P. (1952 –).
    Molecular Biology / by G.P. Jeyanthi. –
Chennai : MJP Publishers, 2009
    xviii, 450 p. ; 23 cm.
    Includes Glossary, References and Index.
    ISBN 978-81-8094-057-6 (pbk.)
    1. Biology, Molecular I. Title.
    572.8 JEY MJP 053

ISBN 978-81-8094-057-6
© Publishers, 2009
All rights reserved

MJP PUBLISHERS
47, Nallathambi Street
Triplicane
Chennai 600 005

Publisher : J.C. Pillai
Managing Editor : C. Sajeesh Kumar
Project Editor : P. Parvath Radha
Acquisitions Editor : C. Janarthanan
Editoral Team : B. Ramalakshmi, N. Pushpa Bharathi,
L. Mohanapriya, N. Yamuna Devi,
Lissy John, M. Gnanasoundari
CIP Data : Prof. K. Hariharan

This book has been published in good faith that the work of the author is original. All efforts have been taken to make the material error-free. However, the author and publisher disclaim responsibility for any inadvertent errors.

PREFACE

The term molecular biology was first used in 1945 by William Astbury. He defined this branch of science as the study of the chemical and physical structures of biological macromolecules. After several breakthroughs in this study, the concept of molecular biology was modified. It can be defined as the biochemistry of genes and of the specific products of genes. It deals with the molecular basis of biological or genetic specificity. It is concerned with the fine structure, function and genesis of cells, organelles and macromolecules of living organisms. It emerges out as a complex field with the components of genetics, biochemistry, cell biology, physics, organic chemistry and biophysical chemistry.

Flow of genetic information from generation to generation and expression of the inherited genetic information in an individual is the fundamental central dogma of molecular genetics. However, the strategies of implementation differ among living organisms—both prokaryotes and eukaryotes.

This book on Molecular Biology, in which these features have been incorporated, has been presented in thirteen chapters. Chapter 1 deals with the basic concepts of the genetic material and the identification of the nature of the genetic material. The second, third and fourth chapters deal with the chemistry of nucleic acids, and the features and properties of DNA.

The distinctive features and organization of the prokaryotic and eukaryotic chromosomes are discussed in chapter 5.

Synthesis of DNA in the cells, requirements for replication, fidelity of replication, regulation and inhibition are the highlights of chapter 6. Common lesions that occur in DNA and DNA repair mechanisms are also dealt within this chapter. Mutations and their types and the molecular mechanism of mutagenesis are explained in chapter 8.

The chemistry of RNA, their types and structures, RNA synthesis, genetic code and protein biosynthesis with elaborate discussion on regulation of gene expression in prokaryotes and eukaryotes are discussed in the chapters 9, 10, 11 and 12. Selected analytical techniques used in the study of DNA are discussed in chapter 13.

The contents have been prepared with utmost care to enable the students to understand the concepts clearly, to prepare for their academic and also competitive examinations. The materials presented are based on the undergraduate and postgraduate course syllabi offered by colleges and universities all over india for the major subjects biochemistry, biotechnology, bioinformatics, life sciences, molecular genetics and medical genetics.

Another special feature of this textbook is a section with challenging questions/problems with worked out solutions for each chapter and another section with review questions. This will help the students to have a strong insight into the concepts discussed and to prepare for NET and other competitive examinations. Inserts in boxes in the text highlight the current concepts, practical applicability and striking features relevant to the chapter contents.

I acknowledge with gratitude the support rendered by Thiru. T.K. Shanmuganandham, Chancellor, Avinashilingam University for Women, Coimbatore and Dr. Saroja Prabhakaran, Vice Chancellor, Avinashilingam University for Women, Coimbatore, in the completion of this work.

Suggestions and comments are welcome for improvement of the contents of this book.

G.P. Jeyanthi

CONTENTS

1

GENETIC MATERIAL

NATURE OF THE GENETIC MATERIAL

For a long period, proteins were thought to be the genetic material. Although they are not the genetic material, they are very much a part of the expression of genetic information encoded in DNA. The genes in DNA carry the information that is expressed in proteins. Both DNA and proteins are molecules composed of linear chains of subunits. The subunits of DNA are nucleotides and the subunits of proteins are amino acids. Once it became clear that DNA is the inherited material and that genes consist of DNA, it soon became apparent that the linear sequence of nucleotides in the DNA molecule must be related to the linear sequence of amino acids in the proteins. Because DNA resides in the cell nucleus, and proteins are synthesized outside the nucleus, it seemed that there must be an intermediate substance that transferred the information from DNA to protein.

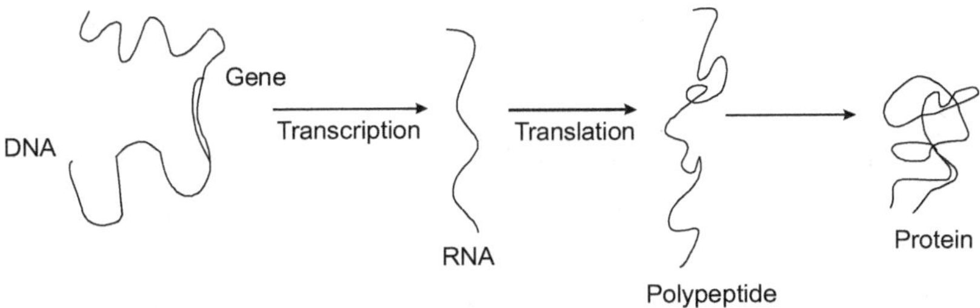

Figure 1.1   The central dogma of molecular genetics

BOX 1.1   HISTORY OF MOLECULAR BIOLOGY

Molecular biology is the study of biology at molecular level. The field overlaps with other areas of biology and chemistry, particularly genetics and biochemistry. Molecular biology includes interaction between DNA, RNA and protein biosynthesis in the cell and their regulation. The history of molecular biology began in the year 1930. The name "molecular biology" was coined by Warren Weaver of the Rockefeller Foundation in 1938. Following the advent of the Mendelian chromosome theory of heredity in the 1910s and the maturation of atomic theory and quantum mechanics in the 1920s, researches were attempted at the intersection of biology, chemistry and physics while prominent physicists Niels Bohr and Erwin Schrodinger turned their attention to biological speculation. In the 1930s and 1940s work in colloid chemistry, biophysics and radiation biology, crystallography and other emerging fields all seemed promising. In 1940, George Beadle and Edward Tatum demonstrated the relationship between genes and proteins. In the course of their experiments connecting genetics with biochemistry, they switched from *Drosophila* model to *Neurospora*. In 1944, Oswald Avery working at the Rockefeller Institute of New York demonstrated that genes are made of DNA. In 1952, Alfred Hershey and Martha Chase confirmed that the genetic material in bacteriophages is made up of DNA. In 1953, James Watson and Francis Crick discovered the double helical structure of the DNA molecule. In 1961, Francois Jacob and Jacques Monad hypothesized the existence of an intermediary between DNA and its protein product, which they called messenger RNA. Between 1961 and 1965, the genetic code was determined to be the link between DNA and proteins. At the beginning of the 1960s, Monad and Jacob also showed the presence of regulatory proteins which were responsible for the regulation of gene expression. The main discoveries of molecular biology took place in a period of only about twenty five years. Another fifteen years were required to discover the advanced findings like isolation and characterization of gene, manipulation of gene, and recombinant DNA technology which were possible with sophisticated techniques.

By the late 1950s, the idea of an RNA intermediate between DNA and protein was well-accepted. This understanding led to the concept of the Central Dogma of molecular genetics (Figure 1.1). That is, a gene composed of DNA is transcribed into RNA, which is then translated into a polypeptide, and is then processed to become a protein.

Genes were first detected and analysed by Johann Gregor Mendel and later by many other scientists, by following their patterns of transmission from generation to generation. However, they could not predict the structure or molecular composition of genes. Subsequent studies established the relation between the patterns of transmission of genes from generation to generation and the behaviour of chromosomes during sexual reproduction. Through several experiments, strong evidences were provided to show that genes are located on the chromosomes. Then a question arose regarding the biochemical composition of chromosomes. Whatever its chemical composition, it was clear even in Mendel's time that the genetic material had to fulfil two important requirements:

1. The genotype function or replication—the genetic material must be capable of storing the genetic information and must transmit the information faithfully from parents to progeny, i.e., generation after generation.

2. The phenotype function or gene expression—the genetic material must dictate the phenotype of the organism, whether it is a virus, a bacterium, a plant, or an animal such as a human being.

Chromosomes are composed of two types of large organic molecules (macromolecules), namely proteins and nucleic acids. The nucleic acids are again classified into two types deoxyribonucleic acids (DNA) and ribonucleic acids (RNA). For many years, scientists disagreed as to which of these three macromolecules carries the genetic information.

EXPERIMENTAL EVIDENCES FOR DNA AS THE GENETIC MATERIAL

Griffith Experiment

In 1928, Fred Griffith performed the first experiment towards the identification of the nature of the genetic material. Griffith selected

Streptococcus pneumoniae (pneumococci) for his experiment. Pneumococci exhibit genetic variability that can be recognized by the existence of different phenotypes. Two such phenotypic characteristics in Griffith's experiment were (i) the presence or absence of a surrounding polysaccharide capsule and (ii) the type of capsule having specific molecular composition of the polysaccharides present in the capsules. The pneumococci that form large and smooth colonies when grown on appropriate media like blood agar are designated as type S. Such encapsulated pneumococci are pathogenic and cause pneumonia in humans. These virulent, disease-causing type S pneumococci mutate to a nonvirulent (nonpathogenic) form that has no polysaccharide capsule. Such non-encapsulated, nonvirulent pneumococci form small and rough-surfaced colonies when grown on blood agar medium and are designated as type R. The capsulated pneumococci may be of several different antigenic types, depending on the specific molecular composition of the polysaccharides, which is due to the genotype of the cell. Based on this fact, the organisms are said to belong to different types of strains like type II and type III. The characteristics of *S. pneumoniae* strains when grown on blood agar medium are as shown in Table 1.1.

Table 1.1   Four types of *S. pneumoniae* used by Griffith for his experiments

Type	Appearance	Size	Capsule	Virulence
II R	Rough	Small	Absent	Nonvirulent
II S	Smooth	Large	Present	Virulent
III R	Rough	Small	Absent	Nonvirulent
III S	Smooth	Large	Present	Virulent

When heat-killed encapsulated type III *S. pneumoniae* were injected into mice, the mice did not develop pneumonia. Similarly, when live non-encapsulated type II R cells were injected into mice, the mice showed no ill effects. But an injection of live type III S pneumococci resulted in severe pneumonia and caused death of the animals. Surprisingly, injection of heat-killed type III S cells (virulent if alive) together with live type II R cells (nonvirulent) caused the death of the mice that succumbed to pneumonia. Live type III S cells were recovered from the carcasses (Figure 1.2).

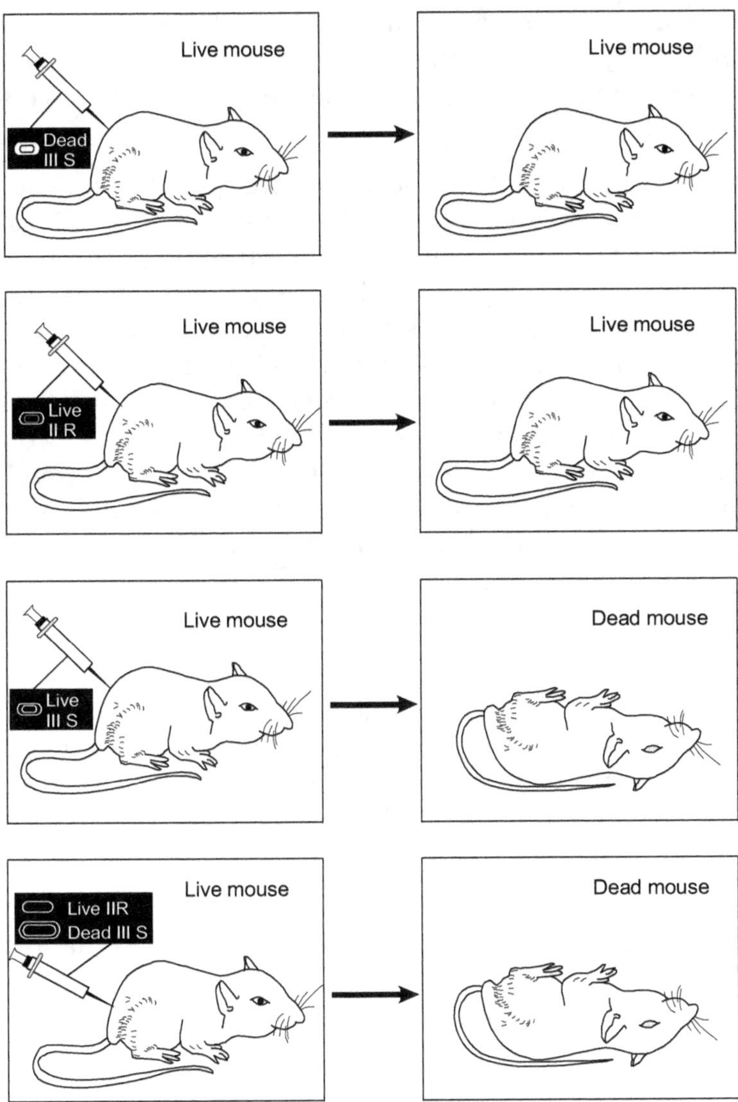

Figure 1.2   The four experiments performed by Griffith

When mice were injected with heat-killed type III S cells alone, none of the mice died. The observed virulence in the fourth experiment was therefore not due to a few type III S cells that survived the heat treatment. It cannot also be argued that the unexpected observation could be due to a mutation of live II R to live II S cells because the colonies identified in the dead animal were not of II S type but of III S. Thus, the "transformation" of nonvirulent type II R cells to virulent type III S cells indicates that some component of the dead type III S cells referred by Griffith as the "transforming principle" must have converted the type II R cells to type III S cells.

Subsequent experiments showed that the phenomenon described by Griffith, now called transformation, was not mediated in any way by a living host. The same phenomenon occurred in the test tube when live type II R cells were grown in the presence of dead type III cells or extracts of type III S cells. It was clearly shown that the II R phenotype is altered to III S phenotype because of a change in the genotype as a result of transformation. The hereditary character responsible for causing pneumonia has been transformed to live II R cells from dead III S cells. This transformation had set the stage for determining the chemical basis of heredity in pneumococcus. Yet another problem was to determine what component of the cell extract was responsible for transformation.

Avery Experiment

The "transforming principle" was shown to be DNA in 1944 by Oswald Avery, Colin MacLeod and Maclyn McCarty. They showed that if highly purified DNA from type III S pneumococci was present with type II R pneumococci, some of the type II R cells were transformed to type III S (Figure 1.3).

But proving the complete purity of any macromolecular substance is extremely difficult. May be the DNA preparation contained a few molecules of protein and these contaminating proteins might be responsible for the observed transformation. Therefore, in their experimental proof that DNA was the transforming principle, they used enzymes that specifically degrade DNA, RNA or protein. Therefore, they treated the highly purified DNA from type III S cells with (1) deoxyribonuclease (DNase) which degrades DNA, (2) ribonuclease (RNase) which degrades RNA, (3) protease which degrades proteins and then tested for its ability to transform type II R cells to type III S.

Only the DNase had an effect on the transforming activity of the DNA preparation. It totally eliminated all transforming activity. The results obtained by Avery and his co-workers clearly established that the genetic information in pneumococcus was present only in the DNA. The later experiments done to investigate the molecular mechanism of transformation proved that the segment of the DNA in the chromosome of pneumococcus that carries the genetic information specifying the synthesis of a type III capsule is physically integrated into the chromosome of the type II R recipient cell by a specific recombination process.

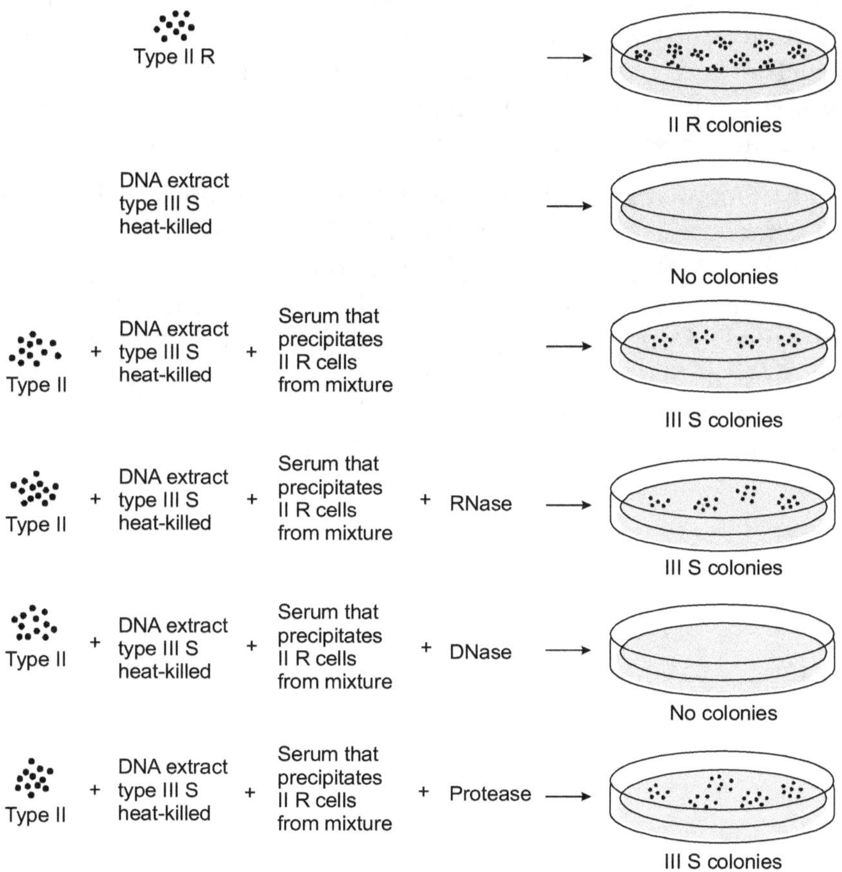

Figure 1.3 The experiment performed by Avery and co-workers

Hershey–Chase Experiment

In 1952, A.D. Hershey (1969 Nobel Prize winner) and M. Chase proved DNA to be the genetic material in the bacterial virus (bacteriophage T2).

Their results, although less definitive than the results of Avery, Macleod and McCarty, had a great impact on the acceptance of DNA as the genetic material by the scientists.

Viruses are the smallest living organisms. Their reproduction is controlled by the genetic information stored in nucleic acids in the same way as in cellular organisms. However, viruses can reproduce only in appropriate host cells. Their reproduction is totally dependent on the metabolic machinery of the host. Viruses have been extremely useful in studying many genetic processes because of their simple structure and chemical composition (most of the viruses contain only proteins and nucleic acids) and their very rapid rate of reproduction (15–20 minutes for some bacterial viruses under optimal conditions).

Bacteriophage T2 (Figure 1.4), which infects the common colon bacillus *Escherichia coli*, is composed of about 50% DNA and about 50% protein. Experiments done prior to 1952 had shown that bacteriophage T2 reproduction takes place within the *E. coli* cells. They showed that only the DNA of the virus particle entered the cell, whereas most of the proteins of the virus remained adsorbed to the outside of the cell. This strongly implied that the genetic information necessary for viral reproduction was present only in the DNA.

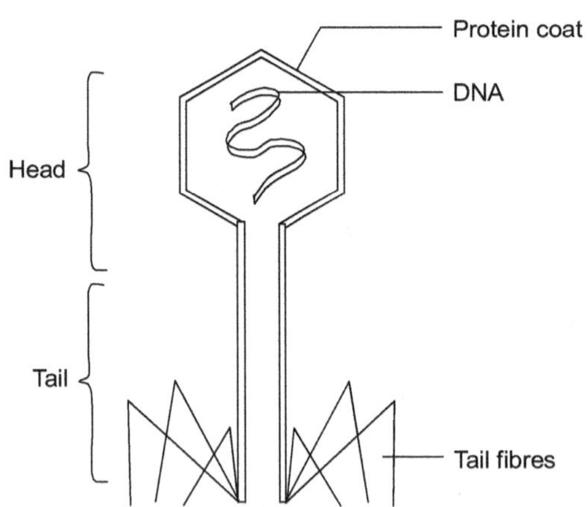

Figure 1.4   Structure of bacteriophage T2

The basis for the Hershey–Chase experiment is that DNA contains phosphorus but not sulphur, whereas proteins contain sulphur but not phosphorus. Therefore, Hershey and Chase specifically labelled the phage DNA with a radioactive isotope of phosphorus (^{32}P) by growing in a medium containing ^{32}P in place of the normal isotope ^{31}P and the phage protein coat with the radioactive isotope of sulphur ^{35}S by growing in a medium containing ^{35}S in place of the normal isotope ^{32}S (Figure 1.5).

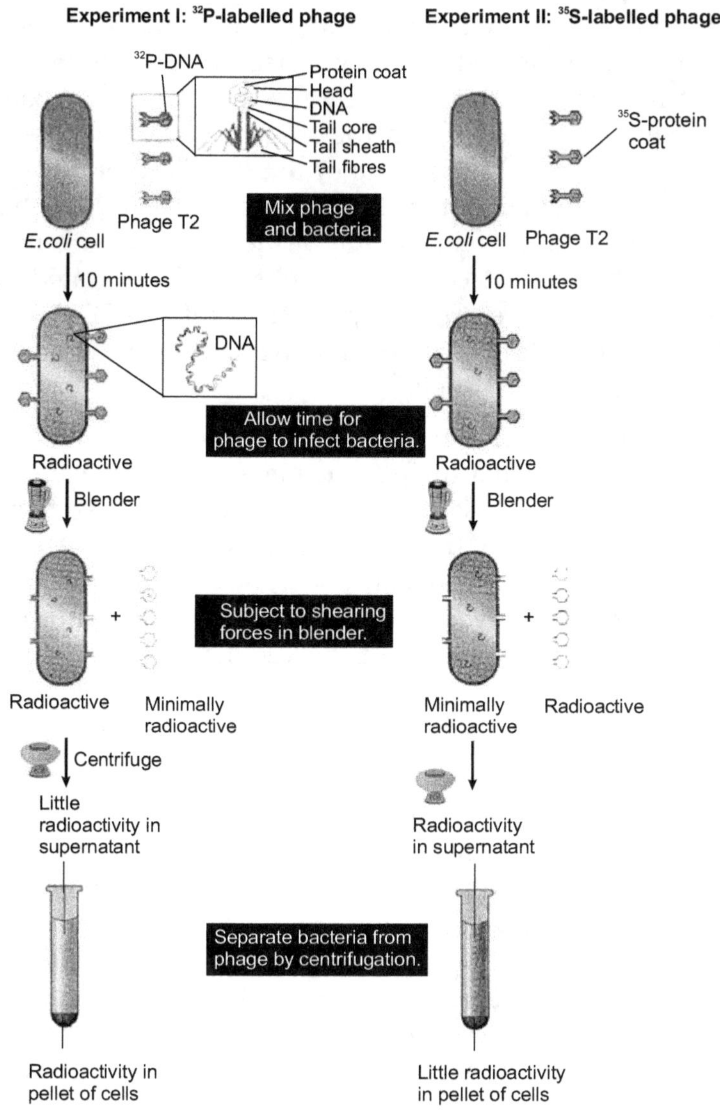

Figure 1.5   The experiment performed by Hershey and Chase

When T2 phage particles labelled with ^{35}S were mixed with *E. coli* cells for a few minutes and were then subjected to shearing forces by placing the infected cells in a Waring blender, it was found that most of the radioactivity (the protein fraction of the phage) could be removed from the cells without affecting the progeny phage production. When T2 phage DNA labelled with ^{32}P were used, the total radioactivity was found inside the cells, that is, it was not removed by shearing in a blender. The sheared-off phage coats were separated out as a supernatant from the infected cells by low-speed centrifugation which pellets (sediments) the cells. Also, the progeny particles were shown to contain some of the ^{32}P, but none of the ^{35}S of the parental phage. These results indicated that the DNA of the virus enters the host cell, whereas the protein coat remains outside the cell. Since progeny viruses are produced inside the cell, their results clearly indicated that the genetic information directing the synthesis of both the DNA molecules and the protein coats of the progeny viruses must be present in the parental DNA.

However, the Hershey–Chase experiment did not provide unambiguous proof that the genetic material of phage T2 is DNA. It was found that a significant amount of ^{35}S (and thus protein coat of phage) was injected into the host cells with the phage DNA during viral infection. Thus, one could always argue that this small fraction of the phage protein very well contained the genetic information. Later, the transfection experiments proved that DNA and not protein is the genetic material.

Transfection Experiments

Additional researches using bacterial viruses following the Hershey–Chase experiment, provided even more solid proofs that DNA is the genetic material. In 1957, it was reported that if *E. coli* was treated with the enzyme lysozyme, the outer wall of the cell could be removed without destroying the bacterium. Such naked cells have only the cell membrane as their outer boundary and are called spheroplasts or protoplasts. Such spheroplasts could be infected with disrupted T2 particles, that is, for infection of spheroplasts, the virus need not necessarily be intact. In 1960, George Guthrie and Robert Sinsheimer purified DNA from bacteriophage ϕX174 (a small phage that contains a single-stranded circular DNA). When this purified DNA was added to *E. coli* protoplasts, complete φX174 bacteriophages were produced. This process of infection by only the viral nucleic acid, called transfection, proved that φX174 DNA alone contains all the necessary information for the production of mature viruses. Thus, it strengthened the evidence supporting the conclusion that DNA serves as the genetic material. However, all these evidences were based on bacterial and viral studies.

INDIRECT EVIDENCES SUPPORTING DNA AS THE GENETIC MATERIAL IN EUKARYOTES

Distribution of DNA

The genetic material should be present where it functions, i.e., in the nucleus as part of chromosomes. Both the DNA and proteins fit this criterion. But protein is also found abundantly in the cytoplasm, whereas DNA is not present. Both mitochondria and chloroplasts are known to perform genetic functions and DNA is also present in these organelles. These observations are consistent with the interpretation favouring DNA over proteins as the genetic material.

Earlier it was established that the chromosomes present in the nucleus contain the genetic material. Hence a correlation was expected to exist between the ploidy of a cell (i.e., n or $2n$) and the amount of the molecule serving as the genetic material. A comparison between the amount of DNA and proteins present in the gametes (egg and sperm) and in the somatic cells, indicate that diploid ($2n$) somatic cells contain twice the amount of DNA as present in haploid gametes (n). No such correlation is observed in the protein content of the gametes and the somatic cells. This provides an evidence favouring DNA over proteins as the genetic material in eukaryotes.

Mutagenesis

Ultraviolet light is one of the agents capable of inducing mutations in the genetic material. Any organism can be irradiated with various wavelengths of ultraviolet light, and the effectiveness of each wavelength can be measured by the number of mutations it induces. When the data are plotted, an action spectrum of UV light as a mutagenic agent is obtained (Figure 1.6a). This action spectrum can then be compared with the absorption spectrum of any molecule (Figure 1.6b) suspected to be the genetic material.

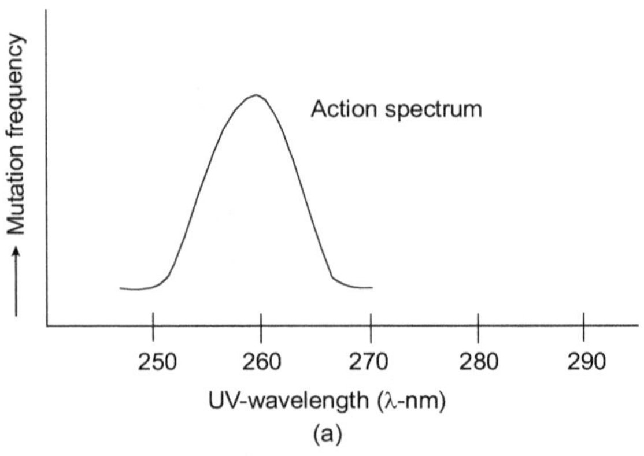

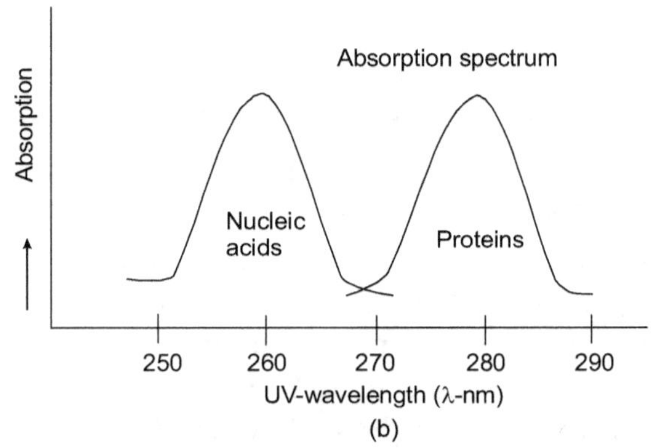

Figure 1.6   Action and absorption spectra of the suspected genetic material

The molecule serving as the genetic material is expected to absorb UV light at the wavelengths found to be mutagenic. UV light is mutagenic mostly at the wavelength of 260 nm. Both DNA and RNA absorb UV light most strongly at 260 nm. On the other hand, protein absorbs most strongly at 280 nm, yet no significant mutagenic effects are observed at this wavelength. This indirect evidence supports the idea that a nucleic acid is the genetic material and not the protein.

BOX 1.2   THE HUMAN GENOME PROJECT

The US Human Genome Project was started in 1990 and was a 13-year effort coordinated by the U.S. Department of Energy and the National Institute of Health. The project originally was planned to last 15 years. But rapid technological advances accelerated the completion date to 2003. The project goals were to

1. identify all the approximately 20,000–25,000 genes in human DNA

2. determine the sequence of the 3 billion chemical base pairs that make up human DNA

3. store this information in databases

4. improve tools for data analysis

5. transfer related technologies to the private sector and

6. address the ethical, legal and social issues (ELSI) that may arise from the project.

To help achieve these goals, researchers also studied the genetic make-up of several non-human organisms. These include the common human gut bacterium (*Escherichia coli*), the fruit fly (*Drosophila melanogaster*) and the laboratory mouse.

Landmark papers detailing sequence and analysis of the human genome were published in February 2001 and April 2003 issues of *Nature* and *Science*.

DIRECT EVIDENCES SUPPORTING DNA AS THE GENETIC MATERIAL

Recombinant DNA Studies

The strongest direct evidence that DNA is the genetic material has been provided by the application of molecular analysis in the form of recombinant DNA technology. In rDNA technology, segments of eukaryotic DNA corresponding to specific genes are isolated and literally spliced into bacterial DNA. The expression of such genes are monitored in the bacterial cell and the corresponding gene products—the proteins—are identified which gives a direct proof that DNA is the genetic material.

RNA AS GENETIC MATERIAL

Some viruses contain an RNA core rather than DNA. In these viruses, it would appear that RNA might serve as the genetic material (Figure 1.7). In 1956, it was demonstrated that when purified RNA from tobacco mosaic virus (TMV) was spread on tobacco leaves, the characteristic lesions of TMV appeared later on the leaves. This concluded that RNA is the genetic material of this virus.

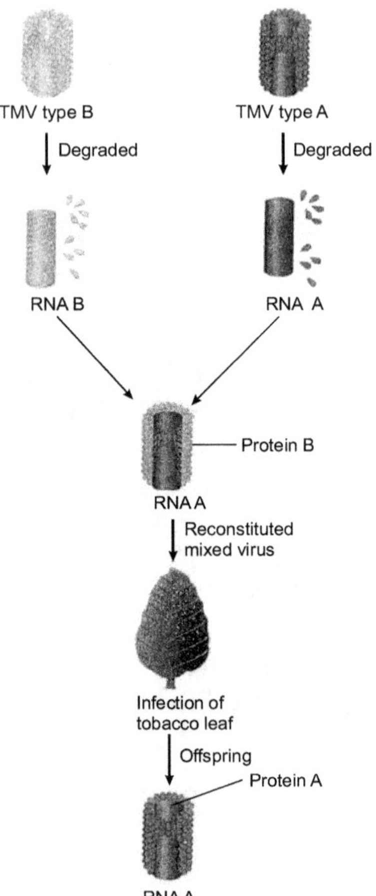

Figure 1.7   Experiment to prove that RNA is the genetic material

Heinz Fraenkel-Conrat and B. Singer demonstrated an experiment to prove that RNA could serve as the genetic material in some viruses. In their experiment, RNA and coat proteins were separated and isolated from tobacco mosaic virus strains A and B. Then they mixed the RNA of one strain and the protein of the other and produced a "reconstituted virus". When this hybrid virus was spread on tobacco leaves, the lesions that were formed on the tobacco leaves were similar to those formed by the original virus from which the RNA fraction was obtained, i.e., hybrid viruses with type A RNA and type B protein coats produced lesions on tobacco leaves characteristic of that produced by type A strain, and hybrid viruses with type B RNA and type A protein coats produced lesions characteristic of that produced by type B strain. Thus they concluded that RNA serves as the genetic material in these viruses.

Spicy Questions and Answers

1. **Before identifying that DNA is the genetic material, why did scientists think that protein is the genetic material?**

    It was found out that the chromosome is the site of genetic information which was proved to contain both DNA and protein. Protein was characterized earlier than DNA. The sequence of amino acids and structural organization were elucidated and was thought that it was more complex than DNA. Hence scientists guessed that protein may carry the genetic information.

2. **Why was the work of Avery *et al.* not accepted as a conclusive proof that DNA is the genetic material?**

    Though Avery and his co-workers proved that addition of DNase enzyme only prevented transformation and not protease or RNase, chemical analysis of the purified transforming DNA showed the presence of a small amount of protein which raised a doubt whether this small fraction of protein would be the genetic material. It was also argued that DNA acting as a mutating agent would have caused a change in the real genetic material creating a new virulent bacteria. That is, DNA would have simply played a catalytic role in capsule synthesis; therefore the cells display smooth type III colonies.

3. **If RNA was the universal genetic material, how would this have affected Avery's experiment and the experiment by Hershey and Chase?**

    In Avery's experiment, when they did the experiment with RNase, they would not have got type III S colonies. Actually, they did not get colonies only when they used DNase which made them confirm that it is DNA that is the genetic material.

If Hershey and Chase had selected an RNA bacteriophage, they would have got the same result, that is ^{32}P would have been incorporated into RNA and the same observation would have been recorded. In that case, they would have concluded that RNA is the genetic material.

4. **Why did Hershey and Chase choose ^{32}P and ^{35}S to label the bacteriophage while proving that DNA is the genetic material?**

    There were doubts whether it was DNA or protein that was the transforming principle, that carries the genetic information. In order to selectively label DNA and protein in two different experiments, ^{32}P labelling was done in one experiment to label phage DNA and ^{35}S labelling was done in the second experiment to label the protein coat of the phage. Because phosphorus is present only in DNA and is absent in the protein, and sulphur is present only in the protein, and is absent in DNA, these two radioisotopes were used to distinguish DNA and proteins.

5. **How did the transformation experiments of Griffith differ from those of Avery and his associates?**

    Griffith did the experiment in animal model utilizing the *in vivo* conditions. Avery and his co-workers did the work in *in vitro* conditions, repeating the same experiment in Petri plates in order to identify the exact nature of the genetic material.

6. **What was the contribution of Griffith and Avery *et al.* to molecular genetics?**

    Griffith proved the presence of a transforming principle, but he could not identify its nature which was proved to be DNA by Avery and his co-workers.

7. **Why was Griffith's work not an evidence for DNA as the genetic material whereas the experiment of Avery *et al.* provided direct proof that DNA carried the genetic information?**

    Because Griffith's work was done in mice, he could not identify clearly the reason for the transformation of heat-killed type III S to live type III S colonies causing pneumonia in animals. Avery *et al.* took DNA extract from heat-killed III S cells and performed a series of experiments, every time using an enzyme—RNase, protease or DNase—to find out which enzyme inhibited the transformation event.

8. **Would it have been suitable to use ^{14}C and ^{3}H as radioactive tracers for the Hershey–Chase experiment? Why or why not?**

    No, because carbon and hydrogen are both present in both DNA and protein, use of ^{14}C and ^{3}H will result in getting both DNA and protein radiolabelled and it will not be possible to distinguish them.

9. **How does DNA fulfil the requirements of a genetic material?**

    DNA is present in the chromosome. DNA replication is very accurate and DNA is truly copied from the inherited parental strands. DNA is transcribed to RNA and RNA carries the inherited information in the form of codons which are translated into protein products. These proteins are responsible for the phenotypic expression of the genetic characters inherited from the parents.

Review Questions

1. Is it correct to say that DNA is always the genetic material?

2. Summarize the evidence that RNA can be the genetic material in some organisms.

3. What critical finding in Hershey and Chase experiment enabled them to interpret the results as a proof that the genetic material in bacteriophage T2 is DNA?

4. Briefly explain the contribution of the following scientists to molecular genetics:

    i.   Griffith

    ii. Avery

    iii. Hershey and Chase

    iv. Fraenkel-Conrat and Singer

5. A cell-free extract is prepared from type III S pneumococcal cells. What effect will treatment with this extract have on the capacity of (i) protease (ii) RNase (iii) DNase to transform recipient type II R to type III S? Why?

6. How could it be demonstrated that the mixing of heat-killed type III S pneumococcus with live type II resulted in transfer of genetic material from type III to II rather than restoration of viability to type III by type II?

7. What macromolecular composition of a bacterial virus such as phage T2 allowed scientists to use them in experiments in determining the nature of the genetic material?

8. What chemical properties do DNA and proteins possess, which allow researchers to specifically label one or the other of these macromolecules with a radioisotope?

9. What evidence suggested that the transforming principle in Griffith's experiment was really a gene or genes?

10. Which experimental approach clearly proved that DNA is the genetic material, that of Hershey–Chase or Avery *et al.*?

11. What are the exceptions to the general rule that DNA is the genetic material in all organisms? What evidence supports these exceptions?

2

CHEMISTRY OF DEOXYRIBONUCLEIC ACID

Historically the search for the chemical identity of genetic material began about a century ago. In 1871, soon after Charles Darwin's work on the theory of evolution and Gregor Mendel's experiments in genetics, Friedrich Miescher (1844–1895) published a method for separating cell nuclei from the cytoplasm. From these cell nuclei he extracted an acid material, nuclein, which proved to contain large amount of phosphorus and contained no sulphur, which excluded the possibility of proteins. In later years, the nuclein, now named nucleic acid, was found to associate with various proteins in combinations called nucleoproteins. When the nucleic acid was separated from the associated protein, several biochemists, especially Levene (1869–1940), showed that it could be broken down further into smaller sections called nucleotides.

CONSTITUENTS OF NUCLEIC ACIDS

Each nucleotide consists of distinct chemical components: a sugar, a phosphate group and a nitrogen-containing portion called a **base**. The sugar is five carbon containing ribose, whose carbon atoms are numbered as C1′ to C5′. When ribose lacks one oxygen atom on the 2′-carbon, it is called **deoxyribose** (Figure 2.1). A particular nucleic acid does not contain these two types of sugars at the same time. Hence there are two types of nucleic acids—**ribonucleic acid (RNA)** which contains ribose sugar and is found commonly in the cytoplasm and **deoxyribonucleic acid (DNA)** which contains deoxyribose sugar and is found, with few exceptions, only in the nucleus.

In both nucleic acids, a phosphate group is attached to each sugar at its 5′ position (Figure 2.2). The term "acid" is used because nucleotides are esters of phosphoric acid; at cellular pH values, the phosphates are present as salts.

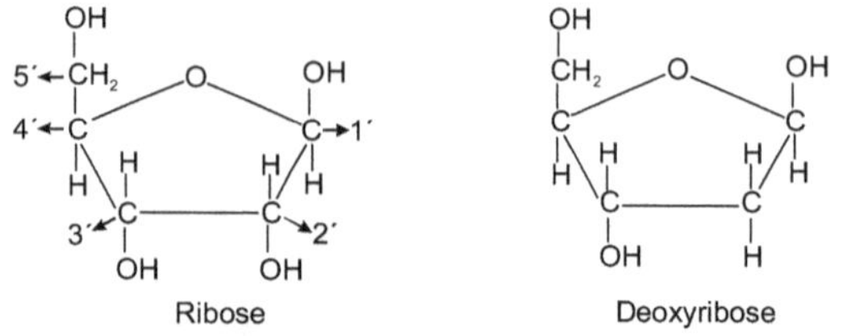

Figure 2.1   Ribose and deoxyribose

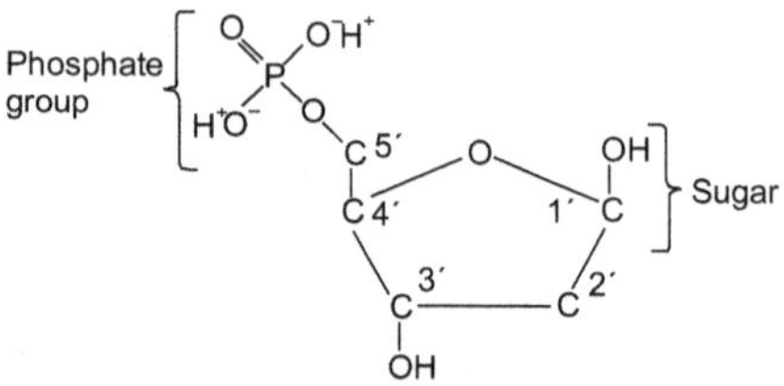

Figure 2.2   Linkage between a phosphate and deoxyribose group yielding a sugar phosphate or phosphorylated sugar

Apart from the sugar and phosphate groups, the nucleotides also contain a nitrogen-containing base associated with each sugar at its number 1 carbon position C1′. The nitrogen unit contains either one or two carbon–nitrogen rings, and is referred to as a base because it can function as a hydrogen-ion acceptor, in contrast to the acidic nature of the phosphate group. Bases containing one carbon–nitrogen ring are the pyrimidines and the two-ring bases are the purines (Figures 2.3, 2.4 and 2.5).

In DNA, the two main pyrimidines found are **cytosine** and **thymine** and RNA has **cytosine** and **uracil**. The difference between thymine and uracil (the presence of a methyl group instead of a hydrogen at position 5 in thymine) is significant, as thymine is not ordinarily found in RNA nor is uracil in DNA. The two main purines, **adenine** and **guanine** are found in both DNA and RNA.

In addition to the difference in ring structures, an amino group (NH_2) is present at positions 4 in cytosine, 6 in adenine and 2 in guanine, while a keto group (C=O) is present at positions 2 and 4 in thymine and uracil, 2 in cytosine and 6 in guanine. A methyl group (—CH$_3$) is present at position 5 in thymine. DNA and RNA can therefore be described as containing two kinds of amino bases and two kinds of keto bases, evenly divided between pyrimidines and purines.

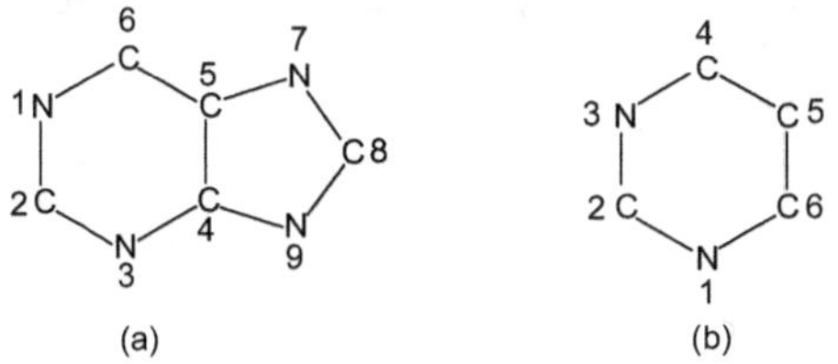

Figure 2.3   (a) Purine ring (b) Pyrimidine ring

In case of purines, the N-9 atom is covalently bonded to the of C1′ the sugar. In case of pyrimidines, the bonding involves N-1 atom.

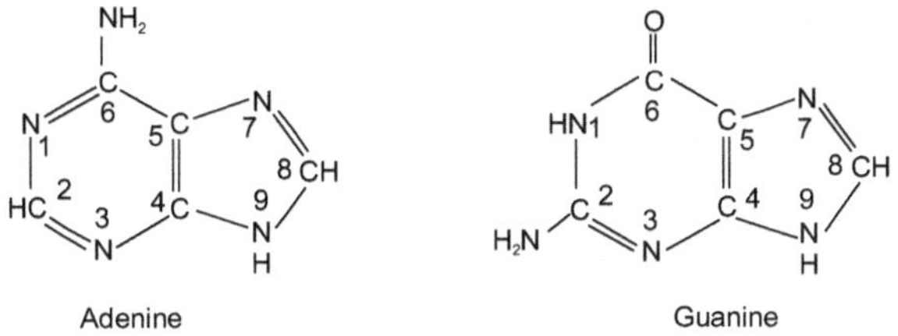

Figure 2.4   Purine bases

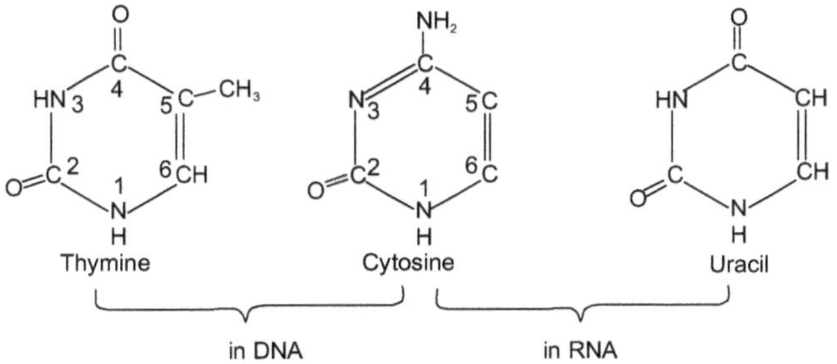

Figure 2.5   Pyrimidine bases

If a molecule is composed of a purine or pyrimidine base and a ribose or deoxyribose sugar, the chemical unit is called a **nucleoside**. If a phosphate group is added to the nucleoside, the molecule is called a **nucleotide**.

Each nucleotide is differentiated by the particular base present in it. There are four different deoxyribonucleotides (Figure 2.6) in DNA and four ribonucleotides in RNA. Thus the combination of phosphate–deoxyribose–adenine yields a nucleotide deoxyadenylic acid or deoxyadenosine 5'-phosphoric acid. The other three DNA nucleotides are deoxycytidylic acid, deoxyguanilic acid and deoxythymidylic acid.

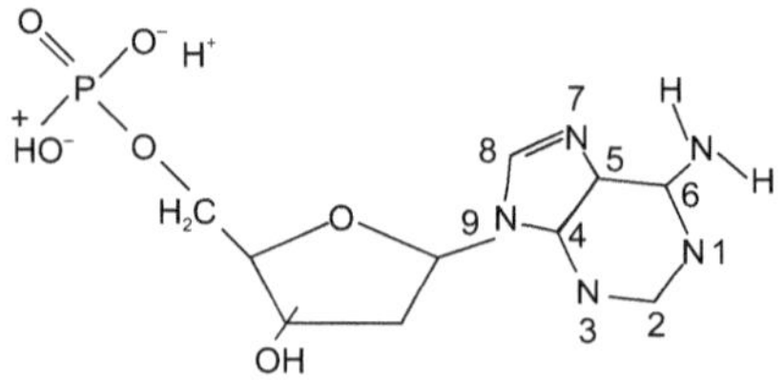

Deoxyadenylic acid (deoxy 5'-monophosphoric acid, dAMP)

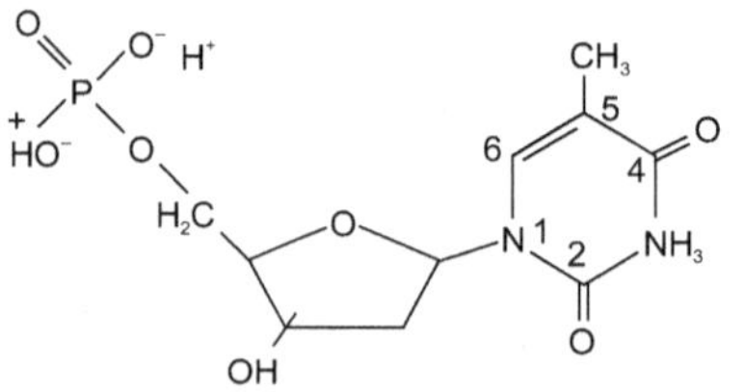

Deoxythymidylic acid (deoxythymidine 5'-monophosphoric acid, dTMP)

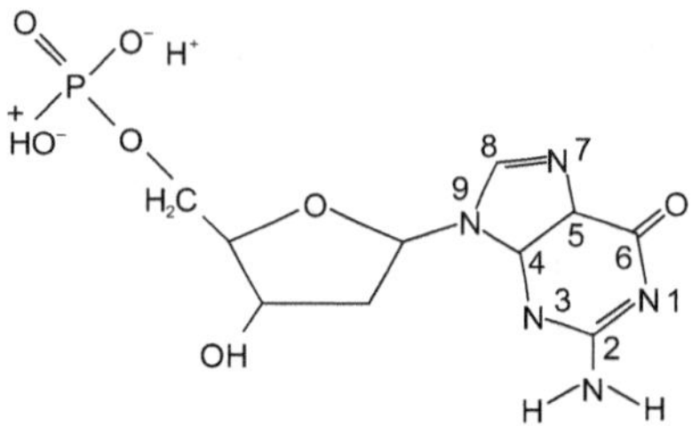

Deoxyguanylic acid (deoxyguanosine 5'-monophosphoric acid, dGMP)

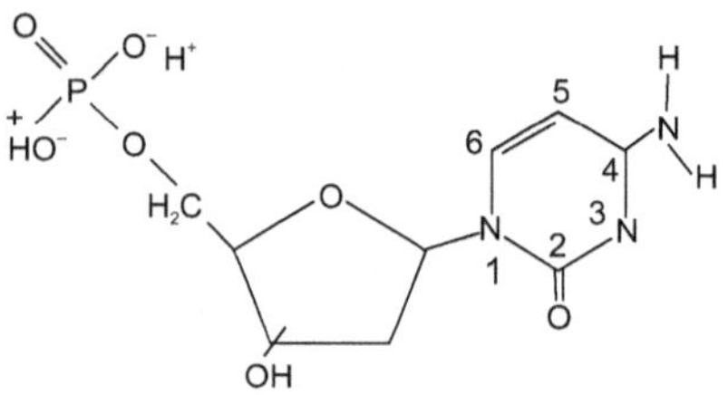

Deoxycytidylic acid (deoxycytidine 5'-monophosphoric acid, dCMP)

Figure 2.6   Four major nucleotides in DNA

The nomenclature of ribo- and deoxyribonucleosides and nucleotides are given in Table 2.1.

Table 2.1   Ribo- and deoxyribonucleosides and nucleotides

Ribonucleosides	Ribonucleotides	Deoxyribonucleosides	Deoxyribonucleotides
Adenosine	Adenylic acid	Deoxyadenosine	Deoxyadenylic acid
Cytidine	Cytidylic acid	Deoxycytidine	Deoxycytidylic acid
Guanosine	Guanylic acid	Deoxyguanosine	Deoxyguanylic acid
Uridine	Uridylic acid	Deoxythymidine	Deoxythymidylic acid

NUCLEOSIDE DIPHOSPHATES AND TRIPHOSPHATES

Nucleotides are also described by the term **nucleoside monophosphate (NMP)**. The addition of one or two more phosphate groups results in **nucleoside diphosphates (NDPs)** and **nucleoside triphosphates (NTPs)**. Those nucleotides which possess deoxyribose moieties are referred to as **dNMP, dNDP** and **dNTP**. The triphosphate form is significant because it serves as the precursor molecule during nucleic acid synthesis within the cell.

POLYNUCLEOTIDES

The linkage between two mononucleotides consists of a phosphate group linked to two sugars. A phosphodiester bond is formed (Figure 2.7), because phosphoric acid is joined to two alcohols (the hydroxyl groups on the two sugars) by an ester linkage on both sides. The mononucleotides which are the monomers in DNA and RNA are linked to one another by the phosphodiester bond.

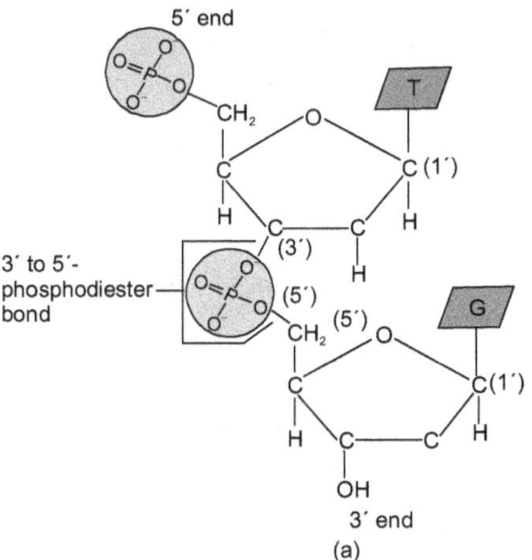

Figure 2.7   Phosphodiester bonds between nucleotides *(Continues)*

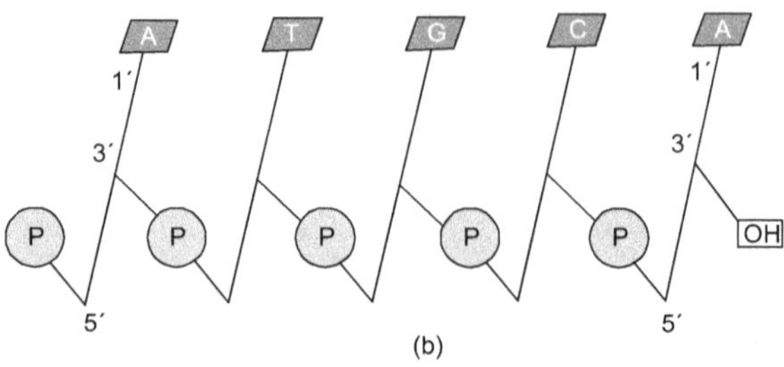

Figure 2.7   Phosphodiester bonds between nucleotides

WHY DNA HAS DEOXYRIBOSE AND NOT RIBOSE SUGAR?

The cell has several nucleotides with ribose as the sugar component. RNA is a nucleic acid that has ribose sugar. In evolution, it is probable that ribonucleotides predated deoxyribonucleotides and RNA predated DNA. The cell requires enormous amount of energy to convert ribonucleotides to deoxyribonucleotides required for DNA synthesis. It is clear that DNA is the repository of genetic information gathered over countless millions of years and it is stored in chemical form in DNA molecules. Chemical molecules usually always have some degree of instability and they tend to break down, whereas DNA molecules have to remain unchanged for a long time, that is for many generations. The presence of 2'-OH group of ribose makes a ribopolynucleotide less stable than DNA. The 2'-OH group is susceptible to a nucleophilic attack in the presence of OH$^-$ ions on the phosphorus atom, thus causing breakage of the phosphodiester link by forming a 2', 3'-cyclic phosphate which is unlikely to happen in DNA that lacks a 2'-OH group. The difference in stability between DNA and RNA is also illustrated by the fact that dilute NaOH will completely destroy RNA at room temperature while DNA is unaffected.

ELUCIDATION OF THE STRUCTURE OF DNA

From 1940 to 1953, many scientists were interested in solving the structure of DNA. Erwin Chargaff, Maurice Wilkins, Rosalind Franklin, Linus Pauling, Francis Crick and James Watson contributed towards prediction of the structure of DNA. All scientific research towards this direction was to find an answer for the question: how does DNA serve as the genetic basis for the living process?

BOX 2.1   BIOINFORMATICS AND DNA STRUCTURE

Bioinformatics involves the use of techniques including applied mathematics, informatics, statistics, computer science, artificial intelligence, chemistry and biochemistry to solve biological problems usually at the molecular level. Major research efforts in the field include sequence alignment, gene finding, genome assembly, protein structure alignment, protein structure prediction, prediction of gene expression protein–protein interactions and the modelling of evolution.

The bioinformatics tools include Biological Database: These biological databases usually contain genomic, proteomic and metabolomic data, but databases are also used in taxonomy. The data include nucleotide sequences of genes or amino acid sequences of proteins. Furthermore, information about function, structure, localization (both cellular and chromosomal), clinical effects of mutations as well as similarities of biological sequences and structures can be found. These tools have three categories:

i.   *Sequence alignment tools* It is a way of arranging the primary sequences of DNA, RNA, or protein to identify regions of similarity that may be a consequence of functional, structural, or evolutionary relationships between the sequences.

ii. *Sequence manipulation tools* They are software programs for analysing and formatting DNA and protein sequences.

iii. *Sequence analysis tools* It implies subjecting a DNA or peptide sequence to sequence alignment, sequence databases, repeated sequence searches, or other bioinformatics methods on a computer. Sequence analysis in molecular biology and bioinformatics is an automated, computer-based examination of characteristic fragments.

In 1953, James Watson and Francis Crick proposed that the structure of DNA is in the form of a double helix. Their proposal was published in a short paper in the journal *Nature*.

The data available to Watson and Crick came primarily from two sources—base composition analysis of hydrolysed samples of DNA and X-ray diffraction studies of DNA. The analytical success of Watson and Crick can be attributed to model building that conformed to the existing data. Watson who entered the University of Chicago at the age of 15, was a 24-year-old postdoctoral fellow in 1953 and Crick was still a graduate student.

Base Composition Studies

Between 1949 and 1953, Erwin Chargaff and his colleagues used chromatographic methods to separate the four bases in DNA samples from various organisms. Later they were quantified, that is, the amounts of the four bases from each source were assessed (Table 2.2).

Table 2.2   Chargaff's DNA base composition data

Molar proportions (Moles of nitrogenous constituent per mole of P)				
Source	**A**	**T**	**G**	**C**
Ox thymus	26	25	21	16
Ox spleen	25	24	20	15
Yeast	24	25	14	13
Avian tubercle bacilli	12	11	28	26
Human sperm	29	31	18	18

Based on the data, the following conclusions were drawn:

1. The amount of adenine residues is proportional to the amount of thymine residues in the DNA of any species. Also the amount of guanine residues is proportional to the amount of cytosine residues.

2. Based on the above proportionality, the sum of the purines (A + G) equals the sum of the pyrimidines (C + T).

3. The percentage of C + G does not necessarily equal the percentage of A + T. The ratio between the two values varies greatly among species.

The base composition and base ratio determined in different sources (Table 2.3) confirm the above facts.

Table 2.3   Base composition of DNA from various sources

Source	A	T	G	C	A/T	G/C	(A+G)/ (C+T)	(A+T)/ (C+G)
Human	30.9	29.4	19.9	19.8	1.05	1.00	1.04	1.52
Sea urchin	32.8	32.1	17.7	17.3	1.02	1.02	1.02	1.58
E. coli	24.7	23.6	26.0	25.7	1.04	1.01	1.03	0.93
Sarcina lutea	13.4	12.4	37.1	37.1	1.08	1.00	1.04	0.35
T7 bacteriophage	26.0	26.0	24.0	24.0	1.00	1.00	1.00	1.08

These conclusions indicate a definite pattern of base composition of DNA molecules. Earlier in 1910, Phoebus A. Levene had proposed the tetranucleotide hypothesis to explain the chemical arrangement of the nucleotides in nucleic acids. He proposed that the four-nucleotide unit AMP, TMP, GMP and CMP linked by phosphodiester bond was repeated in the same order over and over in DNA, assuming a ratio of 1 : 1 : 1 : 1 of these four nucleotides. This discrepancy was ascribed to inadequate analytical technique at that time. DNA base composition data obtained from various sources clearly ruled out the tetranucleotide hypothesis.

X-ray Diffraction Analysis

When macromolecules are subjected to X-ray bombardment, these rays scatter according to the molecule's atomic structure. The pattern of scatter can be captured as spots on photographic film and analysed, particularly for the overall shapes of and regularities within the molecule. This process, X-ray diffraction analysis, was successfully applied to the study of protein structure by Linus Pauling and others. The technique was attempted on DNA in 1938 by William Astbury. By 1947, he detected a periodicity or a repeating unit of 3.4 Å that suggested to him that the bases were stacked like pennies one on top of another.

Between 1950 and 1953, Rosalind Franklin, working in the laboratory of Maurice Wilkins, got an improved X-ray data from more purified samples of DNA. Her reports confirmed Astbury's argument and also indicated some sort of helix nature in DNA structure.

THE WATSON–CRICK MODEL

Watson and Crick proposed the double helical form of DNA with the following major features:

1. Two long polynucleotide chains are coiled around a central axis, forming a right-handed double helix.

2. The two chains are antiparallel. That is, their C-5′ to C-3′ orientations run in opposite directions. They wrap around each other such that they cannot be separated without unwinding the helix, a phenomenon known as **plectonemic coiling**.

3. The bases of both chains are flat structures, which lie perpendicular to the axis. They are stacked on one another, 3.4 Å apart and are located on the inside of the helix.

4. The purine and pyrimidine bases in both the chains are paired to one another by formation of hydrogen bonds (Figure 2.8). A pairs with T by two hydrogen bonds and G with C by three hydrogen bonds.

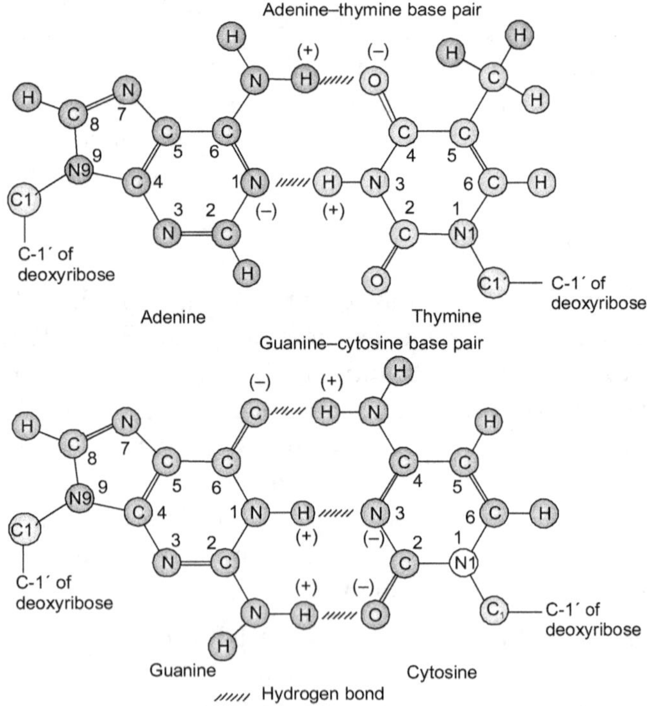

Figure 2.8 Base-pairing in the DNA molecule

5. Each complete turn of the helix is 34 Å long. Thus there are 10 bases in each chain per turn.

6. In any segment of the molecule, there are alternate major and minor grooves (Figure 2.9) along the axis.

7. The double helix measures 20 Å in diameter.

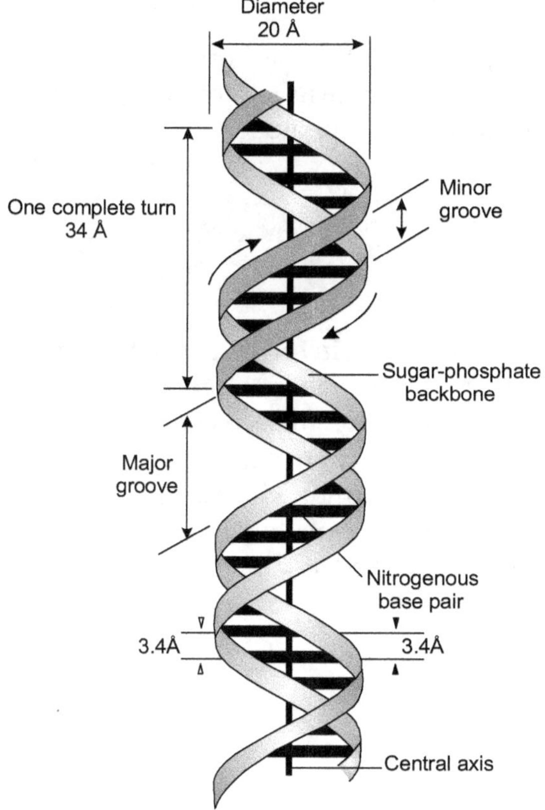

Figure 2.9   Schematic representation of double-helical DNA as proposed by Watson and Crick

CHEMICAL BONDS IMPORTANT IN DNA STRUCTURE

i.   *Covalent bonds*   They are strong chemical bonds formed by sharing of electrons between atoms.

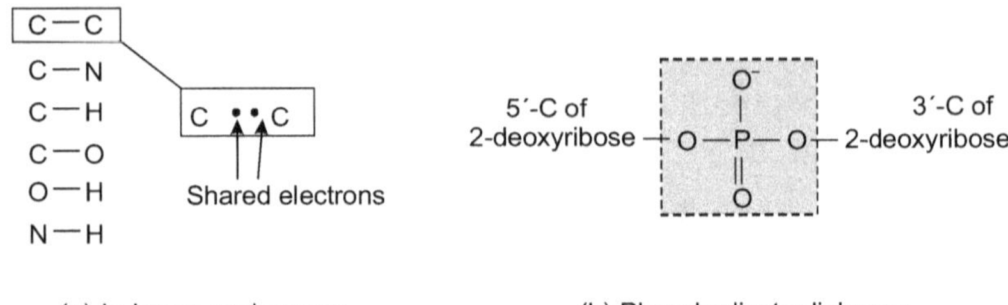

(a) In bases and sugars   (b) Phosphodiester linkages

ii. *Hydrogen bonds*   They are weak bonds between an electronegative atom and a hydrogen atom (electropositive) that is covalently linked to a second electronegative atom.

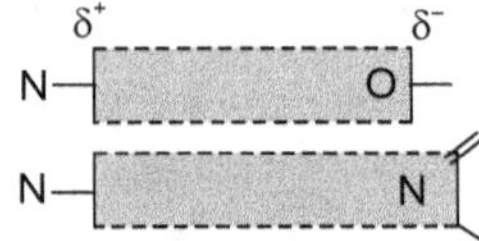

iii. *Hydrophobic bonds*   This involves association of non-polar groups with each other when present in aqueous solutions because of their insolubility in water. Water molecules are highly polar. Compounds which are similarly polar are very soluble in water and are said to be hydrophilic. Compounds which are non-polar have no charged groups and are highly insoluble in water. They are said to be hydrophobic. The stacked base pairs in DNA provide a hydrophobic core.

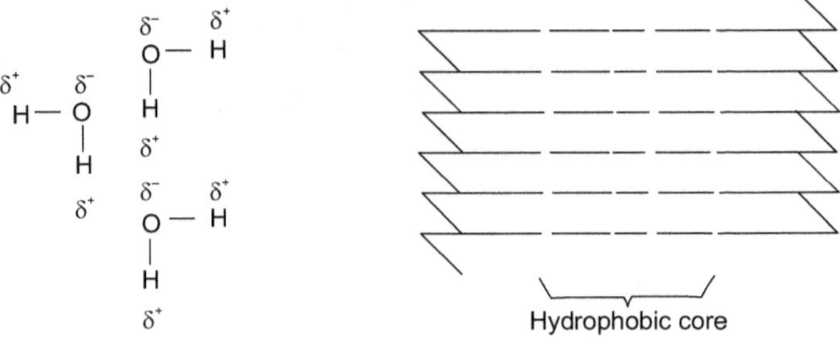

BOX 2.2   GNA NANOTECHNOLOGY

In the Biodesign Institute at Arizona State University, researchers are using DNA to make intricate nano-sized objects. Working at this scale holds great potential for advancing medical and electronic applications. DNA, often thought of as the molecule of life, is an ideal building block for nanotechnology because they self-assemble, snapping together into shapes based on natural chemical rules of attraction. This is a major advantage for Biodesign researchers who rely on the unique chemical and physical properties of DNA to make their complex nanostructures.

While scientists are fully exploring the promise of DNA nanotechnology, Biodesign Institute is working to give researchers brand new materials to aid their designs. In an article recently published in the *Journal of the American Chemical Society*, the research team has made the first self-assembled nanostructures composed entirely of glycerol nucleic acid (GNA), a synthetic analog of DNA.

The DNA helix is made up of just three simple parts: a sugar and a phosphate molecule that form the backbone of the DNA ladder, and one of four nitrogenous bases that make up the rungs. The nitrogenous base-pairing rules in the DNA chemical alphabet fold DNA into a variety of useful shapes for nanotechnology, given that "A" can form a zipper-like chemical bond only with "T", and "G" can pair only with "C."

In the case of GNA, the sugar is the only difference with DNA. The five-carbon sugar commonly found in DNA, called deoxyribose, is substituted by glycerol, which contains just three carbon atoms.

Researchers had a long-standing interest in tinkering with chemical building blocks used to make molecules like proteins and nucleic acids that do not exist in nature. The first self-assembled DNA nanostructure was made at Columbia University in 1998.

In nature, many molecules important to life like DNA and proteins have evolved to exist only as right-handed. The GNA structures, unlike DNA, turned out to be "enantiomeric" molecules, which in chemical terms means both left and right-handed. The research team also found a number of physical and chemical properties that were unique to GNA, including a higher tolerance to heat than DNA nanostructures.

VARIATION OF BASE COMPOSITION OF DNA OF DIFFERENT ORGANISMS

Though it is true that [A] = [T] and [G] = [C] in DNA, there is no rule to govern [G] + [C] / [A] + [T] ratio in a DNA molecule. There is enormous variation in this ratio, ranging from 0.37 to 3.16 for different species of bacteria. The base composition of hundreds of organisms has been determined. G + C content is near 0.50 for higher organisms. For the lower organisms, G + C content varies widely from one gene to another. For example, for bacteria the extremes are 0.27 for the genus *Clostridium* and 0.76 for the genus *Sarcina*. *E. coli* DNA has the value 0.50. A convenient way to measure the base composition is by centrifuging the DNA to equilibrium in caesium chloride, because the density of DNA in CsCl solutions is linearly related to the G + C content.

Spicy Questions and Answers

1. **In a DNA molecule, if guanine content is 22%, can you predict the composition of the other three bases?**

   Yes, we can calculate the composition of the other three bases. If G is 22%, C will also be 22%. The rest, i.e., $100 - 44 = 56\%$ will be A + T or A = 28% and T = 28%.

2. **Double-stranded DNA is isolated from two viruses I and II. DNA from virus I is 1,00,000 base pairs in length and virus II has a molecular weight of 6.6×10^7 Da (i) Which DNA is larger? (ii) What is the size of each DNA molecule in micrometres?**

   i.   Virus II DNA has a molecular weight $= 6.6 \times 10^7$ Da

        Molecular weight of 1 nucleotide pair $= 660$ Da

        Therefore, the number of nucleotide pairs $= 1,00,000$ base pairs

        Both the viral DNAs are of the same size.

   ii. Size of the DNA molecules of both the viruses $= 1,00,000 \times 3.4$ Å
        $= 34\,\mu\text{m}$ long

3. **An animal virus has a circular, double-stranded DNA with 4,00,000 base pairs (i) How many helical turns does this DNA contain? (ii) How many phosphorus atoms does it contain? (iii) How many deoxyribose units does it contain?**

   i.   40,000 turns (B DNA has 10.4 bases per turn; on an average we can have 10 bases)

   ii.   800,000 phosphorus atoms

   iii.   800,000 deoxyribose units

4. What is the reaction of DNA and RNA to alkali treatment?

RNA is sensitive to degradation by alkali because of the 2′-OH on the pentose ring, whereas DNA is more resistant to alkali degradation.

5. Name three co-enzymes that have nucleotide constituents in their overall structures.

NAD, FAD, co-enzyme A.

6. Arrange the various terms listed below and associated with various aspects of mammalian genetic material, in increasing order of complexity: nitrogenous base, nucleotide, dsDNA, genome, nucleoside, chromosome, gene.

Nitrogeneous base, nucleoside, nucleotide, dsDNA, gene, genome, chromosome.

7. Which carbon atoms in deoxyribose and ribose of DNA and RNA are involved in the formation of phosphodiester bond?

3′-carbon and 5′-carbon.

8. Is the Chargaff's rule applicable to the DNA of ϕX174? Why?

No, because it has single-stranded circular DNA.

9. A phage particle is isolated from an *E. coli* cell. The base composition of the phage DNA was found to be G = 15%, C = 20%, A = 35%, T = 30%. What inference is drawn from this data?

Since there is no complementarity of bases, the DNA is not double-stranded and should be circular single-stranded. Single-stranded linear DNA is also not stable.

10. DNA was extracted from a bacterial cell and it was found out that it has 37 cytosine residues. Using this information, can we predict the number of residues of the other three bases? If so, how much? If not, why not?

We can find out the number of guanine residues which will also be 37. But we do not know the total number of nucleotide pairs in the DNA. Therefore we will not be able to predict the number of adenine and thymine residues.

11. **Why are nucleic acids termed as acids?**

Nucleic acids are highly charged polymers. At neutral pH, each phosphate group has a single negative charge. That is why the nucleic acids are termed as acids. They are the anions of strong acids.

Review Questions

1. What are the purine and pyrimidine bases present in DNA and RNA?

2. Distinguish between the structural features and functions of DNA and RNA.

3. Discuss the various bonds involved in the structure of dsDNA.

4. What is Chargaff's rule? How did the rule help in the confirmation of the structure of DNA?

5. Describe the features of Watson–Crick model of dsDNA with a neat sketch of the model.

6. Differentiate between nucleoside and nucleotide with an example.

7. Why does DNA have deoxyribose sugar and thymine base and not ribose sugar and uracil base?

8. Explain the following in the structure of DNA:

    i.   Base pair complementarity

    ii. Major and minor groove

    iii. Phosphodiester bonds

    iv. 5′ and 3′ ends

9. What are polymers? Is DNA a homopolymer or a heteropolymer? What are the monomers present in it?

10. Give the relationship between the length, molecular weight and number of base pairs in a dsDNA.

11. Explain the approach of Watson and Crick in arriving at the structure of DNA.

12. How do covalent bonds differ from hydrogen bonds?

13. What component of the nucleotide is responsible for the absorption of UV light?

14. Draw a short segment of DNA (with three base pairs) at the molecular level and indicate the strand polarity.

15. Why is the bond that holds nucleotides together in nucleic acids called a phosphodiester bond?

16. Draw the enol form of thymine. Can this form a normal base pair with adenine?

17. Draw the chemical structure of the three components of a nucleotide and then link them together.

18. How are the carbon and nitrogen atoms of the sugars and purine/pyrimidine numbered?

19. The chemical name of adenine is 6-aminopurine. How are the other four nucleotides named?

3

STRUCTURAL FEATURES OF DEOXYRIBONUCLEIC ACID

Nucleic acids are regarded as polymers of nucleotides. They are referred to as polynucleotides. Nucleotides are quite strong acids. The primary ionization of the phosphate and protonation of some of the groups on the bases within the nucleotides can be observed at pH values quite close to neutrality as shown in Table 3.1. The bases are also capable of conversion between tautomeric forms.

Table 3.1   Ionization constants of ribonucleotides expressed as pKa values

Ribonucleotide	Phosphate group		Base	
	Primary constant	Secondary constant	Reaction	Constant
5′ AMP	0.9	6.1	Protonation at N-1	3.8
5′ GMP	0.7	6.1	Protonation at N-7	2.4
5′ UMP	1.0	6.4	Loss of proton at N-1	9.4
5′ CMP	0.8	6.3	Loss of proton at N-3	9.5
			Protonation at N-3	4.5

The amino form of adenine and cytosine and keto form of guanine and thymine are the most stable forms. The imino and enol forms of these bases are less stable.

In 1912 Feulgen found out that if DNA is treated with a warm acid, it is subjected to hydrolysis and a reddish purplish staining reaction would occur on adding Schiff's reagent. This reaction was termed Feulgen reaction that was specific for DNA and not for RNA or any other cellular substance. Therefore this reaction formed the basis for the study of chromosomes by cytophotometry involving quantitative studies of measurement of the amount of light transmitted through Feulgen-stained nuclei. The cytophotometric experiments showed a constancy of the amount of DNA for all nuclei of an organism except polyploid conditions and haploids. The amount of Feulgen-stained DNA was found to double during the interphase "S" period of cell growth and was then found to be equally distributed at anaphase to the two daughter nuclei.

SIZE OF DNA MOLECULES

In viruses and bacteria, the total genome is present in a single DNA molecule. In eukaryotes, including the unicells such as algae, yeast and protozoa, the DNA is partitioned into a number of chromosomes, each of which has a single gigantic DNA molecule. The molecular weight, M, of the individual DNA molecules of various organisms is given in Table 3.2.

Table 3.2   Size of various DNA molecules

Organism or particle	Molecular weight, M
Polyoma virus	3.2×10^6
Phage 186	18×10^6
Phage T7	25×10^6
Phage λ	32×10^6
F plasmid	62×10^6
Phage T4	106×10^6
Vaccinia virus	121×10^6
Bacteria	$2.0\text{–}2.6 \times 10^9$
Yeast	6×10^8
Drosophila (fruit fly)	7.9×10^{10}
Human	8×10^{10}

The length of DNA molecules can be obtained from the relation $1\,\mu m = 2 \times 10^6$ molecular weight units.

DIFFERENT TYPES OF DNA MOLECULES

Most of the DNA in most organisms is double-stranded, although some DNA viruses carry single-stranded DNA molecules. DNA molecules may be linear or circular. DNA forms of some organisms are given in Table 3.3.

Table 3.3   Forms of DNA in different organisms

Source	Single-stranded (ss) or double-stranded (ds)	Circular or linear	Number of base pairs (bp)	Length
Simian virus 40	ds	Circular	5243	1.78 μm
φX174 virus	ss	Circular	5386	–
Bacteriophage M13	ss	Circular	6407	–
Cauliflower mosaic virus	ds	Circular	8031	2.73 μm
Adenovirus AD-2	ds	Linear	35,937	12.2 μm
Bacteriophage T2	ds	Linear	$\sim 1.7 \times 10^5$	~58 μm
Bacterium E. coli	ds	Circular	$\sim 4 \times 10^6$	~1.4 μm
Fruit fly	ds	Linear	$\sim 6.5 \times 10^7$	~2 cm

ALTERNATIVE FORMS OF DNA

DNA may exist in different forms. The alternative forms of DNA differ from each other primarily in the degree and direction of helical winding and in the stacking of the bases (Figure 3.1). The most common form of DNA in living cells is a right-handed helix called the B form (or **B-DNA**). Fibres of DNA assume B conformation as indicated by their X-ray diffraction patterns, when the counter ion is an alkali metal such as Na$^+$ and the relative humidity is 92%. B-DNA is regarded as the native form because its X-ray pattern resembles that of the DNA in intact sperm heads. B-DNA helix has 10 bp per turn with a helical twist of 36° per bp. Since the aromatic bases have van der Waal's thickness of 3.4 Å and are partially stacked on each other, the helix has a pitch (rise per turn) of 34 Å.

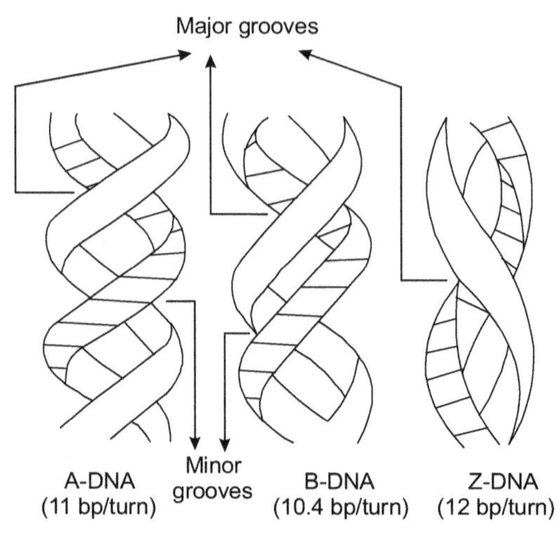

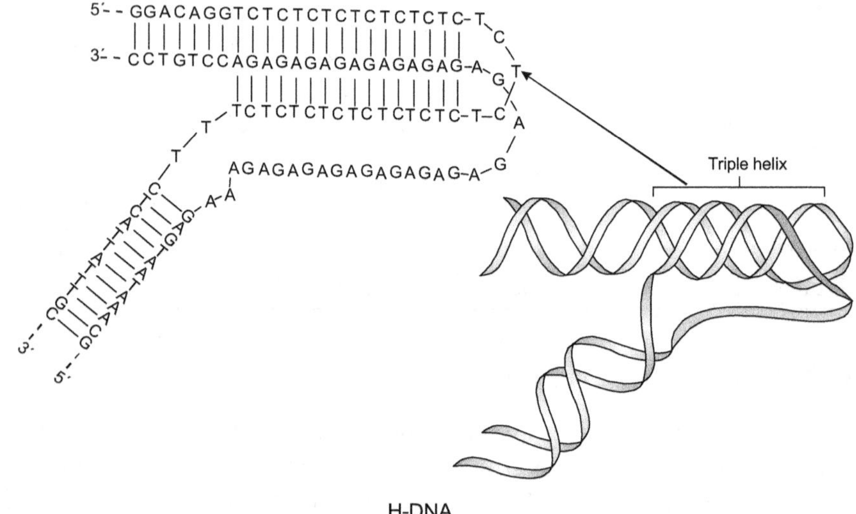

Figure 3.1   Alternative forms of DNA

When the relative humidity is reduced to 75%, B-DNA undergoes a reversible conformational change to the A form. **A-DNA** forms a wider and flatter right-handed helix than does B-DNA. A-DNA has 11 bp per turn and a pitch of 28 Å. The planes of its base pairs are tilted 20°

with respect to the helix axis. Since this helix axis does not pass through its base pairs, A-DNA has a deep major groove and a very shallow minor groove. It can be described as a flat ribbon wound around a 6-Å-diameter cylindrical hole. Most self-complementary oligonucleotides of <10 base pairs, for example, d(GGCCGGCC) and d(GGTATACC) crystallize in A-DNA conformation.

BOX 3.1   BASE PAIR GEOMETRY IN DNA

The geometry of a base pair can be characterized by six parameters—rise, twist, slide, shift, tilt and roll. These values precisely define the location and orientation in space of each base pair in a DNA molecule relative to its predecessor along the axis of the helix. They characterize the helical structure of the molecule. In regions of a DNA molecule where the normal structure is disrupted, these values are used to describe the disruption. Rise is the displacement along the helix axis; twist is the rotation around the helix axis; slide is the displacement along an axis in the base-pair plane directed from one strand to the other and shift is the displacement directed from the minor to the major groove. Tilt is the rotation around the shift axis and roll is the rotation around the slide axis.

Rise and twist determine the handedness and pitch of the helix. Slide and shift are small in B-DNA. Roll and tilt make successive base pairs less parallel and are typically small.

A-DNA has, so far been observed in only limited biological context. Gram-positive bacteria undergoing sporulation contain a high proportion of (20%) small acid-soluble spore proteins (SASPs), some of which induce a B-DNA to assume the A form. The DNA in bacterial spores exhibits a resistance to UV-induced damage that is abolished in mutants that lack these SASPs. This is because the B → A conformation change inhibits the UV-induced covalent cross-linking of pyrimidine bases by increasing the distance between successive pyrimidines.

Z-DNA consists of a helix wound in a left-handed conformation where the sugar–phosphate backbones of Z-DNA form a zigzag pattern instead of a smooth helix. Z-DNA may be present in certain locations within DNA molecules in the cell. Z-DNA is unusual in that its double helix is wound in a left-handed conformation, the molecule is quite narrow and there are more base pairs per helical turn than in B-DNA. The sugar–phosphate backbones of Z-DNA form a zigzag pattern, which is the reason for the Z designation. Although typical cellular conditions do not favour the formation of Z-DNA, certain situations may cause it to appear within limited regions of the DNA molecule. These may include cases where specialized proteins cause the left-handed twisting of the B-DNA molecule or when the nucleotide sequence favours the Z-DNA form. Antibodies specific for Z-DNA bind to a few regions of DNA in cells, suggesting that limited regions of Z-DNA are present.

The left-handed double helix Z-DNA has 12 Watson–Crick base pairs per turn, a pitch of 45 Å and in contrast to A-DNA, a deep minor groove and no discernible major groove. The base pairs in Z-DNA are flipped 180° relative to those in B-DNA. As a consequence, the repeating unit of Z-DNA is a dinucleotide $d(X_pY_p)$, rather than a single nucleotide. The line joining successive phosphate groups on a polynucleotide strand of Z-DNA therefore follows a zigzag path around the helix rather than a smooth curve as in A- and B-DNAs. Fibre diffraction and NMR studies have shown that complementary polynucleotides with alternating purines and pyrimidines, such as poly d(GC)·poly d(CG) or poly d(AC)·poly d(GT), take up the Z-DNA conformation at high salt concentrations. Therefore, Z-DNA conformation is assumed by DNA with alternating purine–pyrimidine base sequences. A high salt concentration stabilizes Z-DNA relative to B-DNA by reducing the increased electrostatic repulsions between closest approaching phosphate groups on opposite strands—8 Å in Z-DNA vs 12 Å in B-DNA. Methylation of cytosine residues at C-5, a common biological modification, also promotes Z-DNA formation because a hydrophobic methyl group in C-5 position is less exposed to solvent in Z-DNA than in B-DNA (Figure 3.1).

The reversible conversion of specific segments of B-DNA to Z-DNA under appropriate circumstances acts as a kind of switch in regulating gene expression.

Runs of homopurine–homopyrimidine DNA sequence [poly (dG)·poly(dC)] seem to set up an A-like helix as proved by circular dichroism spectra. It is assumed that within the general B-DNA molecule, specific regions may exist in an A-DNA form. The structural features of A-, B- and Z-DNA are listed in Table 3.4.

T-DNA is an unusual form of DNA with a helix repeat of 8 bp per turn. This has been purified from bacteriophages T2, T4 and T6. The cytosine residues in T-DNA contain a hydroxymethyl group on the 5'-carbon. In addition, a glucose residue is added to the hydroxymethyl group.

H-DNA is a triple-helical form of DNA. A short region of double helical B-DNA may rearrange to become the H-DNA triple helix. Short segments of H-DNA are present in living cells and H-DNA may help control the expression of genes (Figure 3.1).

Table 3.4 Structural features of A-, B- and Z-DNA

	A-DNA	**B-DNA**	**Z-DNA**
Helical sense	Right-handed	Right-handed	Left-handed
Diameter	~26 Å	~20 Å	~18 Å
Base pairs per helical turn	11	10	12 (6 dimers)
Helical twist per base pair	33°	36°	60° (per dimer)
Helix pitch (rise per turn)	28 Å	34 Å	45 Å
Helix rise per base pair	2.6 Å	3.4 Å	3.7 Å
Base tilt normal to the helix axis	20°	6°	7°
Major groove	Narrow and deep	Wide and deep	Flat
Minor groove	Wide and shallow	Narrow and deep	Narrow and deep
Suger pucker	C-3'-endo	C-2'-endo	C-2'-endo for pyrimidine
Glycosidic bond	Anti	Anti	Anti for Pyr, Syn for Pur

Three other right-handed forms of DNA helices have been discovered when investigated under laboratory conditions. These have been designated **C-DNA, D-DNA** and **E-DNA**. C-DNA is found under even greater dehydration conditions than those observed during the isolation of A- and B-DNA. It has only 9.3 base pairs per turn and is, thus, less compact. Its helical diameter is 19Å. Like A-DNA, C-DNA does not have its base pairs lying flat. They are tilted relative to the axis of the helix. D-DNA and E-DNA occur in helices lacking guanine in their base composition. They have even fewer base pairs per turn as less as 8 and 7½ respectively.

If DNA is artificially stretched, still another form of the molecule is assumed, referred to as **P-DNA**. Compared to B-DNA, P-DNA is longer, more narrow and the phosphate groups found on the outside of B-DNA are present inside the molecules. The nitrogenous bases present inside the helix in B-DNA are found closer to the external surface of the helix in P-DNA and there are 2.62 bases per turn, in contrast to the 10.4 per turn in B-DNA.

It is assumed that DNA might have to take a structure other than the B form when it has to function as the genetic material. The molecule has to undergo strand separation and become accessible to many enzymes and proteins required for DNA and RNA synthesis which becomes possible with the varied structural forms of DNA.

UNUSUAL STRUCTURES OF DNA SEQUENCES

A number of sequence-dependent structural variations have been detected within larger chromosomes that may affect the function and metabolism of the DNA segments in their immediate vicinity. For example, bends occur in the DNA helix whenever four or more adenosine residues appear sequentially in one strand. Six adenosines in a row produce a bend of about 18°. These types of bending are important in the binding of some proteins to DNA.

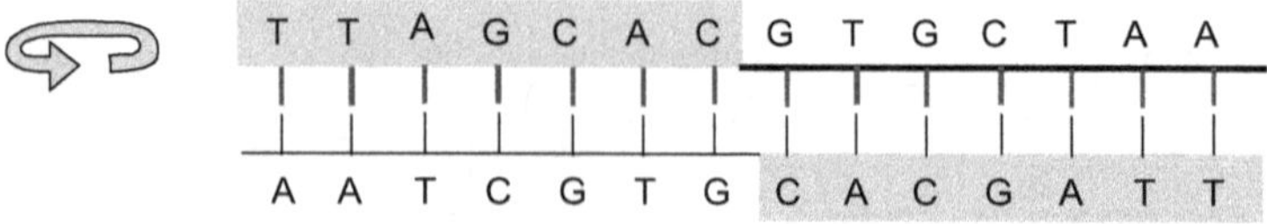

Figure 3.2   Palindrome

A common type of DNA sequence is a **palindrome** (Figure 3.2). A palindrome is a word, phrase or sentence that is spelled identically reading forward or backward, like the words MALAYALAM and ROTATOR and sentences NURSES RUN, ABLE WAS I ERE I SAW ELBA. The term palindrome refers to regions of DNA with inverted repeats of base sequences having twofold symmetry over two strands of DNA. Such sequences are self-complementary within each strand and therefore have the potential to form hairpin (Figure 3.3) or cruciform (cross-shaped) structures (Figure 3.4). When the inverted repeat occurs within each individual strand of the DNA, the sequence is called a **mirror repeat** (Figure 3.5). Mirror repeats do not have complementary sequences within the same strand and cannot form hairpin or cruciform structures.

Some unusual DNA structures involve three or even four strands. They form sites for important events in DNA metabolism. Nucleotides participating in a Watson–Crick base pair can form a number of additional hydrogen bonds, particularly with functional groups in the major groove. For example, a protonated cytidine residue can pair with guanosine residue of a $G \equiv C$ nucleotide pair and a thymidine can pair with adenosine residue of an $A{=}T$ pair. The N-7, O^6, and N^6 of purines, the atoms that participate in the hydrogen-bonding of triplex DNA are referred to as Hoogsteen positions and the non-Watson–Crick pairing is called **Hoogsteen pairing**, after Karst Hoogsteen who first recognized these unusual pairings. Hoogsteen pairing allows the formation of triplex DNAs. A triplex of the type $C \equiv G \cdot C^+$ is highly stable at low pH. Triplexes readily form within long sequences containing only pyrimidines or only purines in a given strand. Some triplex DNAs contain two pyrimidine strands and one purine strand and others contain two purine strands and one pyrimidine strand.

In the DNA of living cells, sites recognized by many sequence-specific DNA-binding proteins are arranged as palindromes and polypyrimidine or polypurine sequences that can form triple helices or even H-DNA are found within regions involved in the regulation of expression of some eukaryotic genes.

Four DNA strands can also pair to form a **tetraplex** called as **quadruplex** (Figure 3.6) when the DNA sequences have very high proportion of G and C residues. Thus quadruplex DNA refers to four-stranded form of DNA. Earlier it was postulated that gene function might be regulated by two double-helix regions of DNA looping back onto themselves to form a four-stranded structure.

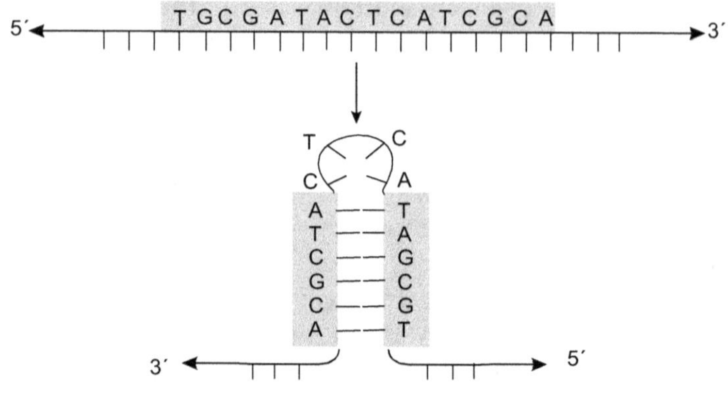

Figure 3.3　Hairpin structure

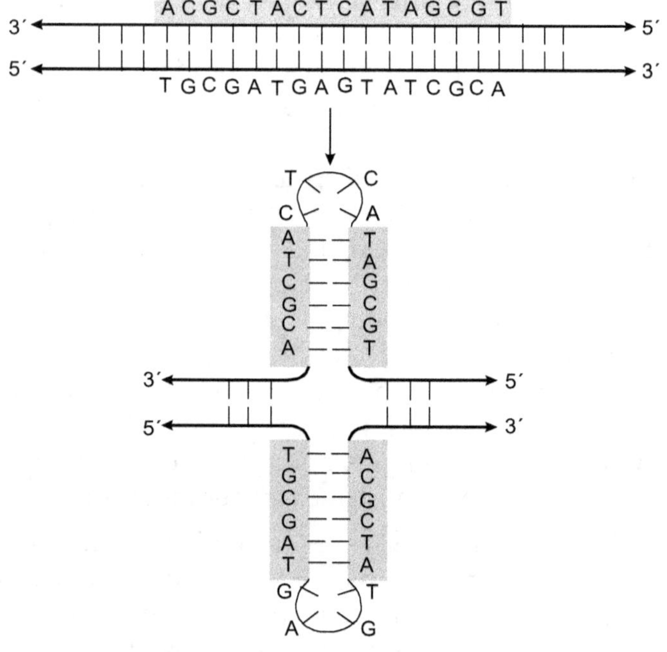

Figure 3.4　Cruciform structure

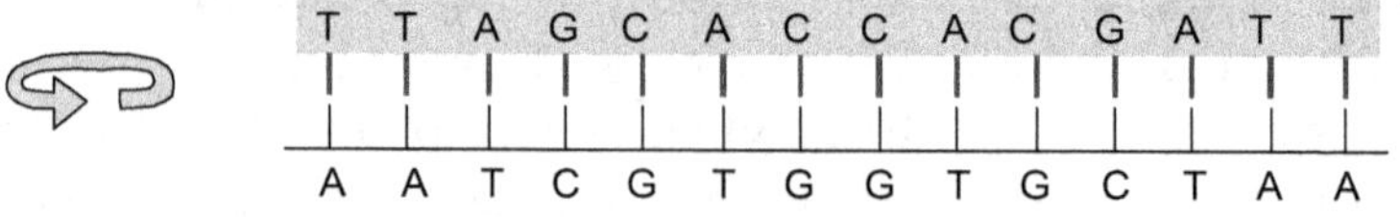

Figure 3.5　Mirror repeat

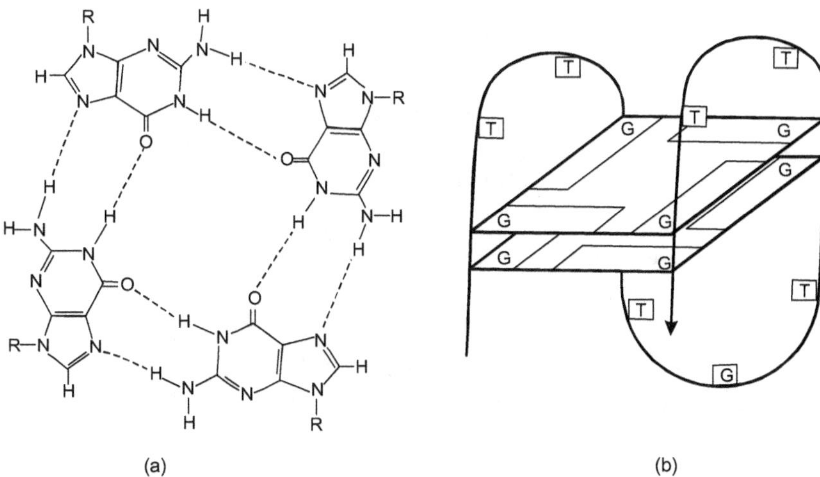

Figure 3.6   G-tetraplex

Using techniques like X-ray crystallography and NMR spectroscopy, such structures were identified in protozoan DNA, telomeres and in a human oncogene. This unique DNA sometimes has chair-shaped structure or boxlike structure with a cavity called G-quadruplexes because they are present in guanine-rich areas in DNA. They may be involved in gene regulation and stabilizing telomeres. Small molecules like porphyrins and anthraquinones bind in the cavity and stabilize the structures. Such molecules may inhibit telomerase activity in continuously extending chromosome ends and thus control cancer.

CIRCULAR AND SUPERHELICAL/SUPERCOIL DNA

The intact DNA molecules of most prokaryotes and viruses are circular. A circular molecule may be a **covalently closed circle** which consists of two unbroken complementary single strands, or it may be a nicked circle, which has one or more interruptions (nicks) in one or both strands. A nicked circle is also called as an **open circle**. With few exceptions, covalently closed circles are twisted. Such a circle is said to be a **superhelix** or a **supercoil** or a **supertwist** (Figure 3.7). The quantitative measure of the intensity of supercoiling is referred to as superhelix density, σ.

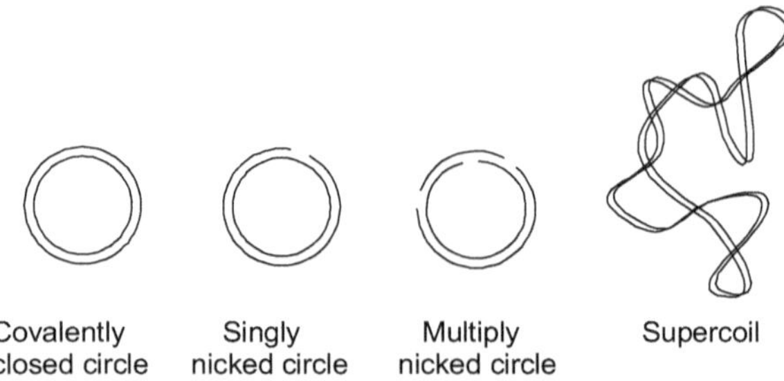

Figure 3.7   Supercoil in DNA

Let us consider a double-helical DNA molecule in which both strands are covalently joined to form a circular duplex molecule (Figure 3.8).

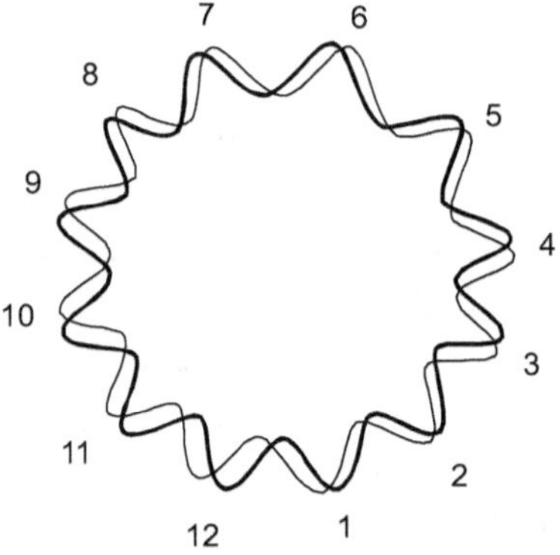

Figure 3.8   Circular DNA

Each strand is joined only to itself because the strands are antiparallel. The number of coils cannot be altered without first cleaving at least one of its polynucleotide strands.

A mathematical expression for a coiled DNA is

$$L = T + W$$

in which, L is the linking number which is defined as the number of times one DNA strand winds about the other, which is an integer quantity. Linking number can be broken into two structural components called writh (W) and twist (T) (Figure 3.9). The **twist** is the number of complete revolutions that one strand makes about the duplex axis in the particular conformation under consideration. By convention, T is positive for right-handed duplex turns. For B-DNA in solution, the twist is normally the number of base pairs divided by 10.4 (the number of base pairs per turn of the B-DNA double helix under physiological conditions).

W is the writhing number, the number of turns the duplex axis makes about the superhelix axis in the conformation of interest. It is a measure of the DNA's superhelicity.

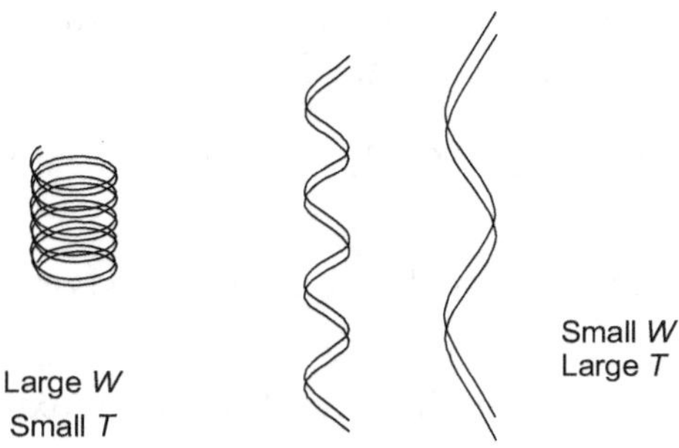

Figure 3.9   Superhelical topology of DNA

The difference between writhing and twisting is shown in Figure 3.9. $W = 0$ when the DNA's duplex axis is constrained to lie in a plane as in Figure 3.8. Then $L = T$, so L may be evaluated by counting the DNA's duplex turns.

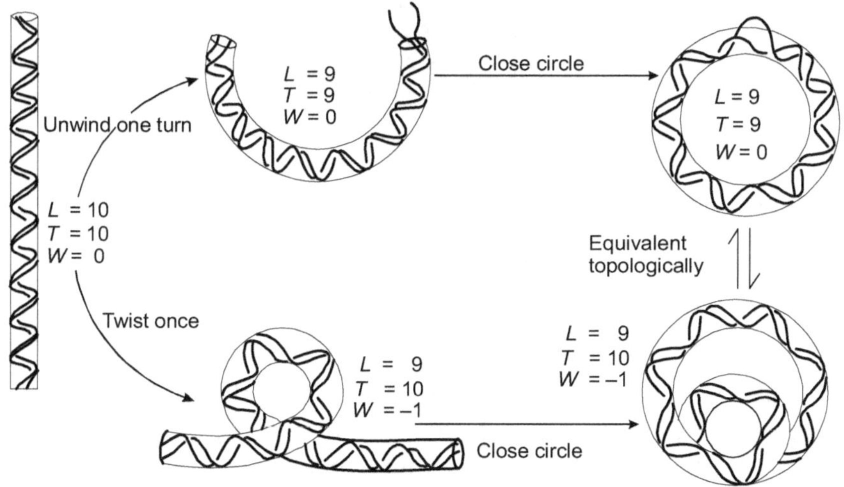

Figure **3.10**   Relation between linking number, twist and writhing number and supercoiling in DNA

The two DNA conformations on the right in Figure 3.10 are topologically equivalent; that is, they have the same linking number, L, but differ in their twists and writhing numbers. T and W need not be integers, only L. Since L is constant in an intact duplex DNA circle, for every new double-helical twist, ΔT, there must be an equal and opposite superhelical twist, that is $\Delta W = -\Delta T$. For example, a closed circular DNA without supercoils (upper right) can be converted to a negatively supercoiled conformation (lower right) by winding the duplex helix the same number of positive (right-handed) turns.

A supercoiled duplex may assume two topologically equivalent forms (Figure 3.11):

1. A toroidal helix, in which the duplex axis is wound as if over a cylinder.

2. An interwound helix, in which the duplex axis is twisted around itself.

These two interconvertible superhelical forms have opposite handedness.

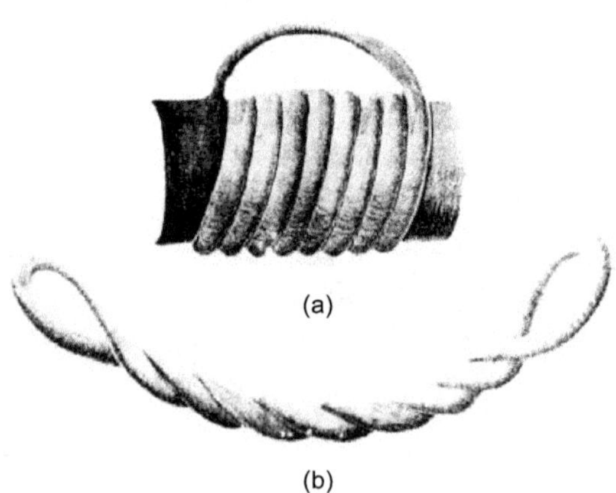

(a)

(b)

Figure 3.11   (a) A toroidal helix (b) An interwound helix

The duplex DNA has two strands in the form of right-handed helices and has an inherent degree of twist, with approximately 10 base pairs per turn. A short piece of linear DNA that is completely free to rotate on its own axis adopts this strain-free configuration and is known as the relaxed state. Instead, if one end of the DNA is clamped and is not free to rotate and if an extra twist is given to the other end, the double-helical coil tightens (Figure 3.12); the number of turns of DNA is increased. In other words, the number of base pairs per turn is decreased. This is known as **positively supercoiled or overwound state**. If the DNA is twisted in the opposite direction, the coil gets opened and the number of turns per unit stretch is reduced or the number of base pairs per turn gets increased. **This is known as negatively supercoiled** or **underwound state**. Both the states are under tension and now the DNA double helix coils upon itself forming a coiled coil or supercoil.

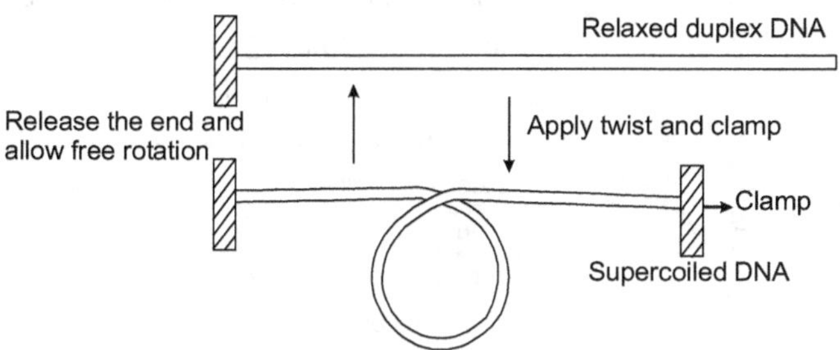

Figure **3.12**   Twisting of a piece of DNA

If the uppermost strand is turning to the left, it is positive and if it turns to the right, it is negative.

BOX 3.2   DNA HELIX GEOMETRIES

DNA is relatively a rigid polymer, modelled as being worm-like. It has three significant degrees of freedom—bending, twisting and compression. Twisting/torsional stiffness is important for the circularization of DNA and the orientation of DNA-bound proteins relative to each other. Bending/axial stiffness is important for DNA wrapping and circularization and protein interactions. Compression/ extension is relatively unimportant in the absence of high tension. The bending stiffness of DNA is measured in terms of "the persistence length" using an atomic force microscope. In aqueous solution, the average persistence length of DNA is 46–50 nm or 140–150 base pairs with a diameter of 2 nm. The persistence length of a section of DNA is somewhat dependent on its sequence which can cause significant variation. The variation is due to base stacking energies and the residues which extend into the minor and major grooves.

The following models (Figure 3.13 and 3.14) explain the supercoiling of DNA with the corresponding linking number, writh and twist, a simplified version of Figures 3.9 and 3.10.

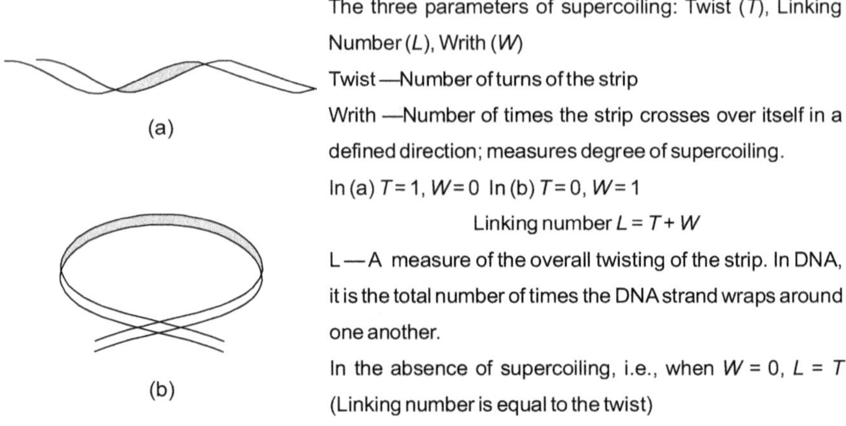

Figure 3.13   DNA supercoiling (a) No supercoil, single twist (b) Supercoiled, no twist

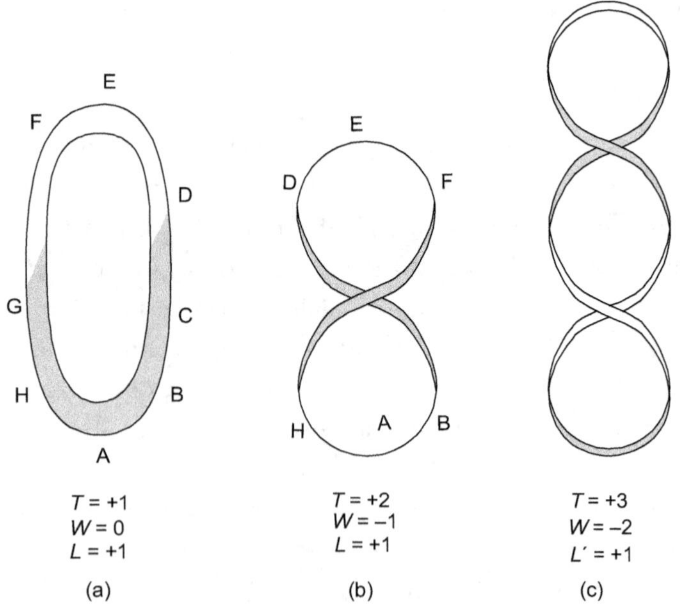

Figure 3.14   (a) Relaxed DNA (b) Right-handed negative and (c) Left-handed positive supercoiling

Spicy Questions and Answers

1. **Following are the base compositions of three DNA molecules. Which one has a higher T_m?**

DNA	Base Composition in %			
	A	G	T	C
Molecule 1	10	40	10	40
Molecule 2	27	23	27	23
Molecule 3	46	4	46	4

T_m value, i.e., the melting temperature of DNA depends on its base sequence. The bond between G and C in the two complementary strands is stronger with 3 hydrogen bonds. This requires more energy to break the bond. Therefore, a DNA molecule with higher GC content will have the higher T_m. Among the three molecules 1, 2 and 3, 1 has higher GC content and will have high T_m.

2. **What is the relationship between the number of nucleotide pairs in a DNA molecule, its length and molecular weight?**

Two adjacent nucleotide pairs are stacked on one another 3.4 Å (0.34 nm) apart.

Each complete turn of the helix with an average of 10 base pairs is 34 Å (3.4 nm) long.

The length of the DNA with n no. of nucleotide pairs is $n \times 0.34 \times 10^{-3}$ m.

The molecular weight of one nucleotide pair is 660 daltons on an average.

$$1000 \text{ base pairs} = 6.6 \times 10^5 \text{ daltons}$$

$$6.6 \times 10^5 \text{ Da} = 340 \text{ nm of dsDNA}$$

3. **It is difficult to isolate and purify DNA with molecular weights above 3×10^7 daltons. Why?**

If the weight of the DNA is above 3×10^7 daltons, it tends to fragment into two pieces of about 1×10^7 daltons on purification.

4. **What do you mean by "syn" and "anti" conformation in DNA?**

The purine nucleotides (as in Z-DNA) adopt "syn" conformation in which the purine base lies directly above the deoxyribose ring whereas the pyrimidine nucleotides in Z-DNA and all nucleotides in the A- and

B-DNA adopt the "anti" conformation (180° rotation around the glycosidic bond). The glycosidic bond rotates 180° about C1 of the sugar and N9 or N1 of the purine/pyrimidine resulting in "syn" or "anti" conformation.

5.   **How can the DNA double helix have a constant diameter even though each strand has a purine and pyrimidine of different size?**

The diameter of the dsDNA is determined by outside dimensions of the base pairs and the angles of the bases with respect to the phosphodiester backbone. AT and GC base pairs have nearly identical widths. Also they are all positioned in the same way between the sugar residues of the backbone.

Review Questions

1.   What is the basis for determining the base composition in DNA using density gradient centrifugation?

2.   You are given a nucleic acid sample extracted from a virus. How will you determine whether the virus has an RNA or DNA genome and if it has DNA, whether it is ss or dsDNA?

3.   What is the difference between a right-handed and a left-handed helix?

4.   Compare and contrast A-, B-, and Z-DNA.

5.   Describe triple helical DNA.

6.   What is meant by sugar pucker effect? Explain with reference to the DNA molecule.

7.   What are palindromes and mirror repeats?

8.   Differentiate between hairpin and cruciform structures in DNA molecule.

9.   Write a note on G-quadruplex.

10.   Describe supercoiling of DNA. What is the difference between positive and negative supercoiling?

11.   Define:

   i.   Superhelical density

   ii.   Linking number

   iii.   Writh number

   iv.   Twist

4

PROPERTIES OF DEOXYRIBONUCLEIC ACID

ABSORPTION OF ULTRAVIOLET LIGHT

Nucleic acids exhibit characteristic absorption in the ultraviolet region. Absorbance of DNA at 260 nm, A_{260}, is proportional to the concentration of the molecule, with a value of 0.02 units per μg DNA per ml. The amount of UV light absorbed by nucleic acids is dependent on the structure of the molecule. The more ordered the structure, the less light is absorbed. Therefore, free nucleotides absorb more light than a single-stranded polymer of DNA or RNA and these in turn absorb more light than a double-stranded DNA molecule. For example, three solutions of double-stranded DNA, single-stranded DNA and free bases each at 50 μg/ml have the following A_{260} values:

Double-stranded DNA $A_{260} = 1.00$

Single-stranded DNA $A_{260} = 1.37$

Free bases $A_{260} = 1.60$

Therefore, double-stranded DNA is said to be hypochromic and the bases are said to be hyperchromic.

DENATURATION OF DNA MOLECULES

The ordered state of DNA which is originally present in nature is called the **native form**. Any macromolecule in a disrupted state, in which the molecule is in a nearly random coil conformation is said to be the

denatured state. A transition from the native to the denatured state is called **denaturation**. When a double-stranded DNA or native DNA is heated, the bonding forces between the strands are disrupted and the two strands separate. Thus, denatured DNA is single-stranded. Denaturation of DNA molecule can be studied by measuring some properties of the molecule that change as denaturation proceeds, as for example, absorption of ultraviolet light. Denaturation of DNA is accomplished by heating a DNA solution or by exposure to chemicals like urea and formamide. A graph of a varying property as a function of temperature is called a melting curve. If a DNA solution is slowly heated and the A_{260} is measured at various temperatures, a melting curve is obtained (Figure 4.1).

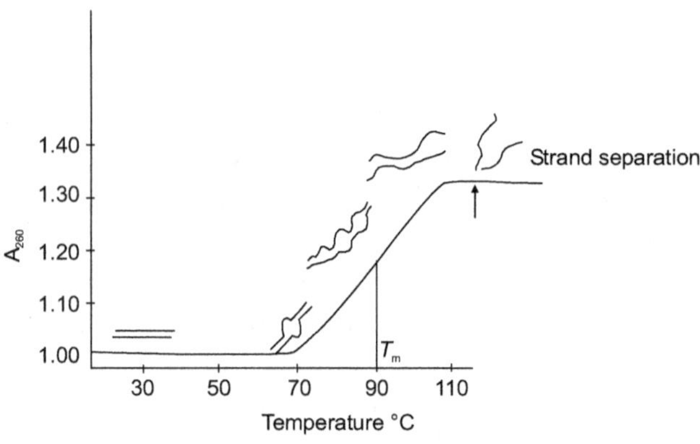

Figure 4.1   Melting curve of DNA

As long as the double-stranded DNA is in native form, there is no change in A_{260}, which starts increasing as the temperature is gradually increased. When both the strands are completely separated at a particular higher temperature, there is maximum A_{260} that indicates complete denaturation of the molecule. The temperature at which the rise in A_{260} is half complete is called the melting temperature, T_m of the DNA molecule. T_m value of a DNA molecule increases with increase in the G + C content of the DNA, because a higher temperature is required to break the three hydrogen bonds between G and C nucleotides.

Compounds like urea and formamide are capable of hydrogen-bonding with the DNA bases. They maintain the unpaired state of DNA molecules and result in lowered T_m value. Formaldehyde reacts with the NH_2 groups of DNA bases and eliminates their ability to hydrogen-bond. Addition of formaldehyde causes a slow and irreversible denaturation of DNA.

There is always a fluctuation in the structure of DNA. The double-stranded regions frequently open to become single-stranded bubbles. This phenomenon is called **breathing**, which enables specialized proteins to interact with the DNA molecule and to read its encoded information. Breathing occurs more often in regions rich in AT pairs than in regions rich in GC pairs.

Certain changes in the physical properties of DNA solutions that accompany denaturation include decrease in viscosity and the ability to rotate polarized light. This indicates that when the hydrogen bonds and hydrophobic interactions are eliminated, the helical structure of DNA is disrupted and the molecule loses its rigidity.

There are many proteins that can unwind a DNA helix. These are called helix-destabilizing or melting proteins. An example is the protein made by gene 32 of *E. coli* phage T4, commonly called the 32-protein. This protein has two properties that enable it to denature DNA:

1. it binds tightly to the bases of single-stranded DNA and

2. the individual molecules of the 32-protein prefer to line up adjacent to one another along a single strand. Binding of the first molecule of the   is made possible by the breathing of the DNA.

Denaturation of DNA is possible by treatment with alkali when there is increased pH. Since DNA is quite resistant to alkali hydrolysis, this procedure is the method of choice for denaturing DNA, because heat treatment may often break the phosphoester bonds and may result in yielding broken fragments of DNA.

RENATURATION OF DNA MOLECULES

A solution of denatured DNA can be treated in such a way that native DNA re-forms. This process is called renaturation or reannealing and the re-formed DNA is called renatured DNA.

> **BOX 4.1   SURPRISING STABILITY OF DNA IN STAINS AT EXTREME HUMIDITY AND TEMPERATURE**
>
> The stability of DNA in air-dried blood and buccal stains at various conditions of relative humidity and temperature when tested revealed that the DNA samples were stable even if incubated for more than one month at 100% humidity and temperature up to 65°C. In wet stains, DNA was found to degrade rapidly. This suggests that even at extreme humid and hot conditions, simple air-drying is adequate for the preservation of blood stains. Collecting whole blood on filter paper simplifies the processing, transport and storage of specimens used for the diagnosis of HIV and other tests including newborn screening for the identifying of inborn diseases in neonates.

Renaturation has proved to be a valuable tool in molecular biology since it can be used to demonstrate genetic relatedness between different organisms, to detect particular species of RNA, to determine whether certain sequences occur more than once in the DNA of a particular organism and to locate specific base sequence in a DNA molecule.

There are two requirements for renaturation to occur:

1. The salt concentration must be high enough so that the electrostatic repulsion between the phosphates in the two strands is eliminated—usually 0.15 to 0.50 M NaCl is used.

2. The temperature must be high enough to disrupt the random, intrastrand hydrogen bonds. However, the temperature should not be too high, because stable interstrand base-pairing will not occur. The optimum temperature for renaturation is 20–25°C below the value of T_m.

Renaturation is a slow process compared to denaturation. The rate-limiting step is not the actual rewinding of the helix but the precise collision between complementary strands such that base pairs are formed at the correct positions. It is a concentration-dependent process. The kinetics of renaturation can be described by an equation in terms of the initial DNA concentration C_0 (expressed in moles of bases per litre), the concentration C of unrenatured DNA at time t (in minutes) and k, a rate constant that depends on the temperature and the size of the DNA fragments:

$$C/C_0 = 1(1 + kC_0 t)$$

Several values of Co can be chosen and C can be measured as a function of time. The data obtained are plotted as C/C_0 versus $\log C_0 t$. When $C/C_0 = \frac{1}{2}$, then $C_0 t\frac{1}{2} = 1/k$. k is inversely proportional to the number of bases per repeating unit (N) that may be present in the DNA. Thus, the value of $C_0 t\frac{1}{2}$ is directly proportional to N. If there are no repeating sequences, so that the DNA molecule itself represents a unique sequence, then N is the number of base pairs in the complete DNA molecule.

SEQUENCE COMPLEXITY OF DNA

$C_0 t$ analysis of the total DNA of numerous organisms has shown that there is a profound difference between the sequence organization of prokaryotes and eukaryotes.

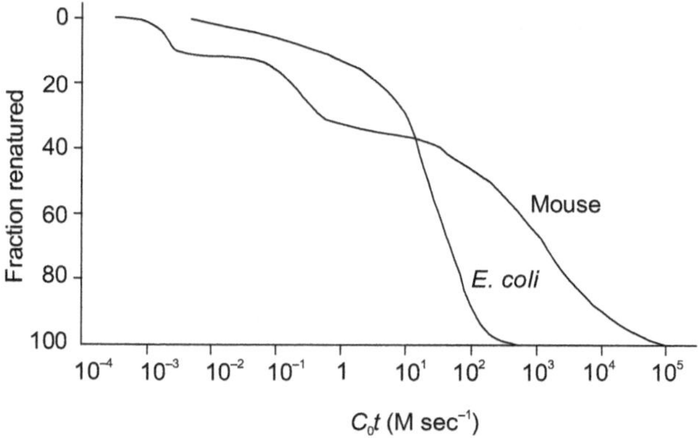

Figure 4.2   DNA $C_0 t$ curve

Figure 4.2 shows the $C_0 t$ curves for bacterial and mouse DNA. The *E. coli* curve is characteristic of all bacteria. It is a smooth one-step curve. This indicates that few sequences exist in multiple copies, that is, each sequence is unique. In eukaryotes, there are four classes of sequences—unique (single copy per genome), slightly repetitive (one to ten copies), middle repetitive (ten to several hundred copies) and highly repetitive (several hundred to several million copies).

If the DNA of a bacterium is isolated and randomly fragmented into hundreds of pieces and subjected to caesium chloride density gradient centrifugation, a single narrow band is obtained. This is because the buoyant density of DNA is proportional to the G + C content of the DNA and the average base composition of segments with several hundred base pairs does not vary much from one section of DNA to the next. If DNA from a crab is analysed, we can find two bands. The major band has a base composition of 58% A + T, whereas the minor band, referred to as **satellite DNA** has 97% A + T, a most unusual base composition. Satellite bands have been observed in the DNA of many organisms (Figure 4.3) and may comprise 1–30% of the total DNA, averaging about 15%. The satellite bands have repeated base sequences of various lengths. For example, in the cow a 1400-bp sequence is repeated again and again which forms the satellite band. But in some monkey species a 172-bp sequence is repeated in the satellite band. In crab, the satellite DNA has only two bases A and T in repeated sequences. *Drosophila virilis* has three satellite DNA bands, each containing a distinct but closely related repeating heptanucleotide namely 5′ A(TC) AAA (TC) T 3′.

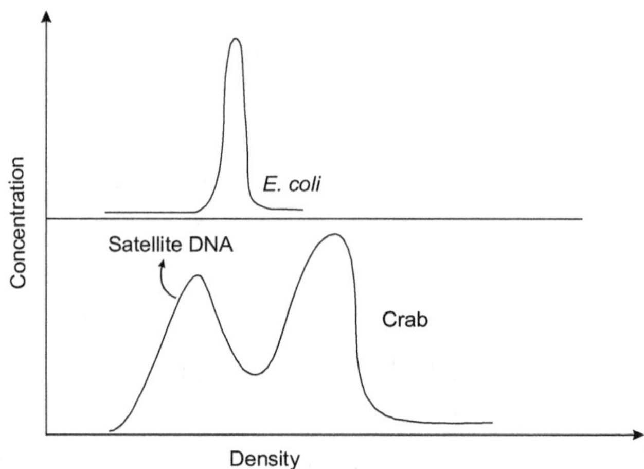

Figure 4.3   Satellite DNA bands

We can analyse the sequence complexity of DNA using a C_0t curve (Figure 4.4). If the C_0t analysis data is as follows: 53 per cent of the sequence having a $C_0t\,\tfrac{1}{2}$ of 10^2, 27 per cent having $C_0t\,\tfrac{1}{2}$ of 1 and 20 per cent having a $C_0t\,\tfrac{1}{2}$ of 10^{-3}, we get a C_0t curve as shown in the figure. From this data, we can determine the number of copies and the size of each sequence. By analysing pure DNA samples of molecules having a unique sequence of known length, a size scale as shown in Figure 4.4b can be obtained. Molecular size cannot be read directly from the observed $C_0t\frac{1}{2}$, because C_0 values used in the X-axis in Figure 4.4a is the total DNA concentration in the renaturation mixture, rather than the concentration of each component.

Therefore, the concentration of each component can be calculated by multiplying each observed $C_0t\frac{1}{2}$, value by the fraction of the total DNA it represents. The real $C_0t\frac{1}{2}$, values are 0.53×10^2, 0.27×1 and 0.20×10^{-3}. The corresponding sizes are 4.2×10^7, 2.2×10^5 and 160 base pairs respectively.

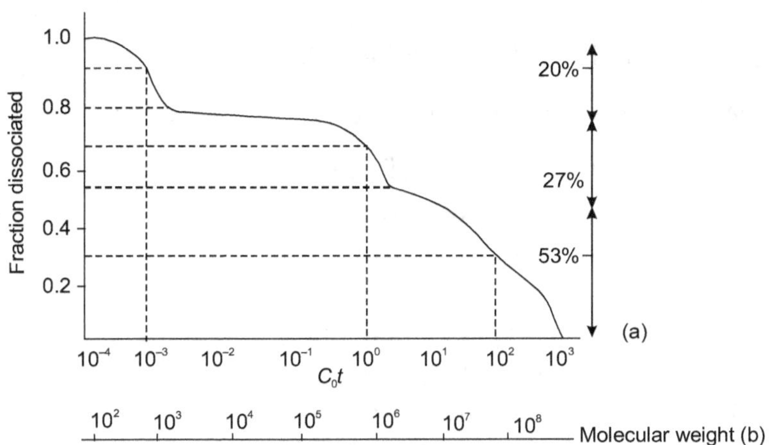

Figure 4.4 C_0t curve analysis

The number of copies of each sequence is inversely proportional to $t\frac{1}{2}$ and hence to the observed (uncorrected) $C_0t\frac{1}{2}$ value. If we assume that the genome contains one copy of the longest sequence, then the cell from which the DNA has been isolated will contain 1, 10^2 and 10^5 copies of sequences having 4.2×10^7, 2.2×10^5 and 160 bp respectively. The total number of base pairs per genome would be

$$4.2 \times 10^7 + 100\,(2.2 \times 10^5) + 10^5\,(160) = 8 \times 10^7$$

The molecular weight of the total cellular DNA would be

$$(8 \times 10^7) \times 660 = 5.3 \times 10^{10}$$

CLEAVAGE OF DEOXYRIBONUCLEIC ACID

Cleavage of nucleic acid structure is mainly required for:

i.   primary structural analysis of DNA and RNA.

ii. cleavage at specific sites for recombinant DNA manipulation. Nucleic acid fragmentation can be achieved by hydrolysis reactions catalysed by acids, bases and enzymes.

RNA withstands treatment with dilute acid, but DNA in 1 M HCl solution undergoes degradation by removal of purine bases. On the other hand, DNA is relatively resistant to alkaline hydrolysis, whereas RNA is hydrolysed under basic conditions at random phosphodiester bonds along the polynucleotide chains to produce polynucleotide fragments of varying lengths.

All cells contain enzymes called nucleases that catalyse the hydrolysis of phosphodiester bonds. They catalyse degradation of damaged or aged nucleic acids by processes that are necessary for general housekeeping in the cell. Nucleases are of two types—deoxyribonucleases (DNase) and ribonucleases (RNase) which act on DNA and RNA respectively. Exonucleases catalyse the hydrolytic removal of terminal nucleotides. Exonucleases are of two types, 5'-exonucleases and 3'-exonucleases, depending on their site of action. Hydrolytic cleavage of internal phosphodiester bonds is carried out by endonucleases. Phosphodiester bonds have two different types of ester bonds: 1) those connecting the 3'-OH group of a nucleotide with the phosphorus and called type *a* and 2) those connecting the 5'-OH group of a nucleotide with the phosphorus and called type *b*. All phosphodiester bridges between nucleotides have one type *a* ester bond and one type *b* ester bond (Figure 4.5).

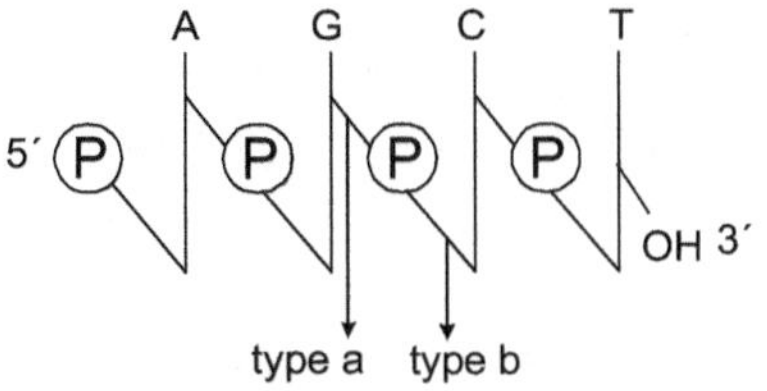

Figure 4.5   Phosphodiester bond with two types of linkages

The most studied nucleases are rattlesnake venom phosphodiesterase, spleen phosphodiesterase, pancreatic ribonuclease and spleen deoxyribonuclease II (Table 4.1).

S1 nuclease, purified from *Aspergillus oryzae*, digests single-stranded DNAs including single-stranded tails or loops or gaps in DNA. At times it will cut at the site of a mismatched base or at sites of chemical modification where normal base-pairing is disrupted. S1 nuclease is used for analysis of cruciforms, Z-DNA and intramolecular triplexes.

Deoxyribonuclease I from bovine pancreas digests DNA to small oligonucleotides of 4 bases in length. It can act on either dsDNA or ssDNA. It is used to introduce random nicks into DNA to permit labelling of DNA with ^{32}P-dNTPs in polymerization reaction. It is also used in "foot printing" studies where nicks are introduced throughout the DNA molecule except where DNA is protected by a tightly bound protein. Exonuclease III, purified from *E. coli* digests only one strand of double-stranded DNA from the 3'-OH end.

Lambda exonuclease, isolated from *E. coli* infected with λ phage, is similar to exo III except that it requires a 5'-PO$_4$ end of DNA. It prefers to digest from a blunt end of DNA. These exonucleases are used to map the sites of covalent chemical modification of DNA and the position of protein-binding sites on DNA.

Phosphatase removes the 5'-phosphate from nucleotides and from the end of a DNA chain to produce a 5'-OH terminus.

Table 4.1   Properties of selected nucleases

Enzyme	Substrate	Type	Specificity
Rattlesnake venom phosphodiesterase	DNA, RNA	exo	Begin at 3' end; no base specificity
Spleen phosphodiesterase	DNA, RNA	exo	Begin at 5' end, no base specificity
Pancreatic ribonuclease A	RNA	endo	Preference for pyrimidines at 3' side
Spleen deoxyribonuclease II	DNA	endo	Internal ester bonds; no base specificity

Restriction endonucleases are specific enzymes that cleave DNA which were discovered in the early 1970s. They are produced by bacterial cells that resist and degrade foreign DNA molecules that enter into the bacterial cells. Restriction enzymes recognize specific base sequences in duplex DNA and catalyse hydrolytic cleavage (Table 4.2). Host DNA is protected from hydrolysis as some of the bases near the cleavage sites are methylated. Many restriction enzymes have been isolated and characterized. The nomenclature for the enzymes consists of a three-letter abbreviation representing the source, a letter representing the strain and a Roman numeral designating the order of discovery. *Eco* RI was the first restriction enzyme to be isolated from *E. coli* (strain R) with cleavage specificity.

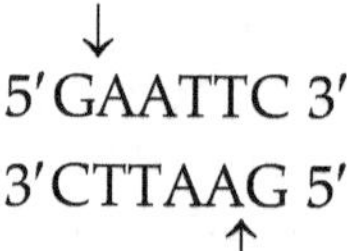

Table 4.2   Some restriction endonucleases and their recognition sequences

Enzyme	Recognition sequence
*Eco*RI	G/AATTC
*Alu*I	AG/CT
Taq I	T/CGA
Hind I	GA/A

Spicy Questions and Answers

1. *E. coli* mutant *cells* with defective DNA ligase are labelled in DNA with ^{3}H-thymine and are allowed to grow in the same medium with ^{3}H isotope of hydrogen-labelled thymine. The new DNA produced is subjected to alkaline sucrose density gradient centrifugation. How does the DNA sediment?

    Two bands will be obtained—one of the template DNA which will be large and another band of the unjoined Okazaki fragments which will be shorter pieces with less molecular weight compared to the other larger strand.

2. It is observed that base ratios vary from one species to another. Suppose, we identify two species with similar base ratios, does this mean that their DNA are identical?

    No. Though their base ratios are similar, the sequence of the nucleotides in DNA may vary.

3. Why is DNA renaturation conducted at about 25°C below the melting temperature of the molecule?

    The denatured DNA has two single strands complementary to one another. Higher temperature is required only for denaturation to disrupt the hydrogen bonds between complementary base pairs. For renaturation, since the two separated strands can easily recognize their complementary bases, a lower temperature is sufficient. At the same time, the renaturation temperature is high enough to disrupt random intra- and inter-strand hydrogen bonds that may interfere with the renaturation process.

4. Why are adenine and thymine not present in equimolar concentrations in ϕX174 DNA?

    Because ϕX174 DNA is single-stranded, there is no chance for complementarity of the AT base pairs.

5. When sodium hydroxide is added to a dsDNA solution, will A_{260} increase or decrease? Why?

    A_{260} will increase because it undergoes denaturation.

6. A renaturation experiment was carried out with genomic DNA of three different bacterial species. They have the following genome size:

    Species I = 2×10^6 bp, Species II = 1×10^8 bp, Species III = 1×10^6 bp.

If we assume that the same total amount of DNA is used in each renaturation experiment and if we draw a C_0t curve for each species, what will be the relative positions of each species in the same graph?

Since the three DNA samples are from bacteria, there will be no complexity in the genome that may be due to the presence of repetitive sequences. The DNA strands from the smallest molecule can undergo rapid renaturation compared to the others. To start with, all DNAs will be single-stranded. This proportion of ssDNA goes on decreasing with renaturation of the molecule. So the smallest genome will have a low C_0t compared to the others.

7. **In a caesium chloride density gradient column, will DNA molecules form bands according to size?**

No. They will form bands according to their density.

8. **Why is dsDNA more stable in solution at higher ionic strength than that at lower ionic strength?**

At high ionic strength, positively charged counter ions reduce the negative charge on the DNA and this lowers the electrostatic repulsion in the DNA structure making it more stable.

9. **What is the effect of acids on nucleic acids?**

In the presence of strong acids like perchloric acid and at increased temperature above 100°C, nucleic acids are hydrolysed to base, sugar and phosphate. In the presence of dilute acids, around pH 3–4, most easily hydrolysed bonds are selectively broken and apurinic/apyrimidinic sites are formed.

10. **What is the effect of alkalis on nucleic acids?**

In the presence of alkalis, bases in the nucleic acids undergo tautomerization. At neutral pH, they are in keto form. If pH is increased, they go to enol form. The hydrogen bonds holding the complementary bases are broken and finally the molecule is denatured and the strands get separated.

11. **Why is DNA more stable than RNA?**

The denaturation of DNA and RNA occurs at a higher pH. But RNA is susceptible to hydrolysis in alkali because of the presence of 2'-OH group in RNA which is perfectly positioned to participate in the cleavage of the backbone structure by intramolecular attack on the phosphate group of the phosphodiester bond. This results in the formation of a free 5′-OH and a 2′,3′-cyclic phosphodiester. This later hydrolyses to either 2′- or 3′- monophosphate. Even at neutral pH, RNA is much more susceptible to hydrolysis than DNA, because of the presence of 2′-OH group. That is why DNA is more stable than RNA.

12. Will ssDNA have a T_m value? If yes, why?

They may have a T_m value if there is complementarity in the bases within the single strand forming hairpin loop structures.

13. What is the relationship between T_m value and GC content of DNA? What is the T_m of *E. coli* DNA having about 50% GC? What is the % GC of DNA from a human kidney cell whose T_m is 85°?

$T_m = 69 + 0.41\ (\%\ GC),\ T_m = 89.5,\ \%\ GC = 39$

Review Questions

1. What is hyperchromic effect? How is it measured?

2. What do you mean by T_m value? How is it related to the base composition of DNA?

3. You are provided with two different samples of DNA of two different GC content. Can you suggest a simple technique to identify the sample with higher GC content?

4. What is meant by denaturation of DNA?

5. Explain renaturation kinetics.

6. Comment on repetitive sequences in DNA and give their biological significance.

7. Write a note on the degradation of DNA by the action of acids, alkalis and enzymes.

8. It is proved that most highly repetitive DNA sequences in eukaryotic chromosomes do not produce any RNA molecules or protein products. What does this indicate about the function of these sequences?

5

PROKARYOTIC AND EUKARYOTIC CHROMOSOMES

The amount of DNA is much greater in eukaryotes than in prokaryotes. However, in both, the DNA must be packed in a systematic manner. Some of the features are common in both the systems—supercoiling, role of proteins in coiling and presence of one DNA molecule per chromosome. Most bacteria and viruses have one circular chromosome with a single, double-helical DNA molecule. In eukaryotes, each chromosome has one double-helical DNA molecule.

A simple analysis helped in finding that a chromosome has one double-helical DNA. In 1974, Ruth Kavenoff and Bruno Zimm determined the molecular weight of the longest chromosome of *Drosophila melanogaster* which was found to be 0.41×10^{11} Da. The haploid content weighed 1.2×10^{11} Da which included four chromosomes. Therefore the haploid content of a single chromosome was found to be 1.2×10^{11} divided by four $= 0.3 \times 10^{11}$ which was close to the value 0.41×10^{11}. This confirmed that the eukaryotic chromosomes have one dsDNA molecule.

PROKARYOTIC CHROMOSOMES

The functional bacterial chromosomes are referred to as nucleoids because they are not bounded by a nuclear membrane. The contour length of the circular DNA molecule of *E. coli* was found to be $1100 \, \mu m$. *E. coli* cell has a diameter of only $1–2 \, \mu m$. This shows that the chromosome must exist in a highly folded or coiled configuration in the cell. The *E. coli* chromosomes were isolated in the absence of ionic detergents to prevent lysis and kept in the presence of a high concentration of cations like polyamines or 1 M salt solution in order to neutralize the negatively charged phosphate groups present in DNA. In this condition, the chromosomes were found to remain in a highly condensed state as in *in vivo* condition (Figure 5.1). This structure is called "folded genome" which is the functional state of the *E. coli* chromosome. In this state, the single DNA molecule of *E. coli* is present as 50 loops or domains. Each of such structure is highly twisted or supercoiled exhibiting positive and negative supercoiled structures.

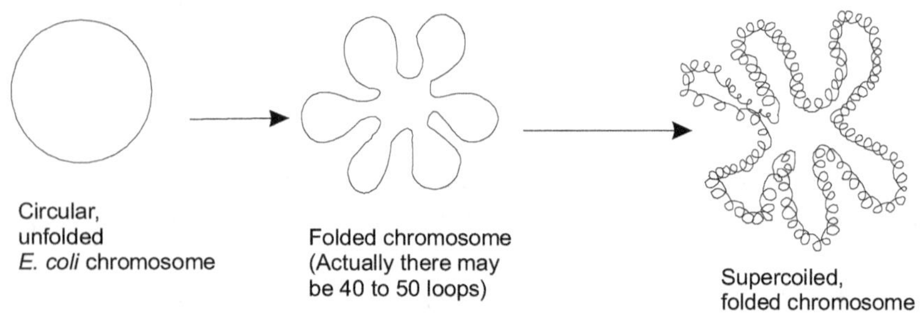

Figure 5.1   Structure of *E. coli* chromosome

ORGANIZATION OF BACTERIAL CHROMOSOME

Bacterial cells possess a single chromosome. Like eukaryotic chromosomes, they exist as a nucleoprotein complex. They are referred to as nucleoids. The nucleoid does not condense prior to cell division and thus does not display the compact structural features of eukaryotic chromosomes. But it is organized into a series of looped domains which reflect functional as well as structural order.

In *E. coli*, the looped domains are approximately 50 kbp in length. There are nearly 100 in the whole genome. Several protein components of the nucleoid have been determined like HU, HI, etc. HU proteins play a central role in holding DNA in a tightly compact complex. The HU protein is a dimer made up of a 9535-Da and a 9225-Da subunit. A single cell has as many as 120,000 copies of this protein. HU is positively charged and binds to the negative charges present in DNA.

A key feature necessary for packaging large, circular prokaryotic chromosomes into small volumes is the twisting of the DNA molecule. Packaging of DNA in prokaryotes and eukaryotes has many of the same features: supercoiling, use of protein to aid in coiling and only one DNA molecule per chromosome. Most bacteria and viruses have one circular chromosome with a single, double-helical molecule of DNA.

The 1-mm long DNA molecule of the *E. coli* chromosome is contained within cells that are only about 2 μm long and 0.5–1 μm wide. A free DNA molecule of this size forms a random coil about 1000 times the volume of an *E. coli* cell. The large volume filled by free DNA is due largely to charge repulsion between the negatively charged phosphate groups. In the cell, this effect is counteracted by association of the DNA with positively charged polyamines such as spermine and spermidine which shield the negative charges of the phosphate groups in DNA.

Numerous small protein molecules are associated with chromosomal DNA, causing it to fold into a more compact structure. The most abundant of these proteins is H-NS which is a dimer of a 15.6-kDa polypeptide. There are about 20,000 H-NS molecules per *E. coli* cell—one H-NS dimer for every 400 bp of DNA.

EUKARYOTIC CHROMOSOMES

The eukaryotic chromosome is uninemic, that is, it contains one double helix of DNA. In 1957, J. Taylor and his colleagues confirmed this with radioactive studies in human and *Drosophila* chromosomes. It was proved that every eukaryotic chromosome contains a single DNA molecule running from end to end, encompassing both arms. The longest molecule was found to be 1.2 cm long with 24–32 $\times$ 10^9 Da molecular weight. The average diploid eukaryotic cell contains many long pieces of DNA. For chromosomes to be properly distributed to each daughter cell during mitosis and meiosis, they must be condensed into structures that are easily manageable. The DNA is wrapped around proteins by which it is coiled and folded in order to get a fully compact chromosome.

HIGHER ORDER CHROMATIN ORGANIZATION

The molecular organization of a eukaryotic chromosome involves several orders of coiling so that the long DNA can be packed in a condensed form.

The DNA of all eukaryotic chromosomes is associated with numerous protein molecules in a stable and ordered aggregate called **chromatin**. Some of the proteins present in chromatin are responsible for the structure of chromosome and for structural changes during cell division. Other proteins play a role in regulating the functions of chromosomes. Chromatin is of two types:

1. **Heterochromatin** These are regions that condense earlier in prophase than the rest of the chromosome, and are highly repetitive, tandomly arranged sequence regions, darkly stained on using certain standard dyes as in Feulgen reaction, and which remain highly condensed (300 Å diameter) throughout the cell cycle.

2. **Euchromatin** It makes up most of the genome and is visible only in the mitotic cycle.

The major heterochromatic regions are adjacent to the centromere and at the ends of telomeric regions and interspersed with the euchromatin (Figure 5.2). In many species, an entire chromosome like the Y chromosome in *Drosophila* and in mammals is almost completely heterochromatin. The number of genes located in heterochromatin is small compared to the number in euchromatin.

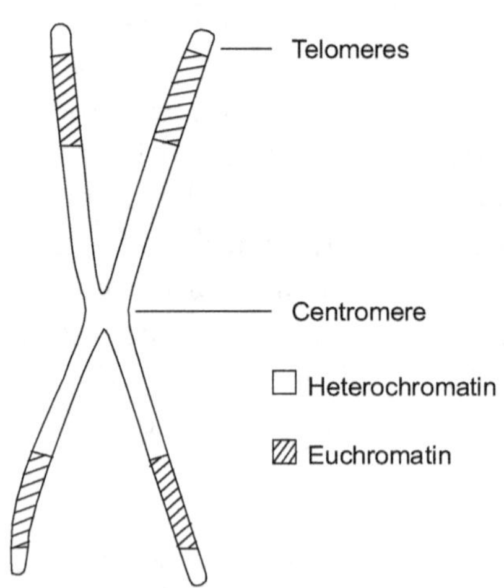

Figure 5.2 Euchromatin and heterochromatin

The winding of DNA to form nucleosomes represents only the first level of structural organization. When nuclei are lysed in a low-salt solution, the characteristic beads-on-a-string structure, termed as 10-nm fibre, represents a packaging ratio of five and lacks one type of chromosomal basic protein, histone H1. At a higher salt concentration, chromatin adopts a more compact structure that includes H1. This forms a coiled 30-nm fibre. The 30-nm fibre is attached at various points to the nuclear matrix (called **nuclear scaffold**) and forms a series of loops containing 30–100 kbp of DNA. This is referred to as the folded fibre model. These chromatin loops are protein-depleted, with their bases attached to scaffold proteins of the nuclear matrix. They form the **chromatin domain** (Figure 5.3).

Interphase chromatin exists in the two distinct forms: the diffuse euchromatin, that comprises of looped 30-nm fibres present in the nucleoplasm and the highly condensed heterochromatin which possesses a high ordered structure and is clustered around the nuclear periphery.

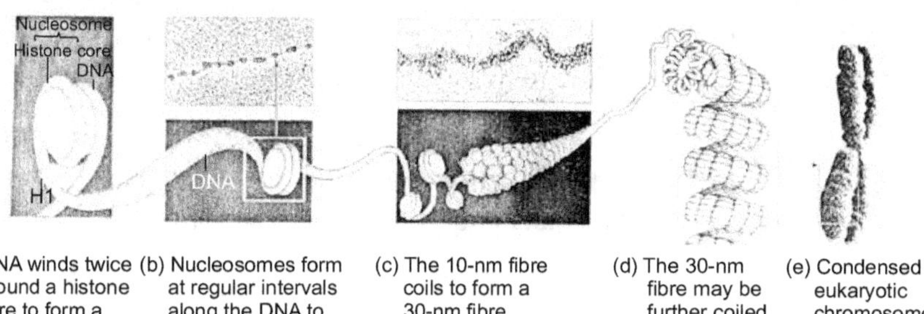

(a) DNA winds twice around a histone core to form a nucleosome.

(b) Nucleosomes form at regular intervals along the DNA to make a 10-nm fibre.

(c) The 10-nm fibre coils to form a 30-nm fibre called a solenoid.

(d) The 30-nm fibre may be further coiled in chromosomes.

(e) Condensed eukaryotic chromosome.

Figure 5.3   Simplified diagram of double-helical DNA organized through solenoid model to the visible structure of the chromosome

Heterochromatin has small number of genes which are almost genetically inert or devoid of function. Heterochromatic blocks can often be rearranged in the genome, duplicated or even deleted without major phenotypic consequences. Heterochromatin is of two types based on their location—**constitutive heterochromatin** that surrounds the centromere, and rich in satellite DNA, and **intercalary heterochromatin** found throughout the chromosome. Constitutive heterochromatin is identified in the C band and the intercalary heterochromatin in G bands during staining of DNA (refer chapter 13, Chromosomal banding techniques). Based on their function, heterochromatin can be said to be **constitutive heterochromatin** that is always inactive and highly condensed and **facultative heterochromatin** which can shift between euchromatic and heterochromatic states, depending on the developmental stage and cell type. The lightly staining euchromatin giving R band, consists of less tightly packed 300-Å fibres. Euchromatin is not visible with light microscope during interphase. Most of the eukaryotic genes are located in euchromatic regions of the chromosomes. Therefore, euchromatin with its less condensed nature is found to be a genetically active region.

Since each eukaryotic chromosome consists of a single, long, double-stranded DNA, an average diploid cell has many of these long pieces of DNA. They are wrapped around specific proteins. This constitutes the first step in a series of coiling and folding processes that finally result in fully compacted chromosome at metaphase. Proteins in chromatin—the complex of DNA and proteins—are of two types: histone proteins and non-histone proteins.

The basic proteins associated with the DNA molecules are rich in arginine and lysine, and are referred to as histones. These positively charged proteins bind to the negatively charged DNA.

There are five types of histone proteins : H1, H2A, H2B, H3 and H4. The relative amounts of arginine and lysine in these five types of histones in all eukaryotic cells is given in Table 5.1. Yeast has no H1 histone.

Table 5.1   Properties of histones

Histone	Lysine (%)	Arginine (%)	Molecular weight (Da)	Relative molar abundance
H1	29	1	22,000	1
H2A	11	9	14,000	2
H2B	16	6	13,700	2
H3	10	13	15,300	2
H4	11	14	11,300	2

All histones occur in equal molar ratios, except for H1, which is present in half the amounts of the others. Histone proteins have nearly the same amino acid sequences in all the species from which they have been isolated and characterized, i.e., they are found to have conserved sequences. This implies that the amino acids play an important role in the histone function. The structure of H1 protein varies slightly between species, but some regions are invariant. Protein H1 varies highly in the cells of some species. It is even given a different name, i.e., H1 is called H5 in avian red blood cells.

In some histone molecules, some of the basic amino acid side chains are modified by post-translational addition of acetyl, phosphate, or methyl groups, neutralizing the positive charges of the side chain or converting it to a negative charge.

The amino acid sequences of four histones, H2A, H2B, H3 and H4 are similar among distantly related species. For example, the sequences of histone H3 from sea urchin tissue and of H3 from calf thymus are identical except for a single amino acid and only four amino acids are different in H3 from the garden pea and that from calf thymus.

Each of the histone proteins making up the nucleosome core contain flexible amino termini of 20–40 residues extending from their globular domains. The N-termini contain several positively charged lysine residues. Some of these interact with the phosphates in DNA of the same nucleosome and some may interact with linker DNA or with neighbouring nucleosomes. These lysines undergo reversible acetylation and deacetylation by specific enzymes that act on specific lysines in the N termini of different histones. In the acetylated form, the positive charge of the lysine ε-amino group is neutralized and its interaction with phosphate group of DNA is removed. Therefore, greater the extent of histone acetylation, less likely are the chromatins to form condensed 30-nm fibres and possibly higher order folded structures.

There is a correlation between histone acetylation and resistance of chromatin DNA to digestion by nucleases. It was found that the active β-globin gene in erythroid precursor cells has acetylated histones and is susceptible to the action of DNase and is less condensed in order to be involved in gene expression. The inactive β-globin gene in non-erythroid cells has unacetylated histones and is highly resistant to DNase action indicating that the chromatin is in highly condensed form.

When the parental strands separate from one another during replication, histones are displaced from DNA. The free histones reassociate immediately with the daughter duplexes. They also get mixed with the newly synthesized histones which accumulate during the G1 stage of cell cycle. It is thought that a molecular chaperone called N1/N2 initiates nucleosome formation by loading a $(H3-H4)_2$ tetramer onto DNA and another chaperone, nucleoplasmin, facilitates the docking of H2A-H2B dimers.

Like replication, transcription also displaces nucleosomes from DNA and reassembly appears to occur following RNA polymerase action. Therefore most transcribed genes retain a nucleosome structure, though the pattern of nucleosome phasing characteristic of nontranscribed genes is lost. The histone octamer reassociates with DNA behind RNA polymerase, perhaps because it remains attached to the nontranscribed strand, or perhaps because it is transiently associated with the enzyme itself.

Chromatin is also composed of non-histone proteins which are not positively charged. They are a heterogeneous group of proteins which help in maintaining the coiled structure of the chromosome. Some are involved in regulation of gene expression or DNA replication like DNA polymerase, DNA ligase, single-strand binding proteins and

RNA polymerase. The content of non-histone proteins in a cell varies depending on the cell type and the organism.

They also include the scaffold proteins which organize higher order chromatin structure and the high mobility group proteins (HMG proteins) which are highly charged and play an important role in gene regulation and structural organization. Their main effect in packaging and transcriptional activation is to introduce sharp bends in the DNA molecule, giving three-dimensional configuration and to bring to close proximity the different regulatory factors which are present at different sites. Protamines are a class of non-histone proteins that facilitate the packaging of DNA into the sperm head. These proteins align the major grooves of adjacent DNA duplexes and fold the DNA into a highly compact array of parallel fibres.

There are specific sequences in DNA called scaffold-associated regions (SARs) or matrix attachment regions (MARs) that are bound to the chromosome scaffold. The scaffold proteins are associated with these regions. SARs are usually found between transcription units. Genes are located within chromatin loops, which are attached at their bases to a chromosome scaffold. In some cases SARs are required for transcription of neighbouring genes. Some SARs can insulate transcription units from each other, so that proteins regulating transcription of one gene do not influence that of a neighbouring gene separated by a SAR.

Other proteins associated with the chromosomes are DNA binding transcription factors, proteins required for DNA replication and some high-mobility group (HMG) proteins with high electrophoretic mobility.

First-Order DNA Coiling

This involves the formation of 11-nm filament of nucleosomes. This filament is produced when about 200 bp of DNA coil around a group of octamers consisting of two each of the histones H2A, H2B, H3 and H4. The DNA–protein unit is called a **nucleosome core particle**. A single molecule of histone H1 is bound to the DNA between the core particles. The H1 linking protein and the core particle together constitute a **nucleosome** (Figure 5.4). With this first order coiling, when chromatin is viewed under an electron microscope, it resembles a thin string of beads about 11 nm in diameter. The coiled core DNA is found to consist of 146 bp holding the histone octamers and 50–60 bp (depending on the species) constituting the linker DNA. Throughout the chromosome, however, genes that are being expressed will not have histones associated with them. The first-order coiling around the histone proteins reduces the length of the chromosomal DNA about fivefold, from about 50 nm of linear DNA to about 10 nm per nucleosome.

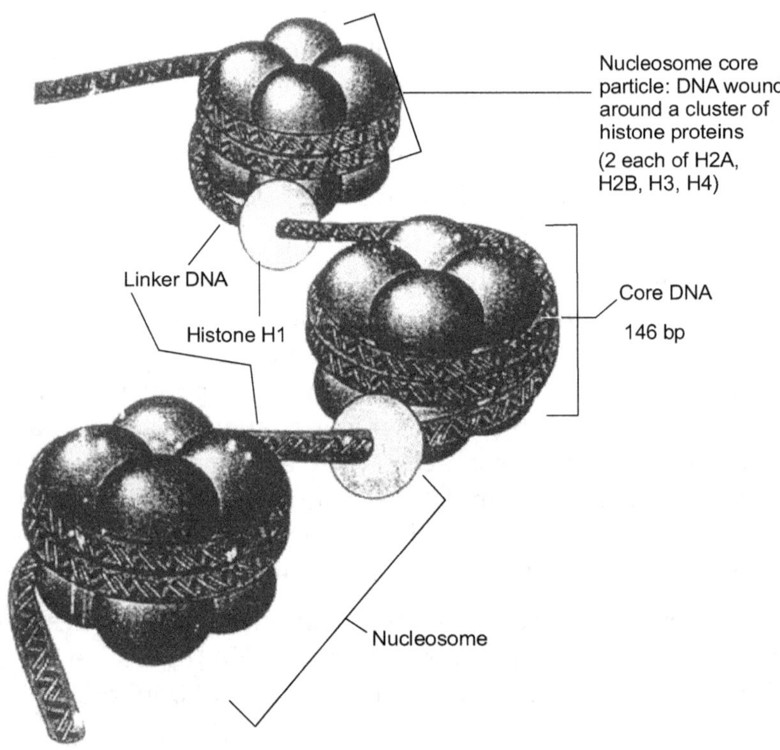

Figure 5.4   Nucleosomes and histones

Second-Order DNA Coiling

The interphase nucleus, instead of showing a beads-on-a-string chromatin structure, presents a filament about 34 nm in diameter. This filament represents a second-order coiling of nucleosomes, known as a **solenoid** which further condenses the chromosome. Each turn of the solenoid contains six to eight nucleosomes. Histone H1 is important in the coiling of the 11-nm filament into the 34-nm solenoid. H1 is an elongated molecule which seals the two turns of the DNA that are coiled around the core histone particle. The two ends of H1 protein have the extended amino terminal and carboxyl terminal arms that attach the adjacent nucleosome core particles. This allows the nucleosomes to arrange in a regular repeating array that facilitates second-order coiling of DNA.

Formation of Coiled Domains

The DNA is separated into loops that are attached to a protein linker every 20,000 to 100,000 bp of double-stranded DNA. The protein linkers and the base to which they are attached are called **scaffolds**. The scaffold represents about 8% of the total protein of the chromosome. The length of the scaffold is approximately the same as metaphase chromosome. This suggests that the scaffold is the primary structure that holds the metaphase chromosome into a compact unit. The chromatin is now composed of 300 nm coiled domains and this represents the final level of coiling. The total diameter of a chromatid is about 700 nm.

The chromosome is not a static body. Coiling and uncoiling occurs continuously throughout the cell division cycle. Different portions of a chromosome can be in quite different states of packaging at the same time. The DNA coils should unwind to be involved in DNA replication and gene expression. The nucleosomes are transiently displaced for these events to take place with minimal structural disturbances. Also, precise regulation of packaging of DNA in eukaryotic chromosomes is very important in regulation of gene expression and cell division.

SPECIFIC REGIONS IN EUKARYOTIC CHROMOSOMES

The organization of genes in eukaryotic DNA is structurally and functionally much more complex than in the prokaryotes. Most bacteria and viruses have a single chromosome, whereas eukaryotic cells usually contain many. A single chromosome may carry thousands of genes. Together all of a cell's genes and intergenic DNA (DNA between genes) form the cellular genome. In humans, the entire length of the chromosomes have different types of DNA sequences.

TYPES OF DNA SEQUENCES IN HUMANS

If DNA is denatured, it will reassociate and this rate will depend on the proportion of unique and repeat sequences present in the molecule.

About 60–70% of the human genome consists of single or low copy number DNA sequences. The remaining 30–40% consists of either moderately or highly repetitive DNA sequences which are not transcribed. This portion consists mainly of satellite DNA and interspersed DNA sequences.

Types of DNA Sequences in Humans

Nuclear DNA ($\sim 3 \times 10^9$ bp)

Genes (50–80000)

Unique single copy

Multigene families

Classical gene families

Gene superfamilies

Extragenic DNA (Unique/low copy number or moderate/highly repetitive)

Tandem repeat

Satellite

Minisatellite

Telomeric

Hypervariable

Microsatellite

Interspersed

Short interspersed nuclear elements

Long interspersed nuclear elements

Mitochondrial DNA (16.6 kb, 37 genes)

Two rRNA genes

22 tRNA genes

13 genes coding for proteins involved in oxidative phosphorylation

---

Nuclear Genes

There are up to 80,000 genes in the nuclear genome, distributed in different chromosomes in varied numbers. Certain parts of the chromosomes like the heterochromatic and centromeric regions contain few genes. Majority of the genes are located in subtelomeric regions.

Unique single copy genes   Most human genes are unique single copy genes coding for enzymes, hormones, receptors and structural and regulatory proteins.

Multigene families   Many genes have similar functions, having arisen through gene duplication with evolutionary divergence resulting in multigene families. Some are present close together in clusters. For example, the α- and β-globin gene clusters are present on chromosomes 16 and 11. Multigene families are of two types—classical gene families which show a high degree of sequence homology and gene superfamilies which have limited sequence homology but are functionally related and have similar structural domains.

Examples of classical gene families include the numerous copies of genes coding for the various ribosomal RNAs which are clustered as tandem arrays at the nucleolar organizing regions on the short arms of the five acrocentric chromosomes and the different tRNA gene families which are dispersed in numerous clusters throughout the human genome.

Examples of gene superfamilies include the HLA genes on chromosome 6 and the T-cell receptor genes which have structural homology with the immunoglobulin genes. It is suggested that these are derived from duplication of a precursor gene.

Gene structure   The original concept of a gene was that it is a sequence of DNA coding for a protein. But analysis of human β-globin gene revealed that the gene was longer than necessary to code for β-globin protein containing non-coding intervening sequences or introns which separate the coding sequences or exons (Figure 5.5).

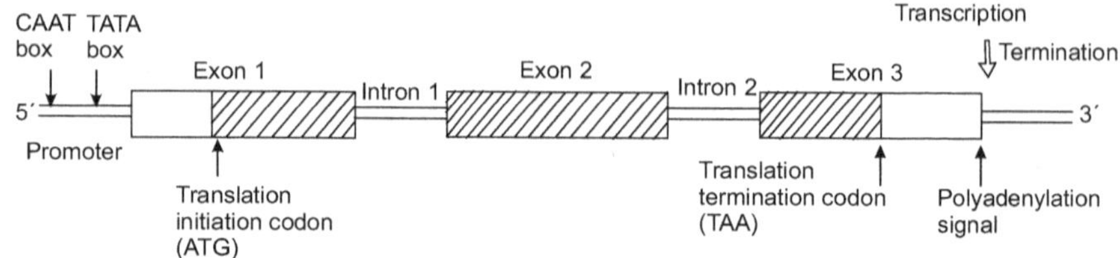

Figure 5.5 Typical human structural gene

The number and size of introns in various genes in humans are extremely variable. Larger the gene, greater is the number and size of the exons. Some introns can be larger than the coding sequence and some have coding sequences for other genes (genes within genes). Usually genes in humans do not overlap and are separated from each other by an average of 30 kb. But some of the genes in the HLA complex are seen overlapping.

Thus many eukaryotic genes have a distinctive structural feature, that is, their nucleotide sequences contain one or more intervening segments of DNA that do not code for the amino acid sequence of the polypeptide product. These non-translated inserts interrupt the otherwise co-linear relationship between the nucleotide sequence of the gene and the amino acid sequence of the polypeptide it encodes. Such non-translated DNA segments in genes are called **intervening sequences or introns**, and the coding segments are called **exons**. Very few prokaryotic genes contain introns. The number of introns, their position and their size are found to be varying in different genes. For example, the egg protein ovalbumin is encoded by a gene in which the introns are much longer that the exons. There are seven introns that make up 85% of this gene's total DNA. In the gene for the β-subunit of hemoglobin, a single intron contains over half the gene's DNA. The gene for the protein conalbumin of the chicken egg has 17 introns and a collagen gene has over 50 introns. Histone genes do not have introns.

Pseudogenes are certain genes that closely resemble known structural genes, but are not functionally expressed, and referred to as pseudogenes. They are obtained either by gene duplication events but made silent because of mutations in the coding or regulatory elements or as a result of the insertion of complementary DNA sequences got by the action of reverse transcriptase on a naturally occurring mRNA transcript, which lack the promoter sequences necessary for expression.

Extragenic DNA

The genes in human chromosomes account only for a maximum of 25% to 33% of the human genome. The rest of the genome is made up of repetitive DNA sequences which are transcriptionally inactive. They do not seem to have any specific function and do not contribute to any phenotypic trait. They are referred to as **selfish DNA**.

Tandemly repeated DNA sequence   They are noncoding DNA sequences and are of three subgroups: satellite, minisatellite and microsatellite DNA.

Satellite DNA   Satellite DNA accounts for approximately 10–15% of the repetitive DNA sequences of the human genome and consists of a very large series of simple or moderately complex short tandemly repeated DNA sequences which are transcriptionally inactive and are clustered around the centromeres of certain chromosomes. These sequences can be separated by density gradient centrifugation as a satellite to the main peak of genomic DNA.

Minisatellite DNA   Minisatellite DNA consists of two families of tandemly repeated short DNA sequences—telomeric minisatellite DNA sequence and hypervariable minisatellite DNA sequence both of which are transcriptionally inactive.

Telomeres are sequences at the ends of eukaryotic chromosomes that help stabilize the chromosome. Yeast telomeres end with about 100 base pairs of repeated sequences of the form

$$5'(TxGy)_n$$
$$3'(AxCy)_n$$

where x and y are between 1 and 4. The number of telomere repeats, n, is in the range of 20 to 100 of the chromosomes of most single-celled eukaryotes and generally over 1500 in mammals. The ends of a linear DNA molecule cannot be routinely replicated by the cellular replication machinery (which may be a reason why bacterial DNA molecules are circular). Repeated telomeric sequences are added to eukaryotic chromosome ends primarily by the enzyme telomerase required for chromosomal integrity.

Two important features of cancer cells are uncontrolled growth and immortality. These are promoted by the action of telomerase. Telomerase is active in cancer cells but turned off in normal cells. In normal cells, telomeres usually get shortened in normal cell cycle (telomerase becomes inactive), chromosomes become unstable, cell division is inhibited and programmed cell death (apoptosis) occurs. The natural, progressive shortening of telomeres may function as a biological clock that limits the number of times a cell can replicate. Telomerase action seems to make cells immortal. Chemical agents that inhibit telomerase activity may serve as potent anticancer drugs.

Inhibition of telomerase can be achieved by the silencing of genes responsible for the formation of the components of telomerase, inhibiting the protein catalytic subunit, TERT of the enzyme, interfering with essential protein factors TRFI and TRF2 and interfering with the action of the RNA template TR.

The latest approach to the enzyme inhibition is by blocking the chromosomal DNA substrate (telomere strands), so that it is not readily available to the enzyme system. X-ray crystallography and NMR studies have shown that single-stranded G-rich regions of telomeres have a tendency to fold back to form G-quadruplex structures, making inaccessible to the telomerase complex. Compounds like anthraquinones, the acridines and tetra (N-methyl pyridyl) porphyrin stabilize the G-quadruplex ends of the telomeres.

Hypervariable minisatellite DNA is made up of highly polymorphic DNA sequences. They consist of short repeats of a common core sequence, the highly variable number of repeat units in different hypervariable minisatellites. This forms the basis of DNA fingerprinting used in forensic and paternity testing. These sequences are referred to as variable number tandem repeats (VNTRs). VNTRs are highly polymorphic when compared to restriction fragment length polymorphisms (RFLPs). RFLP may arise due to variation in even one base pair in DNA nucleotide sequence thereby creating fragments of different length on digestion with restriction enzymes in different individuals.

Microsatellite DNA   Microsatellite DNA consists of tandem single, di-, tri- and tetranucleotide repeat base pair sequences located throughout the genome. These repeats rarely occur within the coding sequences. Trinucleotide repeats in or near genes are associated with certain inherited disorders like Huntington's disease in which there are 9–35 numbers of CAG repeat sequences in the coding region.

Interspersed repetitive DNA sequences   Approximately one-third of the human genome has two classes of short and long repetitive DNA sequences interspersed throughout the genome.

Short interspersed nuclear elements   About 5% of the human genome consists of nearly 750000 copies of short interspersed nuclear elements (SINEs). The most common are sequences of nearly 300 bp which have sequence similarity to a signal recognition particle involved in protein synthesis and referred to as *Alu* repeats which have *Alu* I restriction enzyme recognition site.

Long interspersed nuclear elements   About 5% of the human genome has long interspersed nuclear elements (LINEs). The commonly occurring LINE is LINE-1 (L1 element) with over 100000 copies of 6000 bp sequences that occur once every 50 kb and encodes a reverse transcriptase.

BOX 5.3   DISORDERS CAUSED BY *ALU* SEQUENCES

The most extensively studied SINE sequence in the human genome are the *Alu* sequences. There are approximately 500,000 *Alu* sequences, each about 300 nucleotide pairs long in the haploid human genome. The transposition of an *Alu* sequence into genes and DNA rearrangements caused by recombination between two *Alu* sequences can produce mutant phenotypes. Mutant alleles that cause acholinesterasemia with a deficiency of cholinesterase enzyme resulting in transient paralysis on exposure to muscle relaxants and neurofibromatosis which is a tumour of the nerve tissues under the skin and the central nervous system are found to result from the insertion of *Alu* sequences into the genes. Also hypercholesterolaemia due to deficiency of cholesterol receptors on liver cells and thalassemia due to haemoglobin deficiency are shown to result from gene rearrangements produced by recombination between closely linked *Alu* sequences.

Alu repeat sequences are found to be flanked by short direct repeat sequences and they resemble the transposable elements. Both *Alu* and LINE-1 repeat elements have been implicated as a cause of mutation in inherited human disease.

An overview of the various categories of repetitive DNA is given in Figure 5.6.

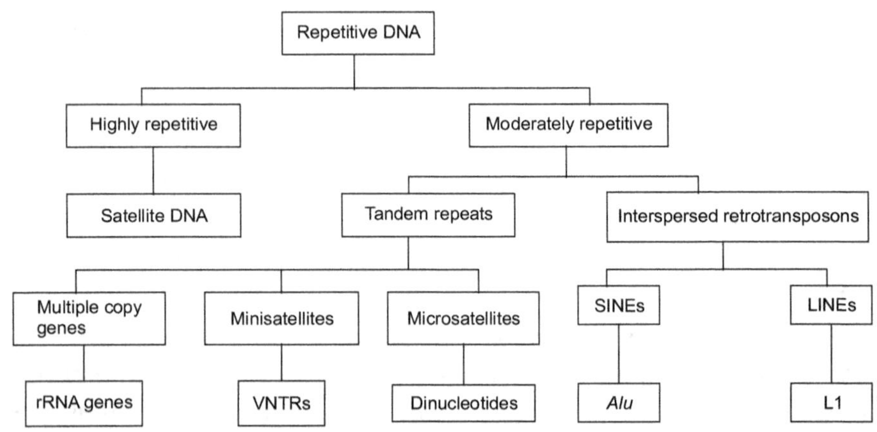

Figure 5.6 An overview of the various categories of repetitive DNA

Mitochondrial DNA

Apart from the nuclear DNA, the thousands of mitochondria of each cell possess their own 16.6 kb circular double-stranded DNA, the mitochondrial DNA. They have little repetitive DNA and code for 37 genes which include two types of rRNA and 22 tRNAs and 13 protein subunits for enzymes such as cytochrome *b* and cytochrome oxidase required for oxidative phosphorylation.

CHROMOSOME MORPHOLOGY

The eukaryotic chromosomes are diploids, because they are derived from one member of each chromosome pair of each parent. Each such pair is referred to as "homologous pair of chromosomes". During replication, each member of the homologous pair undergoes DNA synthesis and this results in the formation of two chromatids.

Each chromosome has two arms—a long *q* arm and a short *p* arm.

A centromere is a highly constricted region of a mitotic or meiotic chromosome where spindle fibres attach (Figure 5.7). Complex sequences of DNA constitute centromeres. The sequences essential to centromere function are about 130-base pairs long and are very rich in AT pairs. The centromeric sequences of higher eukaryotes are much longer and generally contain simple sequence DNA. This consists of thousands of tandem copies of one or a few short sequences of 5 to 10 base pairs in the same orientation. They are protected from endonuclease digestion by various proteins distinct from the histones. They are free from nucleosomes and are decondensed. This accounts for the highly constricted appearance of the centromeric region in condensed chromosomes during mitosis and meiosis.

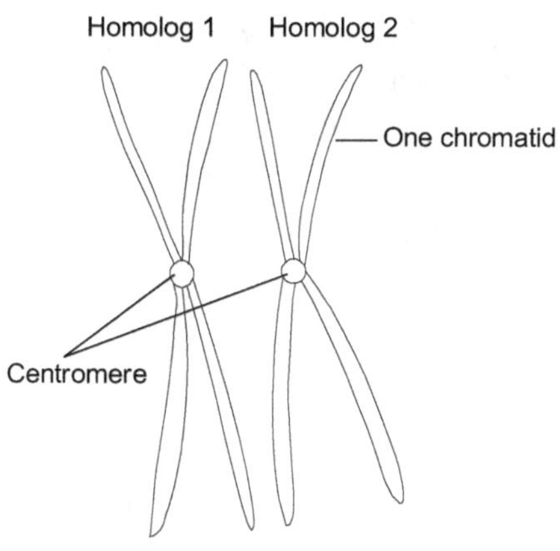

Figure 5.7　Positions of centromeres in the chromosome

Centromeres from budding baker's yeast *Saccharomyces cerevisiae* are about 220 base pairs long. On either side of this sequence of the centromere are endonuclease-sensitive sites whose function may be to promote DNA cutting so that chromatids can separate from each other during anaphase. A chromosome is said to be **metacentric** if the centromere is in or very near the middle. A chromosome having a centromere between the middle and the end is referred to as a **submetacentric** or **acrocentric** chromosome. If the centromere is at or very near the tip, the chromosome is known as **telocentric** chromosome.

Protein complexes associated with the centromeric regions of mitotic and meiotic chromosomes are called kinetochores (Figure 5.8).

Kinetochores bind microtubules of the spindle bundle. In many animal cells, they are disclike structures about 2 μm in diameter that form on one side of each chromatid. Kinetochores in higher plants are ball-like structures that form around the centromere of each chromatid. In many eukaryotes like protozoa, some fungi and insects, no kinetochores are seen.

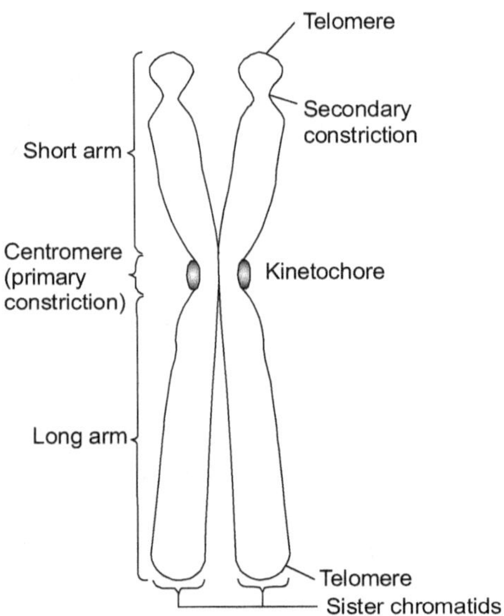

Figure 5.8   Position of kinetochores

CHROMOSOME PARAMETERS

The various parameters characteristic for a chromosome are explained below for a chromosome as shown in Figure 5.9.

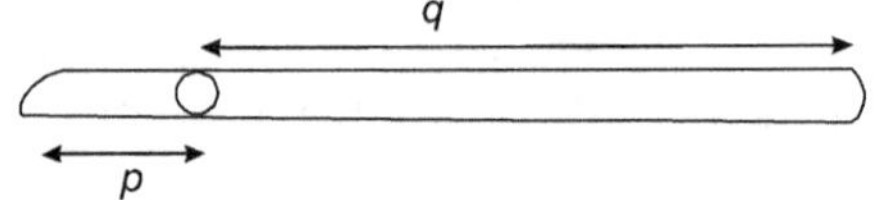

Figure 5.9   Chromosome parameters

***d*-value**   Difference in length between the long and short arms of the chromosome, i.e., $d = q - p$ where q and p are the lengths of the long and short arms respectively.

***r*-value**   The ratio of the lengths of the two chromosome arms, i.e., $r = q/p$.

Centromeric index   The distance from the centromere to the tip of the short arm, expressed as a percentage of the total length of the chromosome, i.e., $i = 100 \times p/(p+q)$.

Monocentric chromosome   Chromosome with a single defined centromere.

Holocentric chromosome   Chromosome with a diffused centromere.

Telocentric chromosome   A monocentric chromosome with a terminal centromere (for a chromosome of length 10 units, $p = 0, q = 10, d = 10, r = \alpha$ and $i = 0$) referred to as T-chromosome or monobrachial chromosome.

Atelocentric chromosome   A monocentric chromosome with a nonterminal centromere (a dibrachial chromosome). This is of the following types:

Metacentric   A monocentric chromosome with a central centromere. When the centromere is exactly at the median point, it is referred to as M-chromosome.

Submetacentric   A monocentric chromosome with a centromere in the submedian region.

Subacrocentric   A monocentric chromosome with a centromere in the subterminal region.

Acrocentric   A monocentric chromosome with its centromere very close to the terminal region.

THE CELL CYCLE

Cell cycle is the sequence of events that take place between successive cell divisions. Many different processes are coordinated during the cell cycle. Some processes occur continuously, like the growth of a cell. Some processes occur discontinuously, like cell division. Cell division must be coordinated with growth and DNA replication so that cell size and DNA content remain constant.

Cell cycle comprises a nuclear or chromosomal cycle (DNA replication and partition) and a cytoplasmic or cell division cycle (doubling and division of cytoplasmic components). In eukaryotes, the two major events, replication and mitosis do not occur simultaneously.

In prokaryotes, both the events are coordinated so that partially replicated chromosomes can segregate during rapid growth. The eukaryotic cell cycle is divided into discrete phases (Tables 5.2 and 5.3) and they proceed in a particular order, whereas the stages of the bacterial cell cycle may overlap.

The Bacterial Cell Cycle

The bacterial chromosome cycle is divided into three phases—the interval phase, the chromosome replication phase and the division phase, represented by the letters I, C and D. This is known as **I + C + D model or Helmstetter–Cooper model**.

DNA replication occurs during the C phase and its duration is fixed which is about 40 minutes in *E. coli*. It is the time taken to replicate the whole chromosome. D phase begins when replication is complete and it ends in cell division. Duration of D phase is also fixed which is about 20 minutes in *E. coli*. During this time, the cellular components required for cell division are synthesized. Because C + D is fixed, any change in the cell doubling time must reflect a change in the duration of I phase, the interval between two successive initiations of replication. The doubling time of *E. coli* can be 20 minutes to 3 hours. During slow growth, I > C + D and replication is completed before cell division. During rapid growth, the doubling time is shorter than the time taken to complete one round of replication and cell division.

Partition of the replicated chromosome marks the culmination of chromosome cycle and is followed by cell division. The cell divides by binary fission referred to as **cytokinesis.**

The Eukaryotic Cell Cycle

The eukaryotic cell cycle (Figure 5.10) is divided into four non-overlapping phases. The main events of the chromosome cycle that includes DNA synthesis and mitosis occur during S phase and M phase respectively. They are separated by G1 and G2 gap phases, when mRNAs and proteins accumulate continuously.

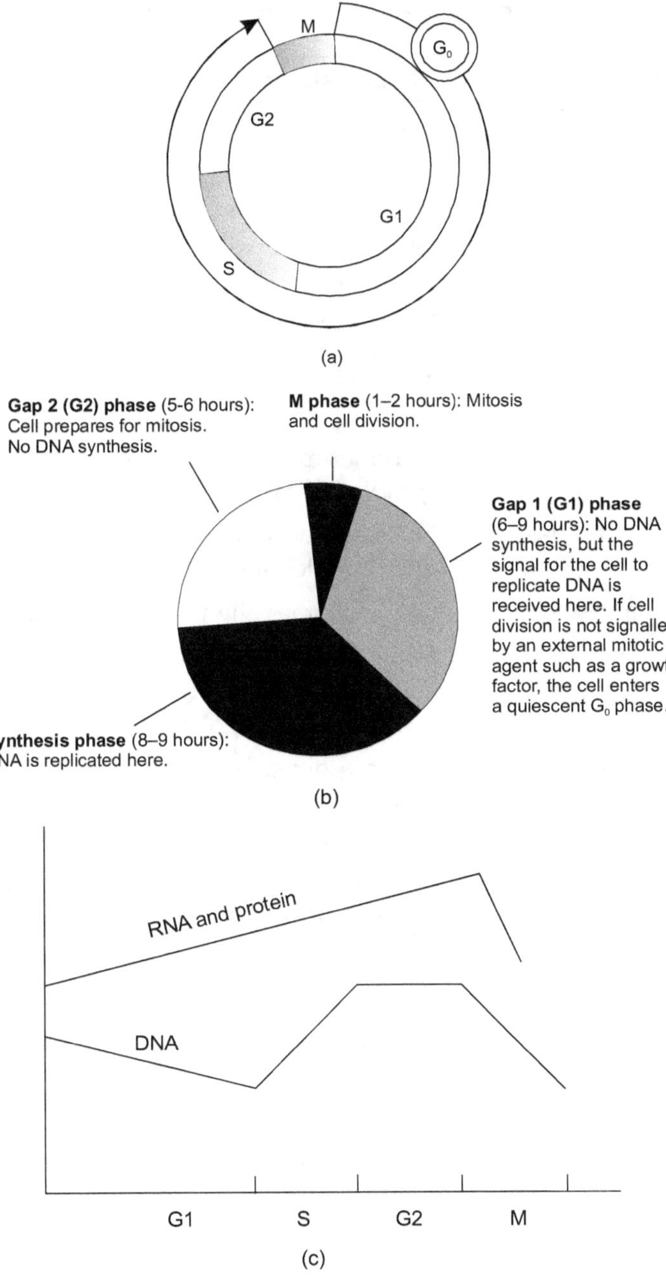

Figure 5.10   Eukaryotic cell cycle (a) Phases of cell cycle (b) Events in the cell cycle phases (c) Cellular constituents in G1, S, G2 and M phases.

Table 5.2   Stages of cell cycle—Mitosis

Stages	Major features
G$_0$ phase	Stable, nondividing period of variable length
Interphase	
G1 phase	Growth and development of the cell; G1/S checkpoint
S phase	DNA synthesis
G2 phase	Cell prepares for division; G2/S checkpoint
M phase	
Prophase	Chromosomes condense and mitotic spindle forms
Prometaphase	Nuclear envelope disintegrates, spindle microtubules anchor to kinetochores
Metaphase	Chromosomes align on the metaphase plate
Anaphase	Sister chromatids separate, individual chromosomes are formed and migrate towards spindle poles
Telophase	Chromosomes reach spindle poles, molecular envelopes re-form, condensed chromosomes relax
Cytokinesis	Division of cytoplasm occurs in plant cells, cell walls form

Table 5.3   Stages in meiosis

Stage		Major features
Meiosis I	Prophase I	Chromosomes condense, synapsis of homologous chromosomes occur, crossing over may occur, nuclear envelope breaks down, mitotic spindle forms
	Metaphase I	Homologous pairs of chromosomes line up on the metaphase plate
	Anaphase I	The two chromosomes (each with two chromatids) of each homologous pair separate and move towards opposite poles

(Contd.)

Table 5.3   (Continued)

Stage		Features
	Telophase I	Chromosomes reach the spindle poles
	Cytokinesis	Two cells are formed with cytoplasm division. Each cell has half the original number of chromosomes
	Interkinesis	In some cells, the spindle breaks down. Chromosomes relax, nuclear envelope re-forms. But there is no DNA synthesis
Meiosis II	Prophase II	Chromosomes condense, spindle forms, nuclear envelope disintegrates
	Metaphase II	Individual chromosomes line up on the metaphase plate
	Anaphase II	Sister chromatids separate and migrate as individual chromosomes towards the spindle poles
	Telophase II	Chromosomes reach the spindle pole, spindle breaks down, nuclear envelope re-forms
	Cytokinesis	Division of cytoplasm occurs

Spicy Questions and Answers

1.   Do prokaryotes have introns?

It may look that introns are just nuisances which should be removed and discarded. Bacteria, with a few exceptions, do not have introns and still they survive well. Therefore it is clear that introns are not essential for gene function. By endosymbiosis, mitochondria and chloroplasts have originated from bacteria and algae. These two cell organelles have introns whereas the prokaryotic genes do not have. This indicates that in prokaryotes, since they lack splicing mechanisms by evolutionary changes, they would have lost the introns. Alternatively, the mitochondrial and chloroplast DNA would have acquired introns from their eukaryotic hosts.

2.   What is the function of exons and introns?

Exons are usually small, which may code for 20–80 amino acids, which is bigger than the size of a stable folded amino acid domain within a protein. It suggests that exons represent functional domains in proteins. Domains are subunits of proteins. They have a definite function or structure. Some proteins have a series of domains. Individual domains need not encode a complete functional protein. They may be just sequences that could be combined with others to form a functional protein. This fact suggests that some exons encode amino acid sequences which may be similar in many structurally similar proteins like β-sheets, α-helices and DNA-binding domains. This suggests that evolution of genes can occur by adding or subtracting domains. Introns represent spaces between domains. Introns may be the regions where recombination events take place without disrupting the domain sequence. They are also referred to as "selfish DNA". Recently, it has been suggested that introns are inserted randomly and do not separate the protein domains.

3.   What is the role of introns in globin gene family?

Globins bind heme group and are involved in the transport of oxygen. There are two members of the globin protein family in hemoglobin— two α- and two β-chains. They are encoded by two different globin genes. These two genes have similar sequences and would have arisen due to duplication and diverged during evolution. There are two introns in the α- and β-globin genes. They are located in positions that divide the gene into a central, heme-binding domain and two flanking domains. Myoglobin gene also belongs to the globin gene family. The protein product is a monomer capable of binding oxygen in muscles. Its sequences show differences compared to α or β globin. Myoglobin gene also has two introns in the same place as the introns of α- and β-globin genes. Leghemoglobin genes in plants also have introns in the same place as globin genes in animals. These suggest that the heme binding domain may have originated as two domains that were combined into one domain during evolution.

4.   Why don't bacteria have introns?

It can be suggested that there is great evolutionary pressure on bacteria to complete replication quickly while competing with other bacterial species. In this race, those with smaller genome size have an advantage. So, this would have led to the shedding off of unwanted nucleotide sequences.

5. **A diploid plant cell has DNA with 2 billion base pairs. i) How many nucleosomes are present in the cell? ii) How many molecules of each type of histone protein are associated with the genomic DNA?**

    Total no. of nucleotide pairs $= 2 \times 10^9$

    No. of nucleotide pairs for 1 nucleosome $= 2 \times 10^2$

    Therefore, total no. of nucleosomes $= 2 \times 10^9 / 2 \times 10^2 = 1 \times 10^7$

    One nucleosome requires 1 H1, 2 H2A, 2 H2B, 2 H3 and 2 H4.

    Therefore, the total no. of histones required are

    $$H1 = 1 \times 10^7$$

    $$H2A, H2B, H3, H4 = 2 \times 10^7 \text{ of each type.}$$

6. **What is the effect of phosphorylation of histone molecules?**

    They will reduce the net positive charge of histones.

REVIEW QUESTIONS

1. What is a genome? How does the complexity of bacterial genes differ from that of the eukaryotic genome?

2. What are microsatellite DNA? What role do these sequences play in human disease? How are they useful in human genetics?

3. Which fraction of the genome contains the most information? Why is this true?

4. Histone proteins were isolated from chromatin and their mass was equal to that of the DNA and the molar ratio of the four types of histones was 1 :1 :1 :1 (H2A : H2B : H3 : H4) and H1 was found in half the amount. Discuss whether or not these data fit the bead- and -string model for the nucleosome.

5. Describe the essential differences or similarities between a centromere and a telomere. What will happen *in vivo* if we remove a telomere from the chromosome? What will happen if the centromere is deleted?

6. Discuss the structural arrangement of the nucleosomes.

7. Write a note on the organization of the prokaryotic chromosomes.

8. Elaborately explain the molecular organization of a eukaryotic chromosome.

9. Write short notes on

    i. Introns and exons

    ii. Euchromatin and heterochromatin

    iii. LINEs and SINEs

    iv. Minisatellites and microsatellites

10. Constitutive heterochromatin is usually found in distinct regions of nearly every eukaryotic chromosome. What structural features of the chromosome do these regions represent?

11. In which of the following would you expect to find the protein coding genes—highly repetitive, moderately repetitive or unique DNA sequences? Explain.

12. Calculate the length (in metres) of the genome of an organism that has a C-value of 1×10^{10} bp of DNA.

13. What is the correlation between the $C_0t\,{1}/{2}$ value of a mixture of DNA molecules and the concentration of individual sequences within a reaction mixture?

14. Explain the levels of coiling of a eukaryotic chromosome and compare this to how the prokaryotic chromosome is packaged.

15. Give an example each of facultative heterochromatin and constitutive heterochromatin.

16. Describe the C-value paradox and explain how it is related to DNA complexity.

17. Pseudogenes often contain many mutations than the functional genes. Why?

18. In every cell why are rRNA genes devoid of nucleosomes?

6

REPLICATION AND REPAIR OF DEOXYRIBONUCLEIC ACID

DNA occupies a unique and central place among all the biological macromolecules because it carries the genetic information. The nucleotide sequences of DNA encode the primary structures of all cellular RNAs and proteins and through enzymes (which are mostly proteins), indirectly affect the synthesis of all other cellular constituents. The information stored in DNA in the form of nucleotide sequences is transmitted from one generation of cells to the next without any error and in an uncorrupted state. DNA metabolism includes the process by which copies of DNA molecules are faithfully formed (DNA replication) and the process that affects the inherent structure of the information (DNA repair and recombination).

MODE OF DNA REPLICATION

When Watson and Crick deduced the double-helix structure of DNA, they recognized that the base-pairing specificity could provide the basis for a simple mechanism for DNA duplication. Three mechanisms were proposed for DNA replication. According to the **conservative mechanism,** the parental double helix is conserved and it directs the synthesis of a new progeny double helix. The **dispersive mechanism** explains how segments of parental and progeny strands are interspersed as a result of the synthesis and rejoining of short segments of DNA. Watson and Crick proposed the semiconservative mode of DNA replication, which suggests that each parental strand serves as template, i.e., a single strand of DNA specifies the nucleotide sequence of a new complementary strand. The parental double helix is therefore half-conserved.

The first critical experiment to identify the correct mechanism of DNA replication was performed by Matthew Meselson and Franklin Stahl in 1958 (Figure 6.1). Their results showed that the chromosome of the common colon bacillus *Escherichia coli*, a prokaryote, replicates semiconservatively.

Meselson and Stahl grew *E. coli* cells for many generations in a medium in which the heavy isotope of nitrogen ^{15}N was substituted for the normal, light isotope, ^{14}N. Nitrogen in the purine and pyrimidine bases in DNA was labelled with ^{15}N, and this DNA will have a greater density than the DNA of cells grown on medium with ^{14}N. DNA molecules with ^{14}N and those with ^{15}N (light DNA and heavy DNA) were separated by equilibrium density gradient centrifugation using caesium chloride density gradient column. The DNA molecules (that had only ^{15}N heavy isotope of nitrogen) which were extracted from *E. coli* cells grown in ^{15}N medium settled near the bottom in the "heavy" region in caesium chloride column.

Meselson and Stahl took cells that had been growing in medium containing ^{15}N for several generations, washed them to remove the medium containing ^{15}N and transferred them to medium containing ^{14}N. After the cells were allowed to grow in the presence of ^{14}N for varying periods of time, the DNAs were extracted and analysed in CsCl equilibrium density gradients. The results of their experiment established the hypothesis of semiconservative mode of replication, excluding both conservative and dispersive models of DNA synthesis. All the DNA isolated from cells after one generation of growth in medium containing ^{14}N had a density halfway between the densities of "heavy" DNA and "light" DNA. This intermediate density is referred to as "hybrid" density. After two generations of growth in medium containing ^{14}N, half of the DNA was of hybrid density and half was light. These results were in agreement with the hypothesis proposed by Watson and Crick—the semiconservative mode of replication.

One generation of semiconservative replication of a parental double helix containing ^{15}N in medium containing only ^{14}N would produce two progeny double helices, both of which had ^{15}N in one strand (the old strand) and ^{14}N in the other strand (the new strand). Such molecules would be of hybrid density.

If conservative replication is the correct mode of replication, no DNA molecules would have been produced with hybrid density. After one generation of replication of heavy DNA in light medium, half of the DNA would have been heavy with ^{15}N in both the strands and the other half would have been light with ^{14}N in both the strands.

If dispersive replication is the correct mode of replication, Meselson and Stahl would have observed a shift of the DNA from heavy towards light in each generation (half heavy or hybrid after one generation, quarter heavy after two generations and so on). These possibilities are clearly inconsistent with the results of Meselson and Stahl's experiment.

The semiconservative mode of DNA replication in eukaryotes was first demonstrated (Figure 6.2) by J. Herbert Taylor, Philip Woods and Walter Hughes in 1956. They did their experiment with the root tip cells of the broad bean, *Vicia faba*. They labelled *Vicia faba* chromosomes by growing the root tips for eight hours (less than one cell generation) in a medium containing radioactive ^{3}H-thymidine. The root tips were then removed from the radioactive medium, washed and transferred to nonradioactive medium containing the alkaloid, colchicine. Colchicine binds to microtubules and prevents the formation of functional spindle fibres. As a result, daughter chromosomes do not undergo their normal anaphase separation. Thus, the number of chromosomes per nucleus will double once per cell cycle in the presence of colchicine. This doubling of the chromosome number each cell generation allowed Taylor and his colleagues to determine how many DNA duplications each cell had undergone subsequent to the incorporation of radioactive thymidine. At the first metaphase in colchicine (c-metaphase), nuclei will contain 12 pairs of chromatids (still joined at the centromeres) and at the second c-metaphase, nuclei will contain 24 pairs and so on.

The radioactivity status of *Vicia faba* chromosomes was examined by autoradiography. Both the chromatids of each pair were similarly labelled at the first c-metaphase. But, at the second c-metaphase, only one of the chromatids of each pair was radioactive, which proved the semiconservative mode of DNA replication.

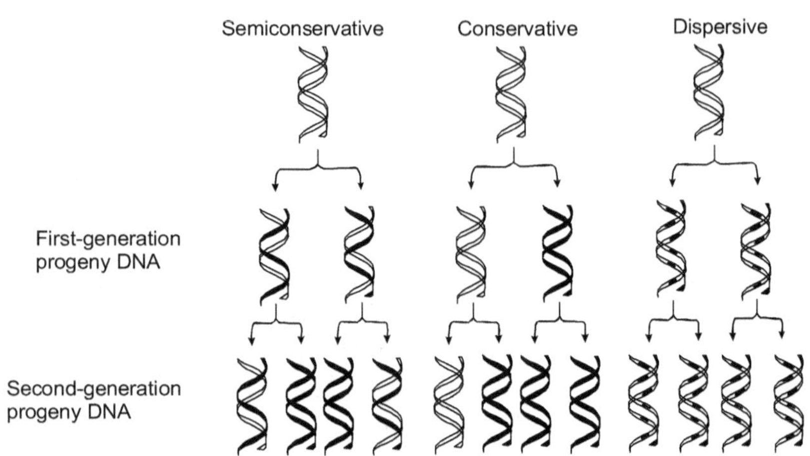

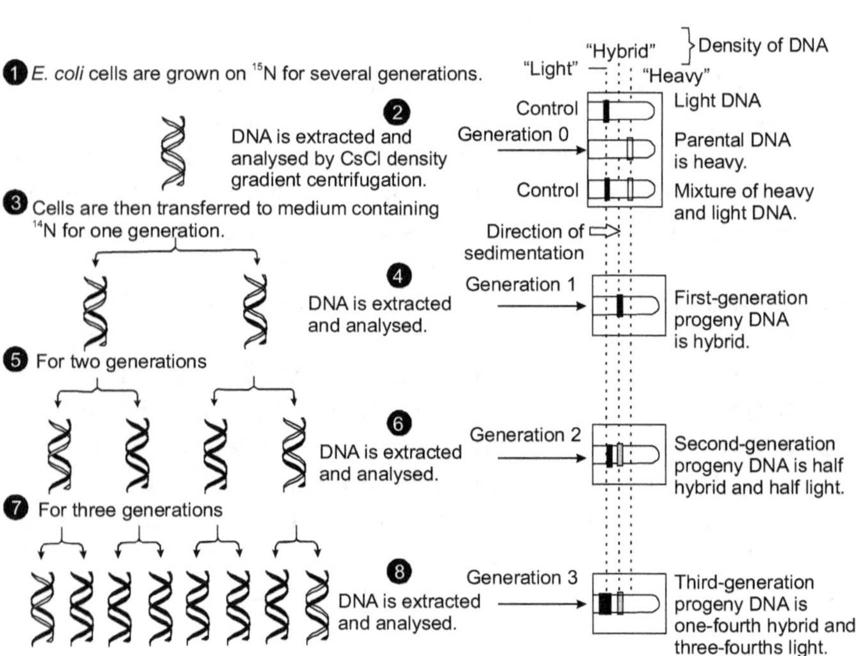

Figure 6.1　Meselson and Stahl experiment to prove the semiconservative mode of DNA replication

Autoradiographs of *Vicia faba* chromosomes

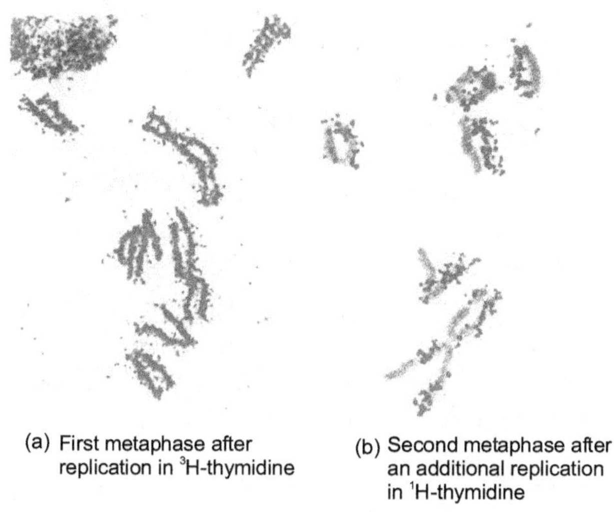

(a) First metaphase after replication in ³H-thymidine

(b) Second metaphase after an additional replication in ¹H-thymidine

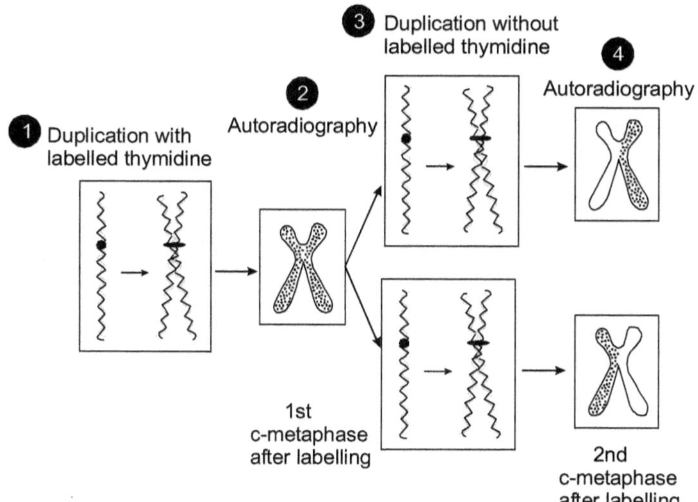

(c) Interpretation of the autoradiograph in terms of semiconservative replication

Figure 6.2   Taylor's experiment to prove the semiconservative mode of DNA replication in eukaryotes

BASIC REQUIREMENTS FOR DNA SYNTHESIS

Basic requirements for DNA synthesis in prokaryotes and eukaryotes are:

1. Parental DNA strands act as templates for the synthesis of new strands of DNA.

2. Free 3'-OH group is required opposite to the template strand, to which deoxyribonucleotides are added.

3. The four important precursor molecules—dATP, dCTP, dTTP and dGTP—collectively referred to as dNTP are incorporated opposite to the template DNA as dNMP monomers.

4. Several enzymes and proteins are required for DNA synthesis—single strand binding protein, DNA helicase, primase, DNA topoisomerases type I and II, DNA polymerase, DNA ligase, telomerase, Dna A, B, and C proteins, etc.

BOX 6.1 DNA SYNTHESIS AND BIOLOGICAL SECURITY

DNA synthesis allows the direct construction of genetic material starting from information and raw chemicals. Like any powerful technology, DNA synthesis has the potential to be purposefully misapplied. Misuse of DNA synthesis technology could give rise to both known and unforseeable threats to our biological safety and security. This is the concern of the International Consortium for Polynucleotide Synthesis (ICPS) and the US Federal Bureau of Investigation (FBI). Continued improvements in DNA synthesis technology are critical for reducing the costs and increasing the pace of basic and applied biological research and enabling the engineering of needed biological technologies. It is imperative that DNA synthesis firms develop and implement effective biological safety and security procedures. Any criminal use of synthetic DNA should be continuously monitored.

STEPS INVOLVED IN DNA SYNTHESIS

One unit of replication is referred to as a replicon which includes a start site, elongation site and a termination site on the DNA. In prokaryotes, the entire single chromosomal DNA is replicated as one replicon with a unique start site and termination site. However, in eukaryotes, there are many replicons with many start and termination sites which are thus replicated simultaneously to complete the replication of the complex eukaryotic chromosomes at a faster rate.

Replication is bidirectional at the start site which is possible by the formation of active replication forks at the site of origin of replication. DNA synthesis always proceeds from 5′ to 3′ direction, i.e., the template DNA is copied from 3′ to 5′ direction. Because the parental DNA strands are antiparallel, and because replication has to proceed only from 5′ to 3′ direction, there is continuous synthesis in one strand in the direction of the replication fork movement and discontinuous synthesis in the other strand against the direction of fork movement.

INITIATION OF REPLICATION

The first event in initiation of replication is the assembly of the replisome in both prokaryotes and eukaryotes. This is an orderly process. It begins at precise sites on the chromosome referred to as site of origin and takes place only at a certain time in the life of the cell.

Prokaryotic Origin of Replication

E. coli replication begins from a fixed origin called oriC and then proceeds in both directions until the forks merge. The first step in the assembly of the replisome is the binding of a protein called Dna A to a specific 13-bp sequence called 'Dna A box' that is repeated five times in OriC (Figure 6.3a and b).

Eukaryotic Origins of Replication

Bacteria like *E. coli* complete a replication cycle from 20–40 minutes, but in eukaryotes, the cycle can vary from 1.4 hours in yeast to 24 hours in cultured animal cells. Eukaryotic chromosomes contain multiple replication origins, because their entire genome should be copied in a reasonable time. Yeast, the simplest eukaryote is found to have 250–400 replicons and mammalian cells have as many as 25,000 replicons. They are referred to as autonomously replicating sequences (ARS). During GI phase ARS sequences are bound by a group of proteins called origin recognition complex (ORC). This results in the formation of a pre-replication complex that is accessible to DNA polymerase.

Replisome and Accessory Proteins at the Replication Fork

DNA replication in both prokaryotes and eukaryotes is by semiconservative mechanism and employs leading and lagging strand synthesis. But the replisome, the replication machinery, consists of enzymes and proteins and its components are different in both the systems. As organisms increase in complexity, the number of replisome components also increases. There are 13 components in *E. coli* replisome and at least 27 in yeast and mammals. Eukaryotic chromosomes exist in the nucleus as chromatin, the basic unit being nucleosomes which consist of DNA wrapped around histone proteins. The replisome has to not only copy the parental strands, but also disassemble the nucleosomes in the parental strands and reassemble them in the daughter molecules. The old histones from the existing nucleosome are randomly distributed to daughter molecules and new histones are delivered in association with a protein called chromatin assembly f a c t o r I (CAF-I) to the replisome. CAF-I binds to histones and targets them to the replication fork, where they can be assembled together with newly synthesized DNA. CAF-I and its associated histones reach the replication fork by binding to the eukaryotic version of the clamp protein called proliferating cell nuclear antigen (PCNA).

Replisome is a nucleoprotein complex. It is a molecular machine with many components. At the replication fork, the catalytic core of DNA pol III is actually part of a much larger complex called the pol III holoenzyme. This consists of two catalytic cores and many accessory proteins. One of the catalytic cores handles the synthesis of the leading strand and the other handles the lagging strand synthesis.

Some of the accessory proteins form a connection that bridges the two catalytic cores and coordinates the synthesis of both the leading and lagging strands. The lagging strand loops around so that the replisome coordinates the synthesis of both the strands and moves in the direction of the replication fork.

An important accessory protein called the sliding clamp encircles the DNA like a doughnut. Its association with the clamp protein keeps pol III attached to the DNA molecule. Thus, pol III is transformed from an enzyme that can add only 10 nucleotides before falling off the template (referred to as distributive enzyme) to an enzyme that stays at the moving fork and adds tens of thousands of nucleotides (a processive enzyme). Therefore, with the help of accessory proteins, both the leading and lagging strands are synthesized in a rapid and highly coordinated manner.

The primase enzyme does not bind to the clamp protein. So it acts only as a distributive enzyme, adding a few nucleotides before getting dissociated from the template.

The replisome has two classes of proteins—helicases and topoisomerases—that open the helix and prevent overwinding. Helicases are enzymes that disrupt the hydrogen bonds that hold the two strands of the double helix together. Like the clamp protein, the helicase fits like a doughnut around the DNA. From this position, it rapidly unzips the double helix ahead of DNA synthesis. The unwound DNA is stabilized by single-strand binding (SSB) proteins, which bind to single-stranded DNA and prevent the duplex from re-forming and may last from 100–200 hours in some cells. Eukaryotes have to solve the problem of coordinating the replication of more than one chromosome, as well as the problem of replicating the complex structure of the chromosome itself.

The origin of replication in simple eukaryotes like yeast are very much like oriC in *E. coli*. They too have AT-rich regions that melt when an initiator protein binds to adjacent binding sites. They are longer with thousands or tens of thousands of nucleotides. Unlike prokaryotic chromosomes, their chromosomes have many replication origins so that the much larger eukaryotic genome is replicated quickly. Approximately 400 replication origins are dispersed throughout the 16 chromosomes of yeast. There are thousands of growing forks in the 23 chromosomes of the humans.

Dna A Protein

A gene called *dna A* produces the Dna A protein which is the prime candidate that interacts with oriC to initiate replication (Figure 6.3). *E. coli* oriC is an ~240-bp DNA segment at the start site for replication of *E. coli* chromosomal DNA. Plasmids or any other circular DNA containing oriC are capable of independent and controlled replication. oriC has a repetitive four 9-bp sequence and three AT-rich, 13-bp sequence referred to as 9-mers and 13-mers respectively.

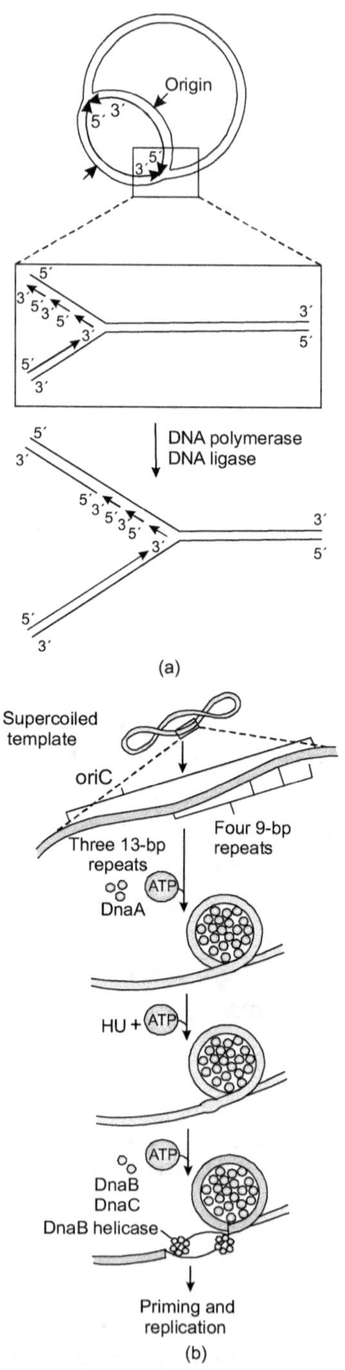

Figure 6.3 (a) Origin of replication (b) Events at origin of replication

Dna A protein binds to the four 9-mers in oriC and forms an initial complex which contains 10–20 protein subunits. Although Dna A can bind to dsDNA of *E. coli* at the origin in the relaxed circle form, it can initiate replication only if the DNA is negatively supercoiled, because only negative supercoils are tightly wound and are easier to melt locally.

Binding of Dna A to the oriC 9-mers facilitates the initial strand separation or melting of *E. coli* duplex DNA which occurs at the oriC 13-mers. This process requires ATP and yields an open complex. In response to this binding, the origin is unwound at a cluster of A and T nucleotides, because it is easier to separate or melt the double helix at stretches of DNA that are rich in AT base pairs.

After unwinding begins, additional Dna A proteins bind to the newly unwound single-stranded regions. Two helicases—Dna B proteins— now bind and slide in a $5' \rightarrow 3'$ direction and unzip the helicase at the fork.

Primases and DNA pol III holoenzymes are now recruited to the replication fork by protein–protein interaction, and DNA synthesis begins.

Dna B Protein

Further melting of the two strands of the *E. coli* chromosome to generate unpaired template strands is mediated by the protein product of dna B locus, a helicase. One molecule of Dna B, a hexamer of identical subunits, clamps around each of the two single strands in the open complex formed between Dna A and oriC. This binding requires ATP and a protein encoded by dna C locus which takes Dna B to Dna A. This generates the prepriming complex.

Helicases are a class of enzymes that can move along a DNA duplex utilizing the energy of ATP hydrolysis to separate the strands. In *E. coli*, the separated strands are prevented from reannealing by single-strand binding protein (SSB protein). Dna B binds to a single-stranded segment of DNA and moves along that strand melting the hydrogen bonds that link this strand to its complementary strand. Like many proteins that bind to DNA, helicases exhibit a directionality with respect to the unwinding reaction. Dna B moves along ssDNA to which it binds in the direction of its free 3' end and unwinds the DNA in the

$5' \rightarrow 3'$ direction. There are other helicases also which can unwind in the opposite direction. Like many proteins that act on DNA, helicase action of Dna B protein is said to possess processivity (the average number of nucleotides added before a polymerase dissociates is known as the **processivity** of the enzyme). Because it forms a clamp around a single-stranded DNA, it does not fall off till it reaches the end of that strand or is unloaded from DNA by another protein.

Primases

Primases catalyse the formation of RNA primers. The gene locus dna G codes for the enzyme primase, an RNA polymerase. In both prokaryotes and eukaryotes, a short RNA molecule is first synthesized opposite to the template strand as a primer molecule that will provide growing 3'-OH ends for the new DNA strand. Primase is usually recruited to a segment of single-stranded DNA by first binding to the Dna B hexamer that is already attached to that site.

Single Strand Binding Proteins

Single strand binding proteins (SSB) rapidly bind to the newly separated single-stranded DNA and maintain it in a single-stranded state until that portion of DNA is replicated. SSBs are also referred to as cooperative tetramers because four protein subunits form a tetramer that binds to ssDNA. Binding of one tetramer facilitates the binding of an adjacent SSB tetramer. This process rapidly continues from one SSB to the next until all the ssDNA is bound with SSBs.

Topoisomerases

In the cell, DNA is not free to rotate on its own long axis. In *E. coli*, the closed circular chromosome clamps the DNA. In eukaryotes, DNA is very long and is arranged in fixed loops. Also it is attached to several proteins. Once again free rotation becomes impossible. But separation of DNA strands demands that the duplex rotates. This causes overwinding, creating positive supercoils ahead of the replication fork and as the helix tightens, further separation is resisted. If unrelieved, a tension is created halting strand separation and DNA replication.

A group of enzymes known as topoisomerases catalyse the removal of positive supercoils ahead of the replicative fork. They act on DNA and isomerize or change its topology.

There are two classes of topoisomerases—types I and II.

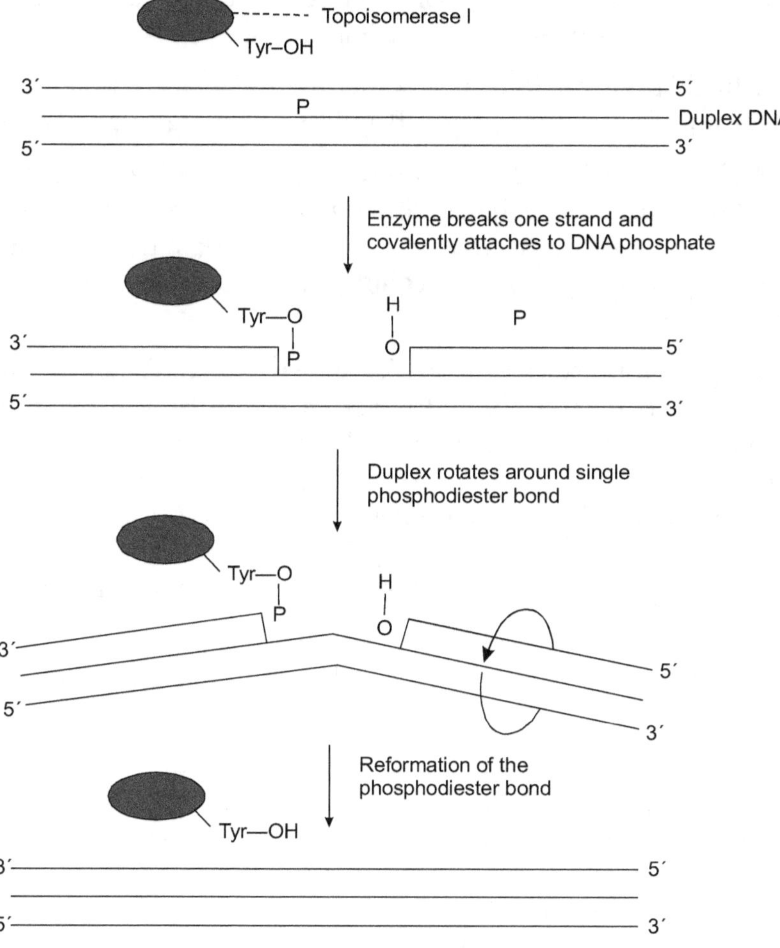

Figure 6.4   Reactions catalysed by topoisomerase I

In type I, the enzyme breaks one strand of a supercoiled duplex, which permits the whole duplex to rotate on the single phosphodiester bond of the partner strand and this introduces a swivel into the DNA. After rotation has occurred, the enzyme reseals the duplex. The enzyme does not hydrolyse the phosphodiester bond it attacks;

it simply transfers the bond from the deoxyribose 3′-OH to the OH of one of its own tyrosine side chains. This process is reversible, because there is no energy loss in the cutting of the chain. This enzyme does not use ATP. A supercoiled DNA molecule, whether it is positively or negatively supercoiled, is in a state of tension and is at a higher energy level than the relaxed state. Topoisomerase I can only relax a supercoiled DNA, it cannot insert supercoiling (Figure 6.4).

Actually there are four topoisomerases in *E. coli*. Type I and III are both grouped as type I, and type II and IV are both grouped as type II.

In *E. coli*, topoisomerase I can relax only negatively supercoiled DNA. So it cannot solve the unwinding problem. The relaxation of the positive supercoiling ahead of the replication fork in *E. coli* is done in a different way—negative supercoiling is inserted by topoisomerase II called DNA gyrase.

A type II topoisomerase breaks two strands of the DNA double helix transferring the bonds to itself, making the breakage of the polynucleotide chains a freely reversible process. The enzyme physically transfers the DNA duplex of a coil through the double gap for which ATP is required. In *E. coli*, topoisomerase II introduces negative supercoil in the DNA. Insertion of negative supercoiling is equivalent to relaxation of positive supercoiling. A positive supercoil is converted into a negative supercoil.

The physical movement of the DNA through the gap is driven by ATP hydrolysis.

In both *E. coli* and eukaryotes, topoisomerase I cuts one DNA strand whereas topoisomerase II cuts two DNA strands which is an ATP-dependent process. Topoisomerase I in *E. coli* relaxes negative supercoils whereas in eukaryotes it relaxes positive and negative supercoils. Topoisomerase II in *E. coli* relaxes positive supercoils and inserts negative supercoils thereby separating interlinked circles. In eukaryotes, topoisomerase II relaxes positive supercoils but cannot insert negative supercoils (Figure 6.5).

The antibiotic nalidixic acid acts on DNA gyrase in *E. coli*. Eukaryotes do not show this inhibition and therefore nalidixic acid is used to treat certain infections in humans. In thermophilic bacteria living at extremely high temperatures, a "reverse gyrase" is seen.

The reason is that positively supercoiled DNA is less likely to undergo strand separation that occurs at high temperatures.

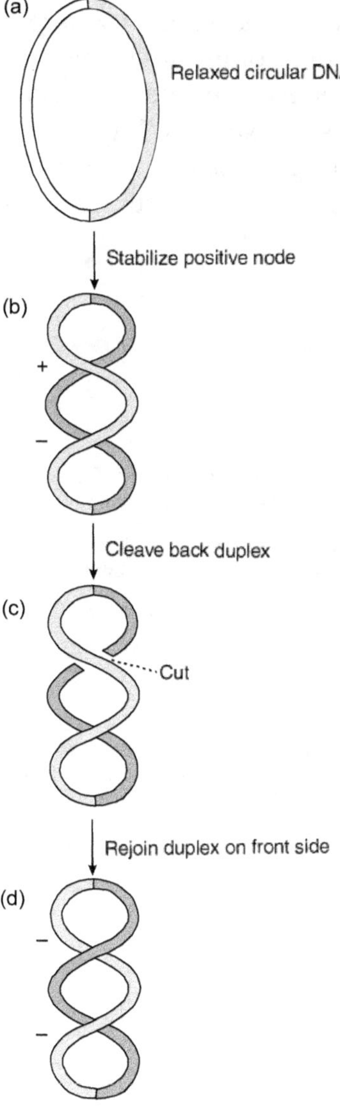

Figure 6.5   Reactions catalysed by *E. coli* DNA gyrase to insert negative supercoils into circular DNA

Eukaryotic DNA, like that of prokaryotes, is underwound or negatively supercoiled in the cell. But eukaryotic topoisomerase cannot insert negative supercoil in DNA. Then how is this achieved? During the assembly of chromatin, DNA is wound around the nucleosomes in such a manner that in the local region where it is in contact with the proteins, it is in an underwound state. This is compensated for by the positive supercoiling elsewhere. The eukaryotic topoisomerases I or II will now relax the positive supercoiling, achieving the insertion of a negative supercoil (Figure 6.6). Since prokaryotes do not have nucleosomes, this type of gyrase activity in eukaryotes differs from that of prokaryotes.

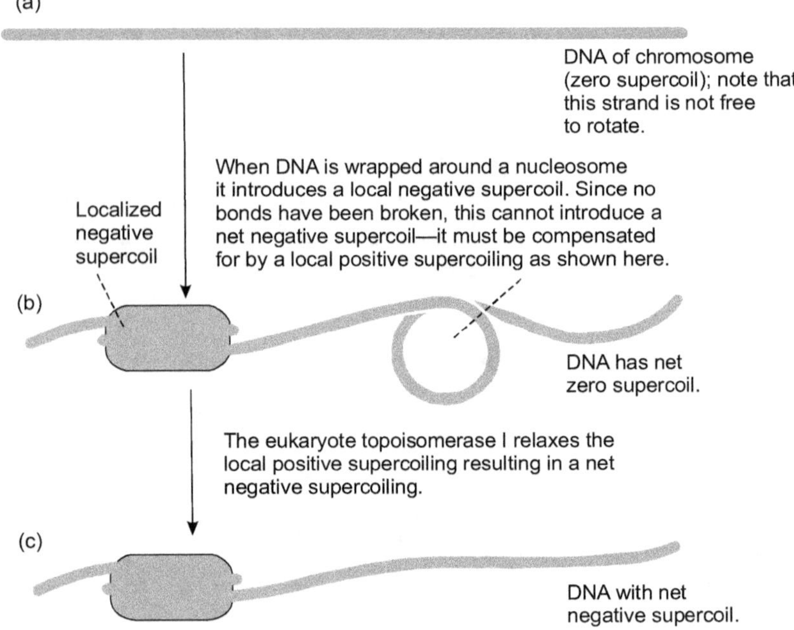

Figure 6.6   Mechanism by which eukaryotic DNA becomes negatively supercoiled despite the absence of any enzyme to insert negative supercoils as prokaryotic gyrase

DNA Polymerases—Key Enzymes for DNA Synthesis

In 1955, Arthur Kornberg and his colleagues purified and characterized an enzyme from *E. coli* cells referred to as DNA polymerase I.

Later four other types, DNA polymerases II–V were identified in *E. coli.* DNA polymerases have three catalytic functions.

Polymerization reaction   The fundamental reaction catalysed by this enzyme is a nucleophilic attack by the 3′-OH group of the nucleotide at the 3′ end of the growing strand on the 5′-α-phosphorus of the incoming deoxynucleoside 5′-triphosphate. Inorganic pyrophosphate is released in the reaction. The reaction is

$$(\text{d NMP})_n + \text{dNTP} \xrightarrow{\text{DNA polymerase}} (\text{d NMP})_{n+1} + \text{PPi}$$

$$\underset{\text{DNA}}{} \qquad\qquad \underset{\text{DNA lengthened by one nucleotide}}{}$$

This reaction has an apparent thermodynamic balance in that one phosphodiester bond is formed and one is destroyed. But, non-covalent base stacking and base-pairing interactions help to stabilize the lengthened DNA product relative to the free nucleotide. Subsequent hydrolysis of the pyrophosphate that is formed by the enzyme pyrophosphatase favours the formation of the product and it favours polymerization.

An important requirement for this reaction is a template. The polymerization reaction is guided by a template (parental) DNA strand which dictates the sequence of bases to be added one by one during synthesis. The second requirement is a primer. A primer is a segment of strand (complementary to the template) with a free 3′-OH group to which a nucleotide can be added. The free 3′ end of the primer is called the primer terminus. All DNA polymerases can only add nucleotides to a pre-existing strand. Primers are often oligonucleotides of RNA. These are synthesized by the specialized enzymes primases whose action has already been discussed.

After adding a nucleotide to a growing DNA strand, DNA polymerase either dissociates or moves along the template and adds another nucleotide. Dissociation and reassociation of the polymerase can limit the overall polymerization rate. This process will be faster if the enzyme adds nucleotides without getting dissociated from the template. DNA polymerases vary in this processivity from a few nucleotides to many thousands.

3′ → 5′ Exonuclease activity of DNA polymerase   All DNA polymerases have 3′ → 5′ exonuclease activity. This double-checks each nucleotide after it is added. If a polymerase has added a wrong nucleotide, its forward movement to add another nucleotide is inhibited when a mismatch base pair is formed. The $3' \rightarrow 5'$ exonuclease activity removes the mispaired nucleotide and the polymerase begins its action again. This is called proofreading or editing function of the enzyme. This function of the polymerase is not simply the reverse of the polymerization reaction because pyrophosphate is not involved in this reaction. In the monomeric DNA polymerase I, the polymerizing and proofreading activities have separate active sites within the same polypeptide.

$5' \rightarrow 3'$ *Exonuclease activity of DNA polymerase*   DNA polymerase activity is essential not only in replication, but also in recombination and DNA repair. The enzyme also has a $5' \rightarrow 3'$ exonuclease activity. This activity is located in a structural domain that can be separated from the enzyme by mild protease treatment. When the $5' \rightarrow 3'$ exonuclear domain is removed, the remaining fragment is called the large or **Klenow fragment**.

DNA synthesis involves two processes—leading strand synthesis and lagging strand synthesis. This is because synthesis should take place in $5' \rightarrow 3'$ direction and at the same time, the two parental strands acting as templates are antiparallel. To solve this problem, one parental strand is copied in one direction, when the new strand is referred to as leading strand and the other parental strand is copied in the opposite direction when the newly synthesized strand is referred to as the lagging strand.

During the synthesis of the lagging strand, polymerase III extends the growing strand until the RNA of the previously synthesized precursor fragment is reached. Where the DNA and RNA segments meet, there is a single-stranded interruption or nick. *E. coli* DNA ligase cannot seal the nick because a triphosphate is present. It can link only a 3'-OH and a 5'-monophosphate. Here, DNA polymerase, with its $5' \rightarrow 3'$ exonuclease activity, removes the nucleotides from the 5' end of a base-paired fragment. This activity is effective both with DNA and RNA acting at the nick and displaces the nick in $5' \rightarrow 3'$ direction by removing RNA nucleotides (primers) one by one and adding DNA nucleotides to the 3' end of the DNA strand. This process is known as nick translation Table 6.1 gives the list of enzymes involved in DNA replication in *E.coli*.

Table 6.1   Enzymes involved in DNA replication in *E. coli*

Enzyme/Protein	Gene locus	Function
DNA polymerase I	*pol A*	Gap filling and primer removal
DNA polymerase II	*pol B*	Damaged template replication
DNA polymerase III		
α subunit	*dna E*	Polymerization $5' \rightarrow 3'$ polymerase
ε subunit	*dna Q*	Polymerization $3' \rightarrow 5'$ exonuclease
θ subunit	*bol E*	Polymerization
β subunit	*dna N*	Processivity clamp (as a dimer)
τ subunit	*dna X*	Preinitiation complex
γ subunit	*dna X*	Preinitiation complex
δ subunit	*bol A*	Processivity
δ' subunit	*bol B*	Processivity
χ subunit	*bol C*	Processivity
ψ subunit	*bol D*	Processivity
Helicase	*dna B*	Primosome, unwinds DNA
Primase	*dna G*	Primosome, creates Okazaki fragments primers
Initiator protein	*dna A*	Binds at oriC
DNA ligase	*lig*	Closes Okazaki fragments
SSB protein	*ssb*	Binds ssDNA
DNA topoisomerase I	*top A*	Relaxes supercoiled DNA
DNA topoisomerase II (DNA gyrase)		
α subunit	*gyr A*	Relaxes supercoiled DNA, ATPase
β subunit	*gyr B*	Relaxes supercoiled DNA
Topoisomerase IV	*par E*	Unconcatenates DNA circles
Termination protein	*tus*	Binds at termination sites
RNase H		Removes RNA primers after they have completed their function
HD protein		Prevents immediate rewinding of unwound DNA

Characteristics of Prokaryotic and Eukaryotic DNA Polymerases

The types and features of prokaryotic DNA polymerases are given in Table 6.2.

Table 6.2   Characteristics of DNA polymerases in *E. coli*

DNA polymerase	$5' \rightarrow 3'$ polymerziation	Exonuclease		Function
		$5' \rightarrow 3'$	$3' \rightarrow 5'$	
I	✓	✓	✓	Removes and replaces primer.
II	✓	✓	✗	DNA repair; restarts replication after damaged DNA halts synthesis.
III	✓	✓	✗	Elongates DNA.
IV	✓	✗	✗	DNA repair.
V	✓	✗	✗	DNA repair, translesion DNA synthesis.

Eukaryotic cells contain a number of different DNA polymerases (Table 6.3) that function in replication, recombination and DNA repair. DNA polymerase α, which contains primase activity, initiates nuclear DNA synthesis by synthesizing an RNA primer followed by a short string of DNA nucleotides. After DNA polymerase δ has laid down from 30 to 40 nucleotides, DNA polymerase δ completes replication on the leading and lagging strands. DNA polymerase β does not participate in replication but is associated with the repair and recombination of nuclear DNA. DNA polymerase γ replicates mitochondrial DNA.

A γ-like polymerase also replicates chloroplast DNA. Similar in structure and function to DNA polymerase δ, DNA polymerase Σ is involved in nuclear replication of both the leading and lagging strands.

There are other DNA polymerases (ζ, η, κ, λ, μ) which allow replication to bypass damaged DNA, referred to as **translesion replication,** or play a role in DNA repair.

Table 6.3 DNA polymerases in eukaryotic cells

DNA polymerase	$5' \rightarrow 3'$ polymerase activity	$3' \rightarrow 5'$ exonuclease activity	Cellular function
α (alpha)	✓	×	Initiation of nuclear DNA synthesis and DNA repair
β (beta)	✓	×	DNA repair and recombination of nuclear DNA
γ (gamma)	✓	✓	Replication of mitochondrial DNA
δ (delta)	✓	✓	Leading and lagging strand synthesis of nuclear DNA, DNA repair and translesion DNA synthesis
Σ (epsilon)	✓	✓	Unknown, probably repair and replication of nuclear DNA
ζ (zeta)	✓	×	Translesion DNA synthesis
η (eta)	✓	×	Translesion DNA synthesis
θ (theta)	✓	×	DNA repair
i (iota)	✓	×	Translesion DNA synthesis
κ (kappa)	✓	×	Translesion DNA synthesis
λ (lambda)	✓	×	DNA repair
μ (mu)	✓	×	DNA repair
σ (sigma)	✓	×	Nuclear DNA replication, DNA repair

LEADING AND LAGGING STRAND SYNTHESIS

Leading Strand Synthesis

At the replication fork, the $3' \rightarrow 5'$ side of parental DNA strand is copied by continuous synthesis initiated by one RNA primer and it proceeds forward in the direction of movement of the replication fork.

Lagging Strand Synthesis

The $5' \rightarrow 3'$ side of the parental DNA strand is copied by discontinuous synthesis. Short stretches of DNA are attached to RNA primers on the lagging strand and are called **Okazaki fragments** after their discoverer Reiji Okazaki. This still leaves a problem of how a DNA polymerase can synthesize DNA backwards while moving forward. This is a physical or topological problem.

Actually, the lagging template strand is looped (Figure 6.7), so that for a short distance it is oriented with the same polarity as the leading strand template. The replicative machinery can therefore proceed in the direction of the fork and synthesize both new strands. Every time a new RNA primer is synthesized, the loop is reformed and is enlarged at regular intervals. This model was explained by Arthur Kornberg.

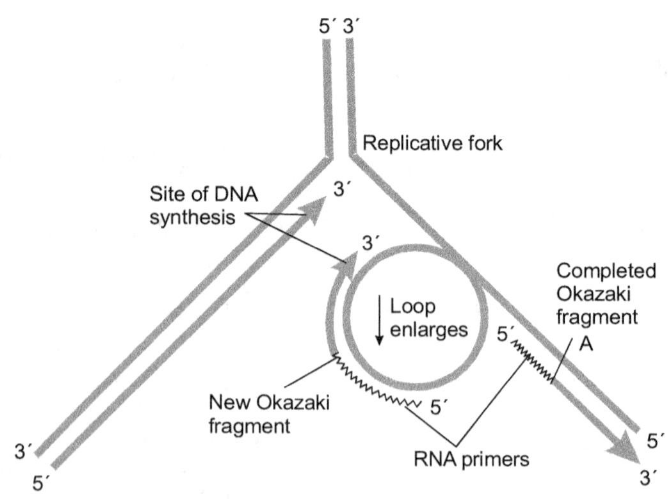

Figure **6.7**   Loop model for Okazaki fragment synthesis

As the loop enlarges and the replicative machinery moves forward, the polymerase will meet the 5′ RNA end of the previous Okazaki fragment. At this point, the polymerase must detach so that the loop falls away and a new one starts.

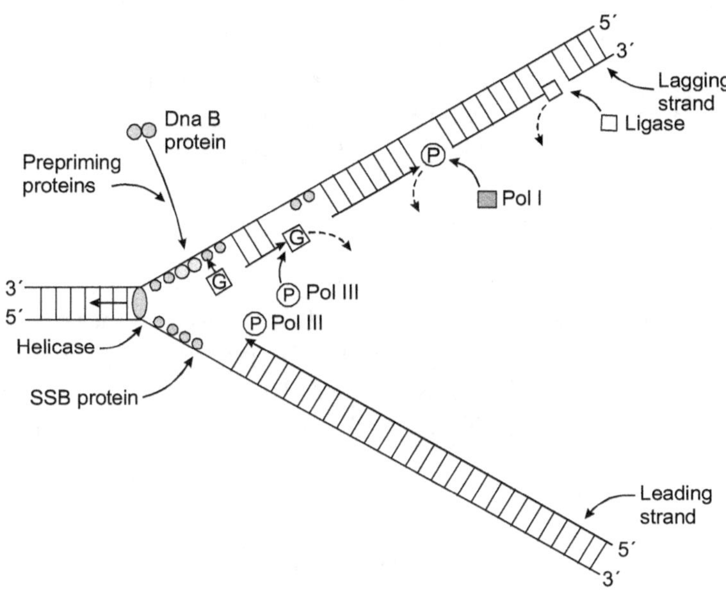

Figure 6.8   Fork model for Okazaki fragment synthesis

In *E. coli*, there are two connected molecules of pol III in the replicative fork (Figure 6.8), one synthesizing the leading strand and the other the lagging strand. They have the same core enzyme, but the holoenzyme dimer is asymmetric with extra subunits present on the lagging strand side. Pol III has high processivity—once it locks onto a DNA strand, it does not fall off but can go on replicating its template strand indefinitely without dissociating. A special mechanism operates to guard against premature dissociation. A ring-shaped protein structure surrounds the DNA that acts as a sliding clamp. The annulus has a hole big enough for double-stranded DNA to slide through it easily but it cannot fall from the DNA. In *E. coli* it is referred to as β protein and in eukaryotes as proliferating cell nuclear antigen (PCNA), a protein in the nucleus required for proliferation. In *E. coli*, β protein is a dimer whereas PCNA has three subunits. An additional protein complex is required to load the clamps onto the DNA present both in *E. coli* and eukaryotes. In *E. coli*, it is known as γ complex. It has a bound ATP molecule and recognizes RNA primer/DNA hybrid region. It attaches a sliding clamp around it. ATP hydrolysis occurs and the loading protein is released.

Because one DNA strand is synthesized continuously and the other discontinuously, DNA replication is said to be **semidiscontinuous synthesis**.

The fragments of DNA formed by discontinuous synthesis of the lagging strand are 1000–2000 nucleotides long in prokaryotes and are much shorter, i.e., 100–200 nucleotides in eukaryotes.

Nick Sealing by DNA ligase

DNA ligase catalyses the formation of a phosphodiester bond between a 3′ hydroxyl at the end of one DNA strand and a 5′ phosphate at the end of the other strand (Figure 6.9). The phosphate group must be activated by adenylation. DNA ligases from viruses and eukaryotes use ATP, and bacteria use NAD^+ as the source of AMP activating group.

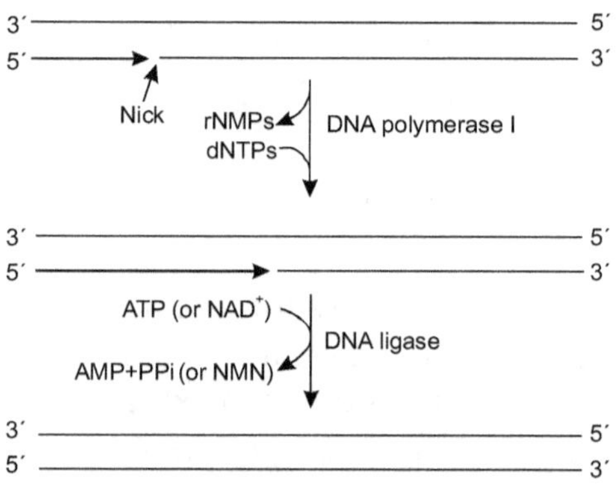

Figure 6.9   Action of DNA ligase

The RNA primers in the lagging strand are removed by the $5' \rightarrow 3'$ exonuclease activity of DNA polymerase I and are replaced with DNA by the same enzyme. The resulting nick is sealed by DNA ligase.

At the nick, the 5′ phosphate is activated. An AMP group (from ATP or NAD^+) is first transferred to a lysine residue of the DNA ligase enzyme and then transferred to the 5′ phosphate at the nick. Then the 3′-hydroxyl group at the other end of the nick, attacks this phosphate and displaces AMP, producing a phosphodiester bond and seals the nick. When the energy source is ATP as in viruses and eukaryotes, during adenylation of the enzyme, DNA ligase pyrophosphate is released. When the energy source is NAD^+ as in prokaryotes, nicotinamide mononucleotide (NMN) is released (Figure 6.10).

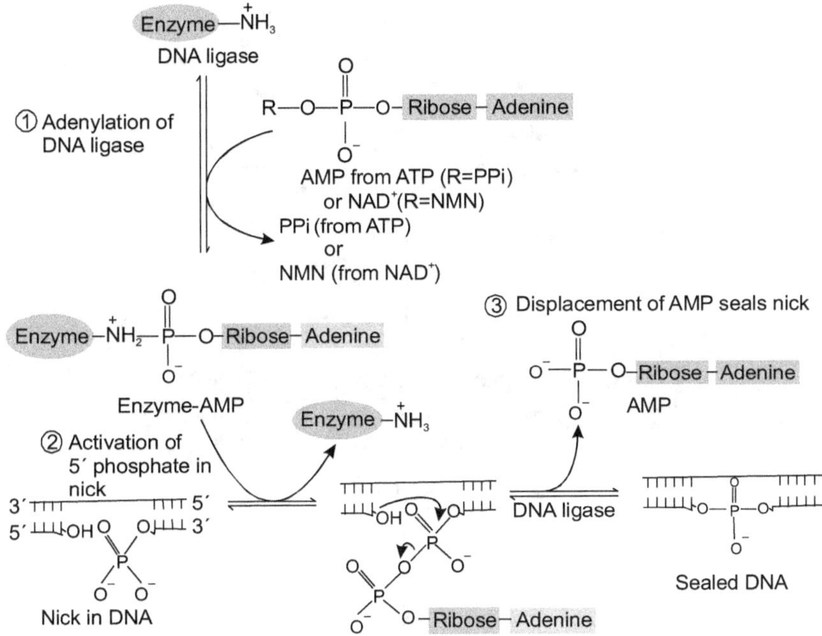

Figure **6.10**   Mechanism of action of DNA ligase

TERMINATION OF REPLICATION

If replication is unidirectional, it terminates at the site where it had originated. It is not so in the case of bidirectional replication. There are two modes of termination:

1. there is a defined termination sequence or

2. the two oppositely moving replication forks meet at a point where termination occurs.

In both the cases, termination occurs halfway around the circle in case of circular DNA molecules. But there is a topological problem in termination. When double-stranded circular DNA replicates semiconservatively, a pair of circular DNA molecules which are interlinked as in a chain is obtained. This structure is called a **catenane**. This molecule is **decatenated** by the enzyme DNA gyrase (Figure 6.11).

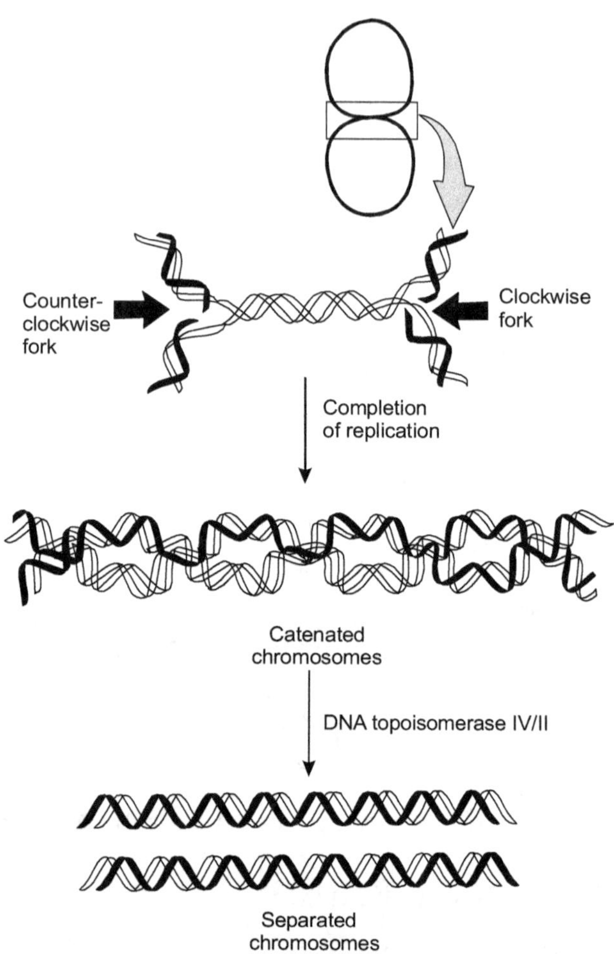

Figure 6.11   Decatenation by DNA gyrase

In SV40 viral DNA, decatenation requires a particular termination site. Such a topological problem does not arise in linear DNA molecules.

Termination of Replication in *E. coli*

In *E. coli* replication terminus is a large region with 350 kb flanked by six identical nonpalindromic ~23-bp terminator sites—Ter E, Ter D and Ter A on one side and Ter F, Ter B and Ter C on the other. A replication fork which moves counterclockwise passes through Ter F, Ter B and Ter C but it stops at Ter A or Ter D or Ter E. In a similar way, a clockwise travelling replication fork transits Ter E, Ter D and Ter A but halts at Ter C or Ter B or Ter F (Figure 6.12). These terminator sites are polar. They act like one-way valves allowing the replication forks to enter the terminus region but not to leave it. Because of this arrangement, the two replication forks which have originated at oriC moving bidirectionally will meet in the terminus site. The arrest of the fork motion at Ter sites require the action of Tus protein, a 309-residue monomer, a product of the *tus* gene (termination utilization substance). Tus protein binds to a Ter site, prevents strand displacement by Dna B helicase, thereby arresting replication fork movement.

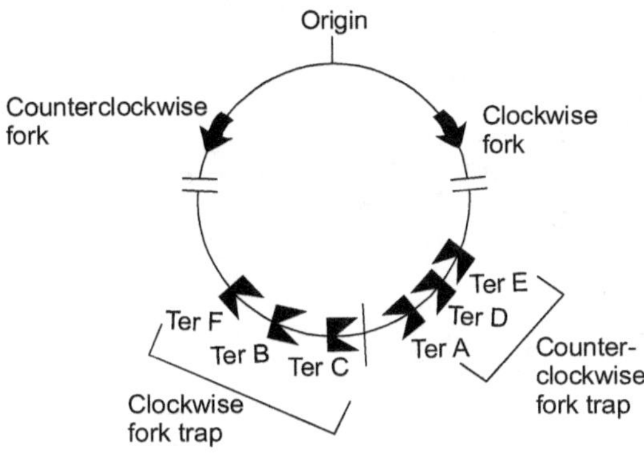

Figure 6.12　Site of origin and termination of replication in *E. coli* chromosome

Termination of Replication in Eukaryotes

Eukaryotic chromosomes are linear and they have specialized ends called telomeres. Telomeres consist of repeated oligomeric sequences TxGy in one strand and CyAx in the complementary strand x and y = 1 to 4). Yeast telomeric repeat sequence is 5′ G$_{(1-3)}$T 3′. This region is very important to overcome the problem of shortening of chromosome length during every round of replication. DNA template strands are synthesized from their 3′ end because replication has to proceed in the 5′ → 3′ direction with one RNA primer in the leading strand and many RNA primers in the lagging strand (Okazaki fragment synthesis). After all the RNA primers are removed, the gaps created by their removal are filled by DNA polymerase because of the availability of 3′-OH groups in the preceding ends (in the lagging strand). However, the gaps obtained by the removal of the single RNA primer in the leading strand and by the removal of the first RNA primer in the lagging strand cannot be filled by DNA polymerase because of the unavailability of a 3'-OH group prior to their starting site (Figure 6.13).

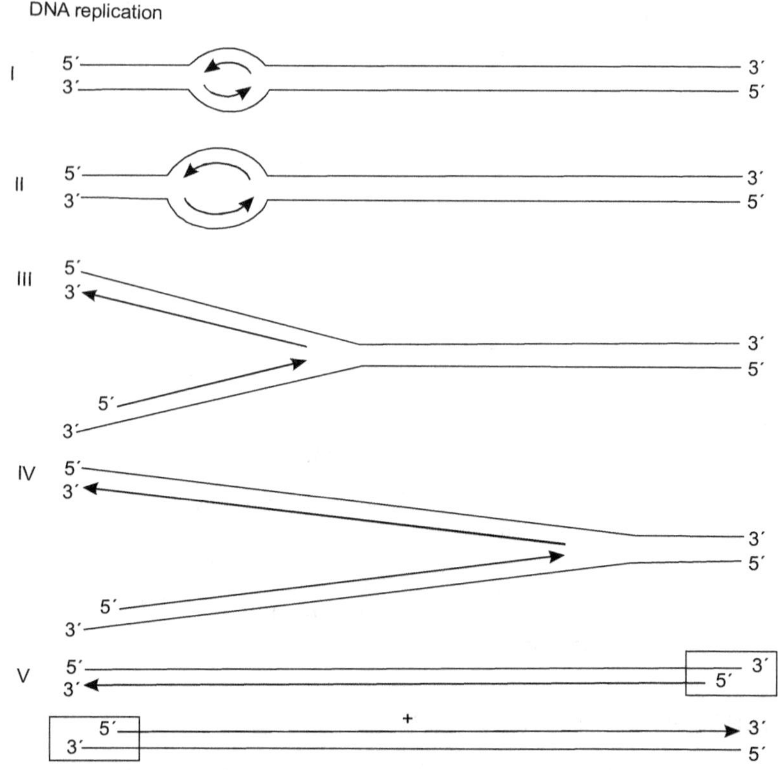

Figure 6.13   Termination in eukaryotes (Gaps denoted by boxes in the 5' end of the new strands)

If this problem is not solved, in the next round of replication, the shortened new DNA strand will act as one of the templates and during replication will lead to further shortening of the new strand synthesized.

This progressive shortening of the chromosome during replication is prevented by a modified reverse transcriptase enzyme called **telomerase** which can fill the gaps in the 5' end of the new daughter DNA strands. This enzyme is a ribozyme, that is, it is a ribonucleoprotein which has an internal RNA template that can base-pair to the 3'-end of the lagging and leading strand templates. The telomerase catalytic site adds dNTPs as dNMPs using the RNA molecule as the template.

The RNA template is about 150 nucleotides long and has about 1.5 copies of the appropriate CyAx telomerase repeat. This region of the RNA acts as a template for the synthesis of the TxGy strand of the telomerase by RNA-dependent DNA synthesis. Telomere synthesis requires the 3′ end of a chromosome as primer and proceeds in the usual 5′ → 3′ direction. After synthesizing one copy of the repeat, the enzyme again repositions to resume extension of the telomere (Figure 6.14).

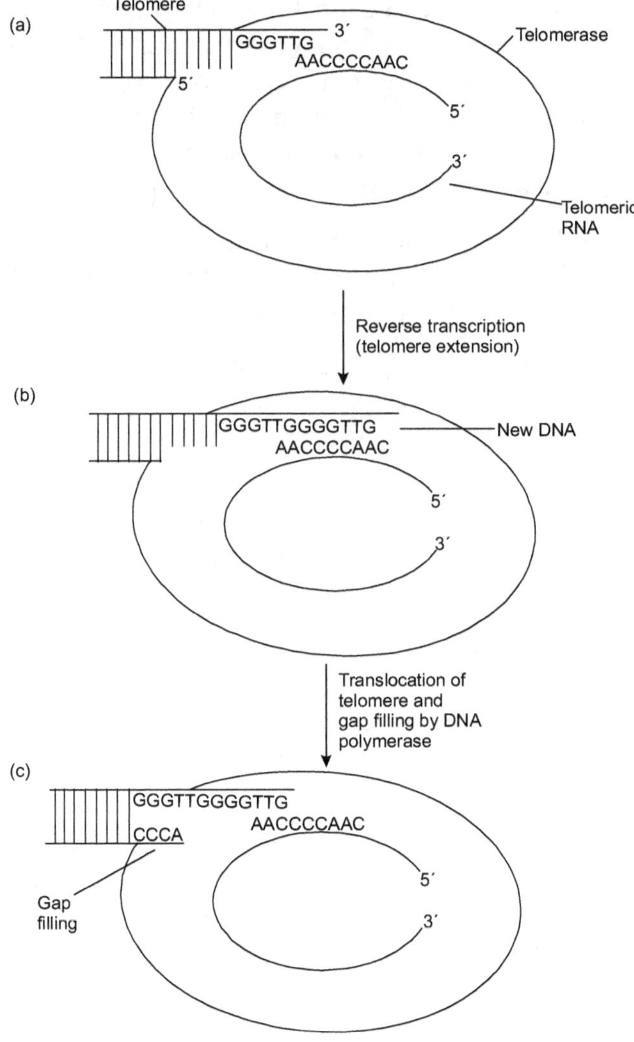

Figure 6.14   Action of telomerase

The complementary C*yAx* strand is synthesized by the DNA polymerase starting with an RNA primer. This single-stranded region is protected from the action of nucleases by specific binding proteins in lower eukaryotes, which have few hundred base pair telomeres. In higher eukaryotes which have many thousand base pair-long telomeres, the single-stranded end forms a T loop structure, folding back and pairing with its complement in the double-stranded portion of the telomere. Formation of T loop involves invasion of the 3′ end of the telomere's single strand into the duplex DNA similar to the initiation of homologous genetic recombination. In mammals, the looped DNA is bound by two proteins, telomere repeat binding factors, TRF 1 and TRF 2. TRF 2 is involved in the formation of T loop. T loop protects the 3′ ends of the chromosomes from nucleases and DNA repair enzymes that act on double-strand breaks.

The extended telomeric ends of the chromosome are not only protected by the novel T loop structure as in mammals, but also by other mechanisms in other organisms.

In some organisms, four guanine residues in the extended region form a complex structure. They form a planar G-tetraplex where the four guanine residues in adjacent positions are hydrogen-bonded to one another.

In the ciliate *Oxytricha nova*, a protein called telomere end-binding protein (TEBP) gets attached to the 3′ ends of telomeres and protects them.

The cells have a special mechanism to keep track of the number of their telomeric repeats. The cells count the number of proteins like TRFI in humans and Rap I in *Saccharomyces cerevisiae* which are bound to the telomeres to determine the number of telomeric repeats to be added.

In humans a link between telomere length and cell senescence (stop of cell division) is observed. In the germ-line cells, which contain telomerase activity, telomere lengths are maintained. In the somatic cells which lack telomerase, the length is not maintained.

In yeast, protozoa and other single-celled organisms, telomerase is active, keeping the ends of the chromosomes at the appropriate lengths. But in most cells of higher organisms, telomerase is not active and the ends of the chromosomes get shorter with each cell division. At a particular telomeric length, there is no further cell division. However, if telomerase becomes active and the ends of the chromosomes lengthen, a signal is conveyed to keep the cells dividing. This may lead to cancerous growth. Therefore telomerase inhibitors are being used in clinical therapeutics to treat cancer. Further, an analysis of normal telomerase shortening acts as a biological clock that may help to assess the aging process and senescence.

DNA REPLICATION AND EUKARYOTIC CELL CYCLE

DNA synthesis takes place only in the S phase of the eukaryotic cell cycle. The onset of DNA synthesis is limited to this single stage through regulation of cell cycle.

In yeast, three proteins are required to begin assembly of the replisome. The origin recognition complex (ORC) first binds to sequences in yeast origin of replication like Dna A protein in *E. coli*. In eukaryotes, the presence of ORC at the origin serves to recruit two other proteins—Cdc 6 and Cdt 1. But these proteins and ORC then recruit the replicative helicase and the other components of the replisome. Binding of helicase "licences" the origin. Origins must be licensed first before they are able to support the assembly of the replisome and begin DNA synthesis.

Replication is linked to the cell cycle through the availability of Cdc 6 and Cdt 1. In yeast, these proteins are synthesized during late mitosis and gap 1 and they are destroyed by proteolytic digestion after the synthesis begins. Therefore, the replisome can assemble only before the S phase. Once replication begins, new replisomes cannot form at the origin because Cdc 6 and Cdt 1 are degraded during the S phase and are no longer available.

LICENSING OF DNA REPLICATION

In eukaryotes, DNA replication starts at many initiation sites. Therefore, the entire genome is replicated in a timely manner. No gene must be left unreplicated and no gene must be replicated more than once. There occurs a precise mechanism to ensure that replication is initiated at thousands of origins only one per cell cycle.

The precise replication of DNA is accomplished by the separation of the initiation of replication into two distinct steps. In the first step, the origins are licensed, that is, they are approved for replication. This step is early in the cell cycle when a replication licensing factor attaches to an origin. In the second step, initiator proteins cause the separation of DNA strands and the initiation of replication at each licensed origin. Initiator proteins function only at licensed origins. As the replication forks move away from the origin, the licensing factor is removed, leaving the origin in an unlicensed state, where replication cannot be initiated again until the license is renewed. To ensure that replication takes place only once each cell cycle, the licensing factor is active only after the cell has completed mitosis and before the initiator proteins become active.

SPEED OF DNA REPLICATION IN *E. COLI*

The time taken for *E. coli* to replicate its chromosome is 40 minutes. Therefore, its genome of about 5 million base pairs must be copied at a rate of about 2000 nucleotides per second. *E. coli* has only two replication forks to copy the entire genome. So each fork must move at a rate of as many as 1000 nucleotides per second.

REPLICATION OF CHROMATIN

In chromatin, DNA is wrapped around a histone octamer and nucleosomes are formed. Before replication starts in eukaryotes, DNA has to be unwrapped from the histone and strand separation should occur. Also, after replication, the newly formed DNA molecules have to get wrapped on histone octamers to form new nucleosomes.

The synthesis of histones occurs simultaneously with DNA replication. Unassociated histone molecules are not found in appreciable amount in the cells. During dissociation, the parental histone octamers do not get dissociated and they do not get mixed with the newly synthesized histone molecules. The parental octamers are attached to the newly synthesized daughter DNA strand, that is, to the leading strand. Nucleoplasmin is a nuclear protein that is required for the assembly of the chromatin.

The other DNA strand has a new octameric histone constructed from histones available in the cellular pool with the help of proteins called chromatin assembly factors.

For example, in fruit flies, a protein complex called the replication-coupling assembly factor is used to assemble new nucleosomes. A protein complex called condensin is required for the condensation of interphase chromosomes to mitotic chromosomes. This complex includes two SMC proteins (structural maintenance of chromosomes) and two non-SMC proteins. SMC proteins also help in mitotic segregation, sister-chromatid adhesion, dosage compensation and recombination.

REPLICATION OF MITOCHONDRIAL AND CHLOROPLAST DNA (D-LOOP MODEL)

Chloroplasts and mitochondria have their own circular DNA molecules. Their replication (Figure 6.15) is slightly different from the replication of circular DNA of the prokaryotes, unlike θ structures or rolling circles.

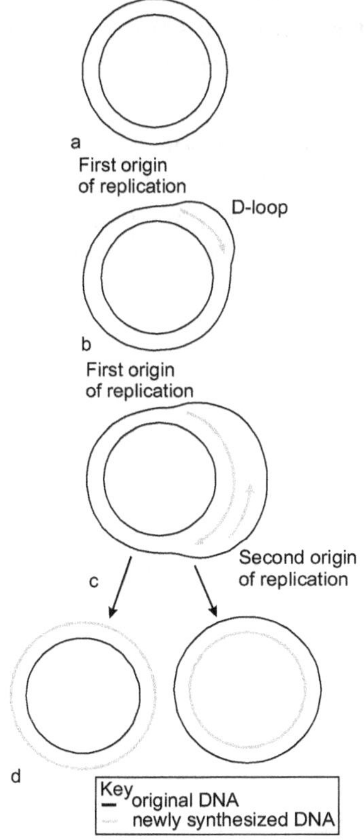

Figure 6.15   Replication of mitochondrial and chloroplast DNA

The origin of replication is at two different points on each of the two parental template strands. Replication begins on one strand and in so doing, displaces the other parental strand. This results in the formation of a displacement loop or D-loop structure. Replication of the first strand continues till the process passes the origin of replication of the other strand. New replication is initiated on the second parental strand in the opposite direction.

REGULATION OF DNA REPLICATION

Initiation of DNA replication in *E. coli* is strictly regulated. Since the chromosome replication in *E. coli* occurs only once per cell division, it is clear that this process is highly controlled. The doubling time of *E. coli* at 37°C varies with growth conditions from <20 min to ~10 h. The constant ~1000 nucleotides rate of movement of each replication fork fixes the 4.7×10^6-bp *E. coli* chromosome's replication time C, at ~40 minutes. Also, the segregation of cellular components and the formation of a septum between them, which must precede cell division requires a constant time, $D = 20$ min., after the completion of the corresponding round of chromosome replication. Cells with doubling times $<C + D = 60$ min. must initiate chromosome replication before the end of the preceding cell division cycle. This results in the formation of multiforked chromosomes for a cell division time of 35 minutes (Figure 6.16).

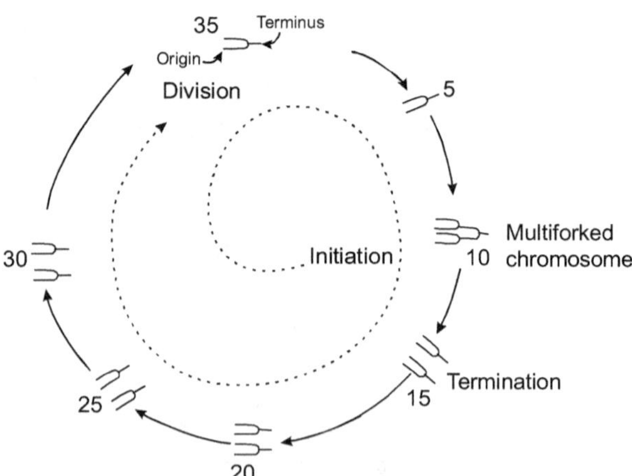

Figure 6.16   An interval of 60 min. between initiation of replication and cell division resulting in the production of multiforked chromosomes

The bacterial or plasmid origin ensures that replication is initiated only once per cell cycle. Initiation is associated with some change that marks the origin so that a replicated origin is distinguished from a nonreplicated origin.

oriC contains eleven copies of ds 5′ GATC 3′ 3′ CTAG 5′ which is a target for methylation (Figure 6.17) at the N^6 position of adenine by the enzyme Dam methylase. Replication of methylated DNA gives hemimethylated DNA, which maintains its state at GATC sites until Dam methylase restores the fully methylated condition.

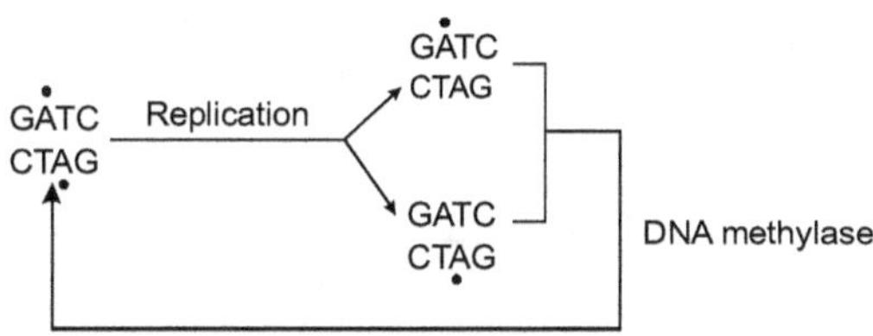

Figure 6.17   Methylation of DNA at oriC

Before replication, the palindromic target site is methylated on the adenines of each strand. Replication inserts the normal (unmodified) bases into the daughter strands. This generates hemimethylated DNA, in which one strand is methylated and one strand is unmethylated. Thus the replication event converts Dam target sites from fully methylated to hemimethylated condition.

Only fully methylated origins can initiate replication. Hemimethylated daughter origins cannot be used again until they have been restored to the fully methylated state (Figure 6.18). The GATC sites at the origin remain hemimethylated for ~13 minutes after replication. This long period is unusual because at typical GATC sites elsewhere in the genome, remethylation begins immediately (<1.5 minutes) following replication. Another region also behaves like oriC: the promoter of the *dna A* gene shows a delay before remethylation begins. While it is hemimethylated, the dna A promoter is repressed, which causes a reduction in the level of Dna A protein. Thus the origin itself is inert and production of the crucial initiator protein is repressed during this period.

The reason for the delay in remethylation at oriC and dna A is that these regions are sequestered in a form in which they are inaccessible to the Dam methylase. *seq A* gene is part of a negative regulatory circuit that prevents origins from being remethylated. Seq A protein binds to hemimethylated DNA more strongly than to fully methylated DNA. It may initiate binding when the DNA becomes hemimethylated, at which point its continued presence prevents formation of an open complex at the origin. Seq A does not have specificity for oriC sequence. This is conferred by Dna A protein, which would explain genetic interactions between *seq A* and *dna A* genes.

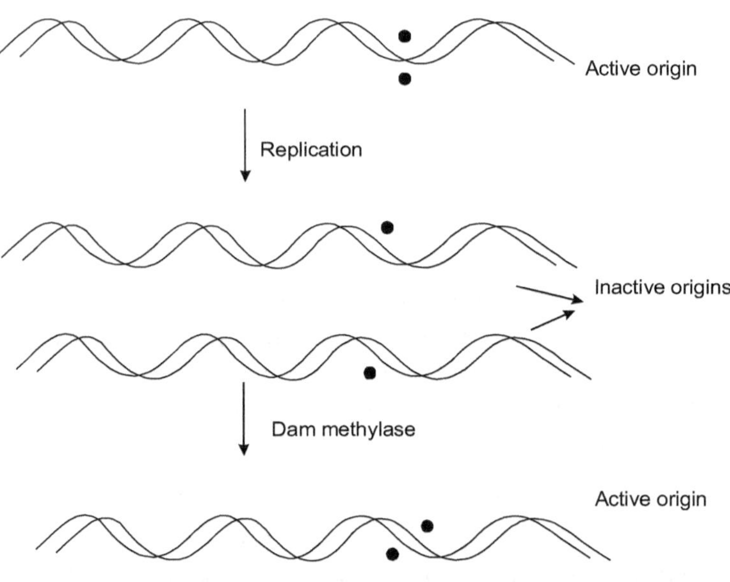

Figure 6.18   Active and inactive origins of replication

Hemimethylation of the GATC sequences in the site of origin of replication favours its association with the cell membrane *in vitro*. Hemimethylated oriC DNA binds to the membranes, but DNA that is fully methylated does not bind. Membrane association may be involved in controlling the activity of the origin of replication.

The properties of the membrane fraction suggest that it includes components that regulate replication. An inhibitor found in this fraction competes with Dna A protein. This inhibitor can prevent initiation of replication only if it is added to an *in vitro* system before Dna A protein (Figure 6.19).

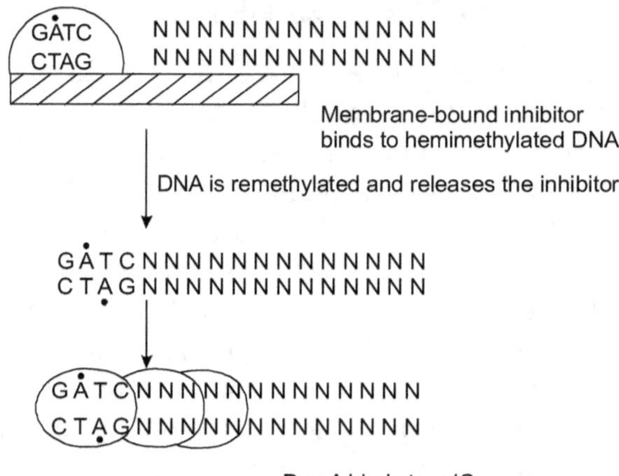

Figure 6.19   Membrane-bound inhibitor and DNA methylation

The inhibitor specifically recognizes hemimethylated DNA and prevents Dna A from binding. When the DNA is remethylated, the inhibitor is released and Dna A is free to initiate replication. If the inhibitor is associated with the membrane, association and dissociation of DNA with the membrane may be involved in the control of replication.

The phospholipids present on membranes interact with Dna A. Phospholipids promote the exchange of ATP with ADP bound to Dna A. To be active in initiating replication, Dna A must be in the ATP-bound form. Dna A has a weak intrinsic activity that converts the ATP to ADP. This activity is enhanced by the β subunit of DNA polymerase III. When the replicase is incorporated into the replication complex, this interaction causes hydrolysis of the ATP bound to Dna A thereby inactivating Dna A and preventing it from starting another replication cycle. This reaction is called **RIDA** (regulatory inactivation of Dna A). It is enhanced by a protein called Hda.

Another factor that controls availability of Dna A at the origin is the competition for binding it to other sites on DNA. A locus called dat has a large concentration of Dna A-binding sites. It binds about eight times more Dna A than the origin. Deletion of dat causes initiation to occur more frequently. This reduces the amount of Dna A available to the origin.

REPLICATION IN BACTERIOPHAGES

Bacteriophages may have either DNA or RNA as the genetic material. The type of genetic material in selected phages is given in Table 6.4. During infection they inject their nucleic acid into the host cells. The phage directs the synthesis of a replicative system that specifically makes copies of phage nucleic acid. Either there is synthesis of phage-specific polymerases or addition of specificity elements to bacterial enzymes. In both the cases many bacterial replication proteins are used. However, RNA-containing phages must encode their own replication enzymes because bacteria do not contain enzymes that replicate RNA. There is a distinct difference in the time of synthesis of phage-specified enzymes (early proteins) and the structural proteins of the phage particle (late proteins). This is accomplished by the timing of mRNA synthesis (of early mRNA and late mRNA). RNA of phages serve as their own mRNA without a need for transcription.

Phages have both essential and non-essential genes. A gene is considered to be non-essential if a mutation in that gene does not prevent plaque formation. If the mutation prevents plaque formation, the gene is essential. Table 6.4 lists the properties of the nucleic acid in some selected phages.

Table 6.4 Properties of nucleic acid in selected phages

Phage	DNA/RNA	Form	Host
φX174	DNA	ss, circ	E
M13, fd, f1	DNA	ss, lin	E
PM 2	DNA	ds, circ	PB
186	DNA	ds, lin	E
B 3	DNA	ds, lin	PA
T 7	DNA	ds, lin	E
λ	DNA	ds, lin	E
P1	DNA	ds, lin	E
T5	DNA	ds, lin	E
Sp01	DNA	ds, lin	B
T2, T4, T6	DNA	ds, lin	E
PBS1	DNA	ds, lin	B
MS2, Qβ, f2	RNA	ss, lin	E
φ6	RNA	ds, lin	PP

E: *E. coli*, B: *Bacillus subtilis*, PA: *Pseudomonas aeruginosa*

PB: *Ps. aeruginosa BAL 31*, PP: *Ps. phasealica*

Specificity in Phage Infection

The ability of a particular phage to infect a bacterium is almost always limited to a single bacterial species and often to a few strains of that species. For example, no phage species that infects the genus *Pseudomonas* can also infect *E. coli*. Also phages that grow in *Ps. fluorescens* fail to grow in *Ps. aerugenosa*. Phage ϕX174 grows well on *E. coli* strain C but fails to grow on most other strains. But there are some exceptions. *E. coli* phage T4 is capable of growing on many strains of *E. coli* and on certain species of the genus *Shigella*.

Several factors contribute to this specificity. One of these is the ability to adsorb, which is determined by the nature of the receptors on the cell wall for phage adsorption and the components of the bacterial cell wall. Another factor is the inability of a "foreign" bacterial RNA polymerase to recognize phage promoters. With most bacteria, there is usually another barrier called host restriction and host modification. The λ.*K*phage that has been grown in *E. coli* strain *k* will form plaques at a low efficiency in strain *B*. Thus λ.*K* is restricted by strain *B*. Similarly λ.B that has been grown in *E. coli* strain *B* is restricted by strain *K*. *E. coli B* contains a restriction endonuclease—*Eco B* nuclease, a site-specific nuclease that cuts DNA strands only within a specific base sequence. Phage λ.*K* has this sequence and is cut by *Eco B* nuclease. *E. coli B* also has this sequence and would destroy its own DNA. But a site-specific methylating enzyme called *Eco B* methylase methylates an adenine in the sequence which now becomes resistant to *Eco B*. Even in some cases of λ.*K* infection in *E. coli B* strain, a few parental phage DNA molecules are methylated before they are restricted.

E. coli phage T4 The DNA of phage T4 is large with 166,000 base pairs. It contains adenine, thymine and guanine but no cytosine. Instead, there is a modified form of cytosine called 5-hydroxymethylcytosine (HMC) which base-pairs with guanine. This base is further modified by glycosylation—a sugar is attached to the OH group of HMC. T4 DNA is terminally redundant, with a sequence of bases repeated at both the ends of the molecule. The terminal redundancy (Figure 6.20) contains about 1600 base pairs.

```
5′ A B C D E F G H I ............ W X Y Z A B C D E F G H I  3′
3′ A′B′C′D′E′F′G′H′I′ ........... W′X′Y′Z′A′B′C′D′E′F′G′H′I′ 5′
```

Figure 6.20 Terminal redundancy in DNA molecule

A remarkable feature of T4 DNA is that although each phage contains one DNA molecule, the molecules differ from phage to phage. A sample of T4 DNA molecule is cyclically permuted (Figure 6.21).

<table>
<tr><td>A B C</td><td>X Y Z A B C</td><td>D E F</td><td>Z A BC D E F</td></tr>
<tr><td>A'B'C'</td><td>X'Y'Z'A'B'C'</td><td>D'E'F'</td><td>Z'A'BC'D'E'F'</td></tr>
</table>

Figure 6.21 Cyclically permuted DNA molecules with redundancy

The nucleotide source for replication of T4 DNA is obtained by the degradation of the host bacterial DNA into dNMP. This is initiated by two endonucleases—the products of the genes *den A* and *den B*. The two enzymes are active only on cytosine-containing DNA. After these enzymes cleave the host DNA, an exonuclease encoded by the phage degrades them to dNMP. From dNMP, dNTPs are obtained by the bacterial enzymes. To ensure an abundant supply of dNTP, five phage-encoded enzymes are also synthesized.

Two phage enzymes T4 hydroxymethylase and HM kinase convert dCMP to dHDP which is converted to dHTP by the *E. coli* enzyme nucleoside phosphate kinase. dHTP is the immediate precursor of HMC in the DNA.

Several enzymes are involved in preventing the incorporation of cytosine into T4 DNA. T4 DNA polymerase cannot distinguish dCTP from dHTP, both of which can hydrogen-bond to guanine. Cytosine should not be incorporated in daughter T4 DNA strands since they will form substrate to T4 endonucleases. Also cytosine-containing T4 DNA cannot act as transcription templates in the later period of the life cycle. Therefore special enzymes are encoded by the phage DNA itself. If cytosine is to be incorporated in a DNA, available dCMP will be converted to dCDP and then to dCTP. A phage enzyme called dCTPase degrades both dCDP and dCTP to dCMP. Another phage enzyme, dCMP deaminase converts dCMP to dUMP. The bacterial and phage dCMP deaminase acting together, increase the amount of dTMP with respect to dCMP. Also if cytosine is present in the progeny phage DNA, the T4 endonucleases will degrade this DNA shortly after synthesis. Since cytosine will be present only on one strand of each daughter duplex strands, this DNA can be repaired by normal repair synthesis. If uracil residues get incorporated, they are removed by the usual uracil N-glycosylase.

The presence of HMC in T4 DNA creates another problem for the phage. *E. coli* has an endonuclease that may attack a DNA with HMC. To avoid this, HMC is glycosylated. Two phage enzymes, α-glycosyltransferase and β-glycosyltransferase transfer a glucose from uridine diphosphoglucose (UDPG) to HMC in the phage DNA. Glycosylation is therefore a post-replication modification.

E. coli phage T7   Phage T7 is a midsized phage and has the usual four nucleotides A, T, G and C in the DNA in equal amounts and has no unusual bases. It has a terminal redundancy of 160 base pairs but it is not cyclically permuted. The complete sequence of 39,936 base pairs of T7 DNA is known. There are 50 closely packed genes and five overlapping genes.

Bidirectional replication of T7 DNA starts at a point about 17 per cent from the left end. Because the two forks move at equal rates, the left end is finished first. The regions at the 3′ end of each template strand remain unreplicated (Figure 6.22).

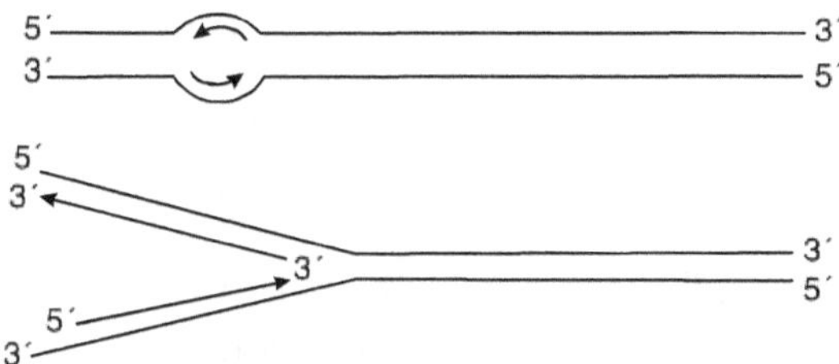

Figure 6.22   Origin and direction of replication of T7 DNA

A hypothesis is proposed for the generation of concatemers from two incomplete daughter molecules. Annealing of the daughter molecules takes place which is possible by the terminally redundant sequence ABC.

The terminal redundancy leads to the formation of a dimer called concatemer. Each unit of the dimer contains a replication origin, so that the dimer can also replicate. The result of this round of replication is a pair of daughter dimers, each of which has complementary single-stranded termini (Figure 6.23). Once again, annealing and trimming occur, to give a tetramer. This process repeats and finally a gigantic linear concatemer is obtained. At any time linear T7 DNA molecules of exact size and terminal nucleotide sequences as in the original parental phages are cut and packed in the viral head.

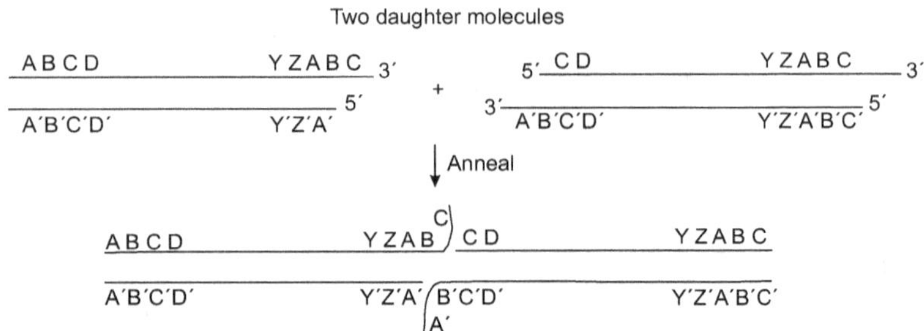

Figure 6.23   Annealing of two daughter DNA molecules in T7 phage

E. coli phage λ   λ is a phage which contains a linear double-stranded DNA molecule with a molecular weight of 31×10^6. It has 48, 514 base pairs. At each end of the molecule the 5′ terminus extends 12 bases beyond the 3′ end and forms cohesive ends (Figure 6.24) which are complementary to one another. This helps in the circularization of the DNA after injection in the host cell.

Figure 6.24   λ DNA with cohesive ends

Replication of λ DNA requires the products of two genes O and P. There are two modes of λ DNA replication—θ and rolling circle replication (Figure 6.25). θ replication increases the number of templates for transcription and further replication. Rolling circle mode provides the DNA for phage progeny. Slowly θ replication stops and rolling circle becomes the predominant form.

In λ DNA, when the molecule circularizes and the sticky ends are joined, the ds region formed is called a cos site (cohesive site). During rolling circle replication like the mechanism of replication and cutting of DNA in T4 and T7, in λ phage also there is a sequence-specific cutting system called the terminase or Ter system and the DNA cutting is called Ter-cutting.

The Ter system requires the presence of Ter proteins which are the components of an empty λ head.

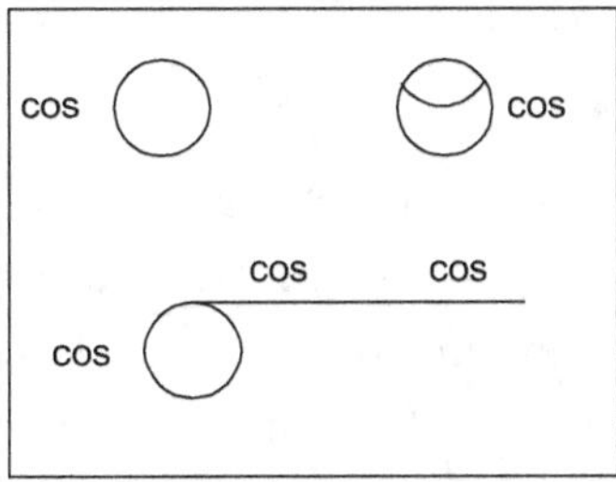

Figure 6.25   Rolling circle replication with cos sites in λ DNA. Three species of λ DNA are shown when maturation begins— circle, θ model and rolling circle

E. coli phage φX174   *E. coli* phage φX174 contains one circular, single-stranded DNA molecule with 5386 nucleotides. The DNA is enclosed in an icosahedral head composed of three coat proteins and one internal protein. The strand contained in the virus particle is referred to as + strand. Replication of φX174 DNA occurs in several steps:

i. *Formation of a double-stranded covalently closed molecule called replicative form (RF) from + strand.* The complementary – strand is the one that is transcribed.

ii. *Formation of many copies of RF.* By transcription of the – strand, a multifunctional protein is produced which is the product of the viral gene A. This protein cuts the + strand between bases 4305 and 4306, i.e., at the origin of replication as it remains covalently linked to the 5′-phosphate terminus. The *E. coli* proteins displace the + strand from the RF, and rolling-circle replication occurs. When one round of replication is completed, the displaced + strand is cleaved from the RF and gets recircularized and used as a template for the synthesis of another strand.

iii. *Synthesis of + strands to be encapsulated.* The packaging system of φX174 captures progeny (+) strands before they can serve as templates for further synthesis of – strands. This occurs when 30–40 copies of RF are formed and the phage heads have been synthesized. When all the + strands are packaged in the viral head, RF synthesis is stopped due to lack of + strand templates.

By genetic approaches, scientists have shown that, in the model bacterium *Bacillus subtilis*, the central carbon metabolism which generates energy from the nutrients and the replication which synthesizes DNA—two key functions in the fields of nutrition and heredity—are tightly linked. It is proposed that the link depends on metabolic signals generated as a function of the activity of the terminal reactions of glycolysis which are sensed directly or indirectly by replication enzymes. This system would then adjust the speed of DNA synthesis and the stability of the replication machinery to the nutritional richness of the environment and thus to the cell's growth rate. That is why, several enzymes of the central carbon metabolism are essential and strongly conserved in living organisms. This metabolism/replication link may be of medical importance, as early events in carcinogenesis (which generally include an up-regulation of glycolysis and a decrease in DNA stability and replication fidelity) may involve perturbations of the metabolism/replication link.

INHIBITORS OF DNA REPLICATION

Study of inhibitors of DNA replication is of therapeutic importance. These inhibitors can be used in pharmacology for treatment of diseases. Identification of the inhibitors that are active separately on prokaryotic and eukaryotic systems may help in formulating medicines which can damage and destroy the infecting organisms without affecting the host cells. These inhibitors can be grouped into four classes.

Compounds that Inhibit Nucleotide Biosynthesis

Methotrexate and fluorodeoxyuridylate interfere with the production of dTTP. Cancer cells duplicate at a very fast rate and divide continuously. Inhibitors of nucleotide biosynthesis will have selective toxicity for such cancers because they cause a depletion or imbalance in the cellular levels of dNTPs required for DNA synthesis. A selective depletion of one of the four dNTPs (i.e., dTTP) by treatment with a drug 5-fluorouracil causes the arrest of DNA synthesis in the cancer cells leading to cell death. However, if dTTP is decreased, the imbalance in the cellular pools of dNTPs may lead to genetic miscoding with fatal mutations.

Methotrexate is a potent inhibitor of dihydrofolate (DHF) reductase which results in the accumulation of DHF in the cell with minor decrease in tetrahydrofolate (THF). Marked decrease in THF may not be seen due to the release of bound THF in methotrexate-treated cells. High levels of DHF are toxic to the cell, inhibiting the reaction catalysed by thymidylate synthase. In leukaemia cells treated with methotrexate, levels of dTTP decrease. There may be less marked decreases in dATP and dGTP resulting from inhibition of amido-phosphoribosyl transferase (PRTase). The consequent imbalance in nucleotide pools results in genetic miscoding and cell death.

Other anticancer drugs which inhibit nucleotide biosynthesis are 5-fluorouracil, hydroxyurea and 6-mercaptopurine. 5-fluorouracil is a very useful anticancer drug which is taken up by the cells and metabolized as follows:

$$FU \rightarrow FUMP \rightarrow FUDP \rightarrow FUTP$$
$$\downarrow$$
$$\rightarrow FdUDP \rightarrow FdUTP \rightarrow FdUMP$$

The primary mechanism of action of FU may be inhibition of thymidylate synthase (dUMP $\rightarrow$ dTMP) by 5-fluorodeoxy UMP. This causes depletion of dTMP and thus dTTP in the cells. However, 6-fluorouracil may also kill cancer cells by two other mechanisms. FUTP which accumulates in cells may be incorporated into RNA causing genetic miscoding, or FdUTP may be incorporated into DNA, again leading to fatal mutations.

Hydroxyurea inhibits ribonucleotide reductase enzyme, accepts the four NDPs (UDP, CDP, ADP and GDP) as substrates and reduces them to the corresponding dNDP. The mechanism of catalysis involves the formation of an unusual tyrosyl radical cation intermediate leading to depletion of all four dNTPs required as substrates for the synthesis of DNA.

6-Mercaptopurine is an anticancer drug that may enter the cells and be metabolized as follows:

$$MP \rightarrow MP - MP \rightarrow MP \cdot DP \rightarrow MP \cdot TP$$
$$\downarrow$$
$$MXMP$$
$$\downarrow$$
$$MGMP \rightarrow MGDP$$
$$\downarrow MdGDP \rightarrow MdGTP$$

6-Mercaptopurine 5′-monophosphate (MP · MP) formed is a potent inhibitor of amino PRTase and thus de novo purine biosynthesis is blocked. The 6-mercapto deoxy GTP (MdGTP) is incorporated into DNA and causes genetic miscoding.

Inhibitors that Interact with the DNA Template

Actinomycin D is a commonly used inhibitor of both DNA and RNA synthesis. It has a planar structure which binds non-covalently between the stacked base pairs of duplex DNA. This is referred to as **intercalation**. In this situation DNA functions as a poor template. Other intercalating agents include acridine and ethidium. Thus they affect the fidelity of DNA replication.

Nucleotide Analog Inhibitors

A number of nucleotide analog compounds function by blocking further chain growth at the replication fork. The 2′, 3′-dideoxy nucleotides (ddNTPs), which lack–OH groups in both 2′ and 3′ positions of ribose, if present in the medium/system can be incorporated in the 3′-OH growing end of DNA chain in bacteria but cannot facilitate the addition of other nucleotides because they lack 3′-OH group. This forms the principle of sequencing of DNA by Sanger's method.

Inhibitors that Bind to Replication Proteins

The arylhydrazinopyrimidines are potent inhibitors of DNA polymerase III from gram-positive bacteria. These compounds form ternary complexes with the polymerase and the DNA template.

Aphidicolin, a tetracyclic diterpenoid is a potent inhibitor of mammalian nuclear DNA polymerase. It does not affect mitochondrial DNA polymerase. It is used to treat herpesvirus infections of the eye.

Nalidixic acid and novobiocin bind to the A and B subunits respectively of *E. coli* DNA gyrase and inhibit its action. Nalidixic acid is used as an antibacterial agent. The antibiotic oxolinic acid is ten times more potent than nalidixic acid.

DNA REPAIR MECHANISMS

All the cells have a number of DNA repair systems. This shows the importance of setting right a damaged DNA for the cell to survive. Many DNA repair processes appear to be extraordinarily inefficient energetically. But when the most important issue of maintaining the integrity of the genetic information is considered, the amount of chemical energy invested in a repair process seems to be almost irrelevant.

DNA repair is possible mainly because the molecule is double-stranded with complementary bases in both the strands. If there is damage in one strand, it can be removed and using the information in the undamaged strand (undamaged strand acting as template), new nucleotide can be synthesized in the damaged strand.

The types of DNA repair systems are mismatch repair, base-excision repair, nucleotide excision repair, direct repair, post-replication repair (recombination repair) and error-prone repair.

Mismatch Repair

In a mismatch damage, if T is in the template strand and if due to an error in replication, G is added in the new strand, mismatch damage occurs and this results in the destabilization of the DNA molecule.

When there is damage in one strand, the damaged region should be replaced and repaired (Figure 6.26) with the information in the template strand that has no damage. Therefore, there should be some way by which the system discriminates between the template strand and the newly synthesized strand (which has the damage). This is done by the cell by tagging the template DNA with methyl groups to distinguish it from the newly synthesized strands.

In bacteria the mismatch repair system includes at least 12 protein components—Dam methylase, Mut H, Mut L, Mut S proteins, DNA helicase II, SSB, DNA polymerase III, exonuclease I, exonuclease VII, Rec J nuclease, exonuclease X and DNA ligase.

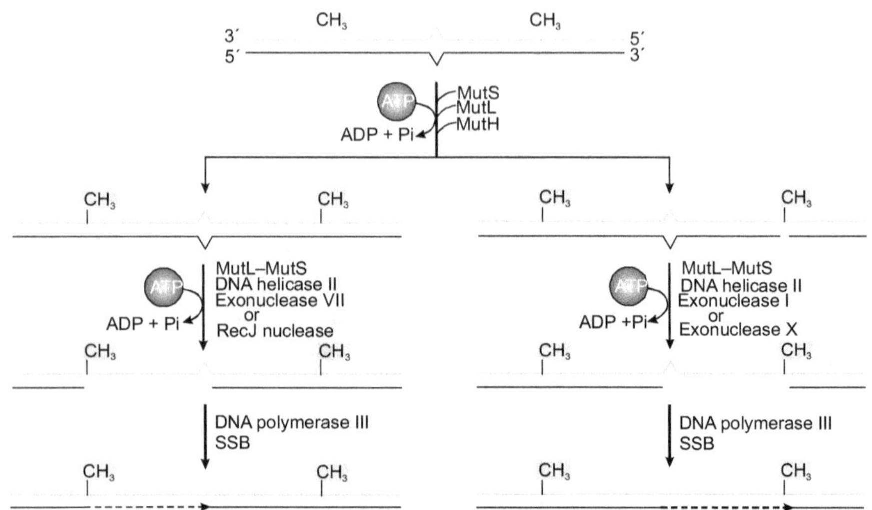

Figure **6.26**   Mismatch repair mechanism

Dam methylase methylates DNA at the N^6 position of all adenines within 5'-GATC sequences. Immediately, after passage of the replication fork, there is a short period of a few seconds when the template strand is methylated but the newly synthesized strand is not methylated. In this short period, when the DNA is hemimethylated the damaged strand that has the mismatch base is repaired.

A protein of the mismatch repair system, Mut L forms a complex with Mut S protein and binds to all mismatched base pairs (except C-C mismatches). Another protein Mut H binds to Mut L and to GATC sequences involving Mut L–Mut S complex. Thus a loop is formed. Mut H has a site-specific endonuclease activity. This activity becomes activated only in the presence of hemimethylated GATC sequence. Mut H catalyses cleavage of the unmethylated strand in the 5' side of the G in GATC. The unmethylated strand now gets unwound and is degraded in the $3' \rightarrow 5'$ direction from the cleavage site and passes through the mismatch region. The gap so created is filled by the synthesis of the new strand. This requires the combined action of DNA helicase 1, SSB and either exonuclease I or exonuclease X (both these enzymes degrade DNA from $3' \rightarrow 5'$ direction), DNA polymerase III and DNA ligase. Alternatively, the unmethylated strand (to be repaired) may also be degraded in $5' \rightarrow 3'$ direction from the cleavage site and pass through the mismatch region. This type of degradation is possible by the action of exonuclease VII or Rec J nuclease.

Mismatch repair is an expensive process in bacteria in terms of the energy required for repair. The mismatch may be 1000s of base pairs from the GATC sequence. The degradation and replacement of a strand segment of this length requires enormous investment in terms of dNTP precursors to repair just a single mismatch base. This shows the importance of maintaining the integrity of the genome.

In eukaryotes, the proteins involved in the mismatch repair are similar to the bacterial Mut S and Mut L proteins. However strand discrimination for repair of the damaged strand may differ in other ways from the methyl-directed system used in bacteria.

Base-Excision Repair

There is a class of enzymes in the cells called DNA glycosylase. When adenine or cytosine is deaminated because of exposure to mutagens, the N-glycosidic bond between the base and the deoxyribose sugar gets weakened. The DNA glycosylases recognize such lesions and remove the affected base by cleavage of the N-glycosidic bond. This results in the formation of apurinic or apyrimidinic site in the DNA referred to as AP site. Each DNA glycosylase is specific for one type of lesion. For example, uracil glycosylase specially removes uracil from DNA that has been formed from spontaneous deamination of cytosine. This enzyme does not remove uracil residues from RNA or thymine residues from DNA. May be this is one of the reasons why DNA has thymine and not uracil. If adenine is deaminated to hypoxanthine, another specific glycosylase recognizes and removes it. Alkylated bases like 3-methyladenine and 7-methylguanine are also removed in a similar manner. Sometimes AP sites may also arise due to slow, spontaneous hydrolysis of the N-glycosidic bonds in DNA.

Once an AP site is formed, another group of enzymes repairs it. The deoxyribose 5′-phosphate in the AP site is removed and replaced with a new nucleotide. This process begins with the action of enzymes called AP endonucleases. These enzymes cut the DNA strand containing the AP site. The position of the cut relative to the AP site both on 5′ side and 3′ side varies with the type of AP endonuclease. Now, a segment of DNA strand including the AP site is then removed, the DNA is replaced by DNA polymerase I and the remaining nick is sealed by DNA ligase (Figure 6.27).

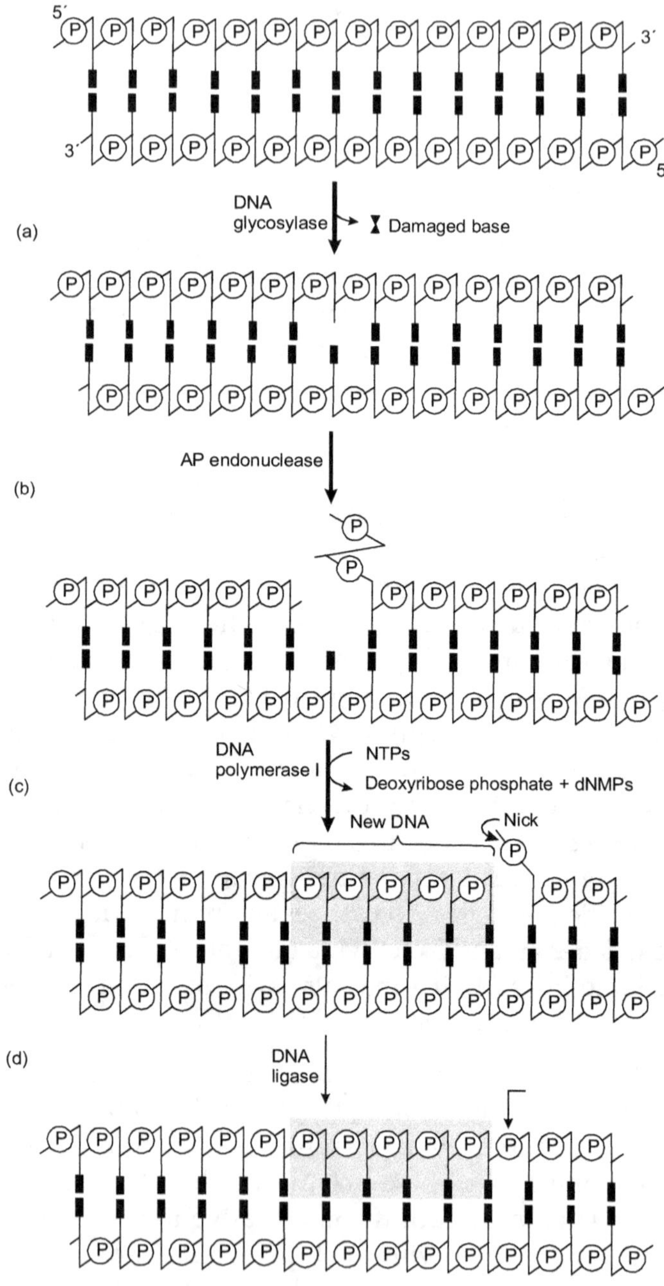

Figure 6.27 Base-excision repair

Nucleotide-Excision Repair

In nucleotide-excision repair, a multisubunit enzyme hydrolyses two phosphodiester bonds, one on either side of the lesion where the nucleotides are damaged. In *E. coli* and other prokaryotes, the enzyme system hydrolyses the 5th phosphodiester bond on the 3'side and the 8th phosphodiester bond on the 5' side and the resulting 12–13 nucleotide fragments are removed. In humans and other eukaryotes, the enzyme hydrolyses the 6th phosphodiester bond on the 3'side and the 22nd phosphodiester bond on the 5'side. Following the dual cut, the excised oligonucleotide is removed and the resulting gap is filled by DNA polymerase I in *E. coli* and DNA polymerase Σ in humans. The nick is sealed by DNA ligase. In *E.coli*, the multisubunit enzyme complex is ABC excinuclease which has three subunits—UvrA, UvrB and UvrC. The nucleolytic activity of ABC excinuclease is novel in the sense that two cuts are made in the DNA. The term "excinuclease" is used to distinguish this activity from that of standard endonucleases.

Uvr A and Uvr B subunits (A_2B) form a complex and bind to the DNA at the site of lesion like the thymine dimers formed due to exposure of DNA to UV rays. Uvr A dimer then dissociates leaving a tight Uvr B–DNA complex. Then Uvr C binds to Uvr B. It is Uvr B that makes the cut in the DNA on the 3'side. This is followed by the action of Uvr C that cuts the DNA on the 5' side. The fragment of 12–13 nucleotides containing DNA strand is removed by Uvr D helicase. The gap is filled and the nick sealed by DNA polymerase I and DNA ligase (Figure 6.28).

This repair mechanism is the primary repair route for many types of lesions like pyrimidine dimers, 6–4 photo products and other base adducts like benzo(a)pyrene-guanine that is formed in DNA by exposure to cigarette smoke. Eukaryotic excinucleases differ structurally from the prokaryotic enzyme but perform the same dual incision.

Direct Repair (Direct Reversal of Damage)

When UV rays of wavelength 200–300 nm promote the formation of a cyclobutyl ring between adjacent thymine residues on the same DNA strand and result in the formation of thymine dimers or cause the formation of thymine–cytosine dimers, the dimers locally distort DNA's

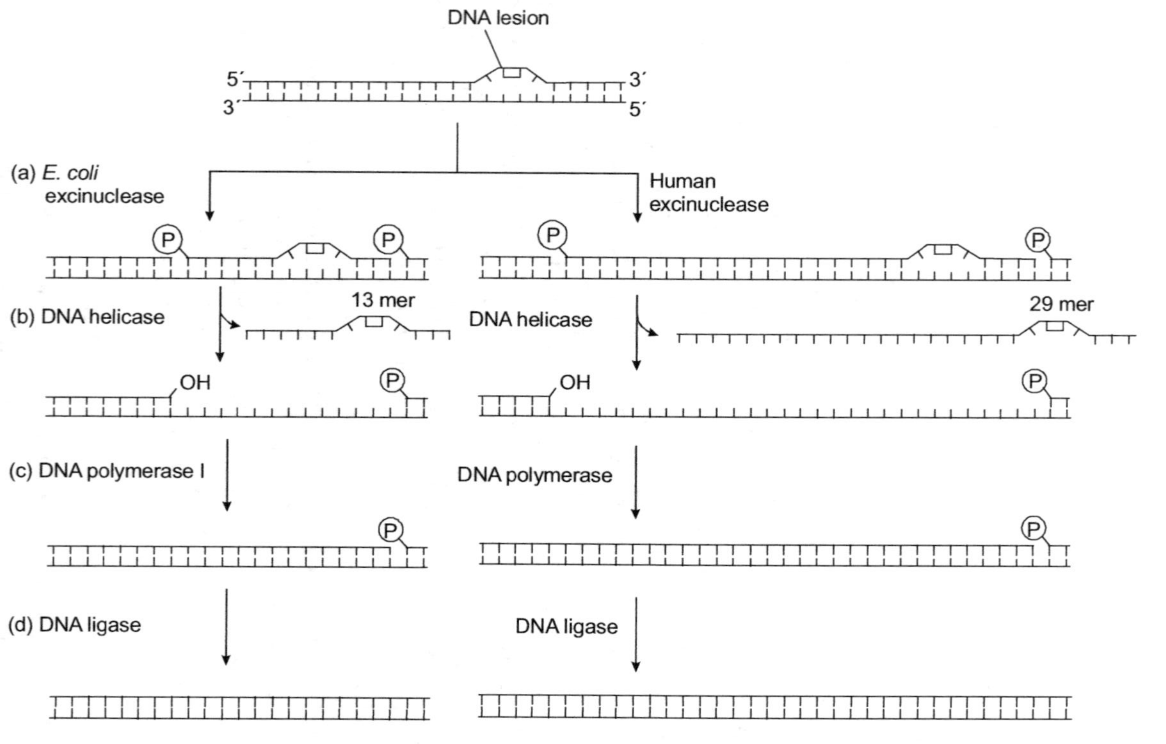

Figure 6.28 Nucleotide-excision repair

base pair structure. In such conditions, DNA strands cannot act as good templates for replication or transcription.

The pyrimidine dimers can be restored to their monomeric forms through the action of light-absorbing enzymes, photoactivating enzymes or DNA photolyases. These enzymes are 55- to 65-kDa monomers. They can bind to pyrimidine dimers in the dark. Pyrimidine dimers result from a light-induced reaction and photolyases use energy derived from absorbed light to reverse the damage. Photolyases generally contain two cofactors that serve as light-absorbing agents or chromophores—FADH is always one chromophore. In *E. coli* and yeast, the other chromophore is a folate—N^5, N^{10} methenyl tetrahydrofolyl polyglutamate (MTHF polyGlu). A blue light of wavelength 300–500 nm is absorbed by this chromophore which functions as a photoantenna. The excitation energy is transferred to $FADH^-$ in the active form of the enzyme. $FADH^-$ is now excited to form $*FADH^-$ which donates an electron to the pyrimidine dimer. Electronic rearrangement restores the monomeric pyrimidines. The electron is transferred back to the flavin radical to once again get back $FADH^-$ (Figure 6.29).

Another example of direct repair is the repair of O^6-methylguanine (O^6-meG), that is formed by alkylation. This is considered as a highly mutagenic lesion. O^6-meG tends to pair with thymine instead of cytosine during replication. This may result in the conversion of GC base pair to AT base pair. Direct repair of O^6-meG is carried out by the enzyme O^6-methylguanine-DNA methyl transferase (Figure 6.30). It is a protein that transfers the methyl group from O^6-meG to one of its own cytosine residues. This is not strictly an enzyme because it accepts a methyl group and becomes permanently methylated and becomes inactive. Therefore an event of methyl group removal from a DNA base and the enzyme getting inactivated is a clear proof to show that the system does not mind wasting one protein molecule in order to maintain the integrity of the cellular DNA.

The methylated enzyme, though not useful for another repair process, functions as a transcriptional activator, increasing the expression of its own gene and other few genes for repair enzymes.

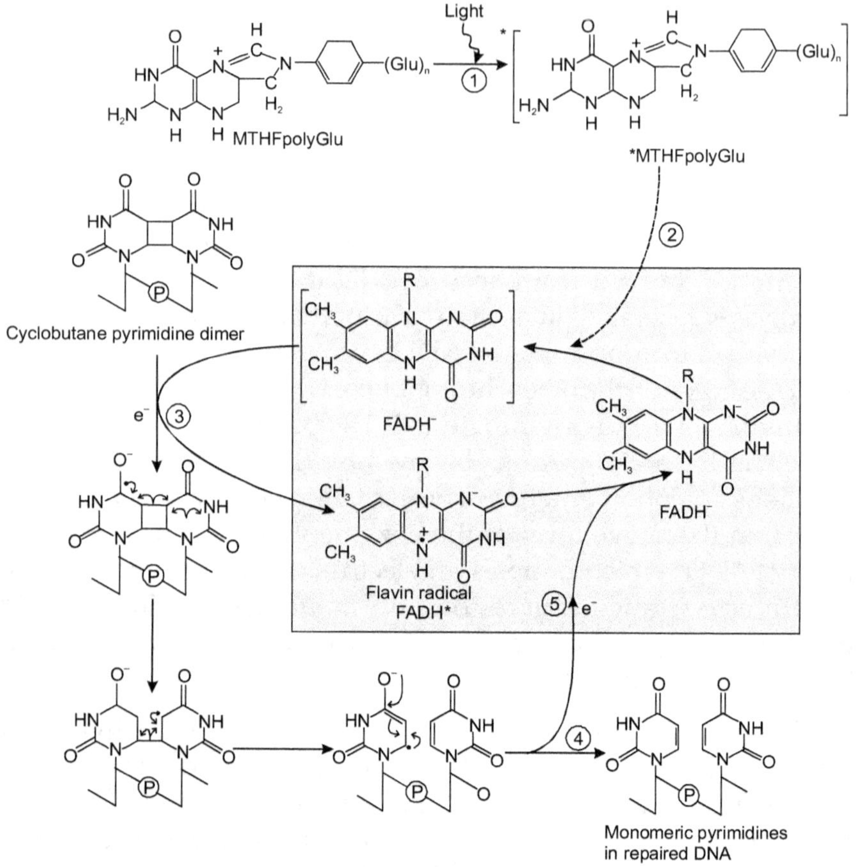

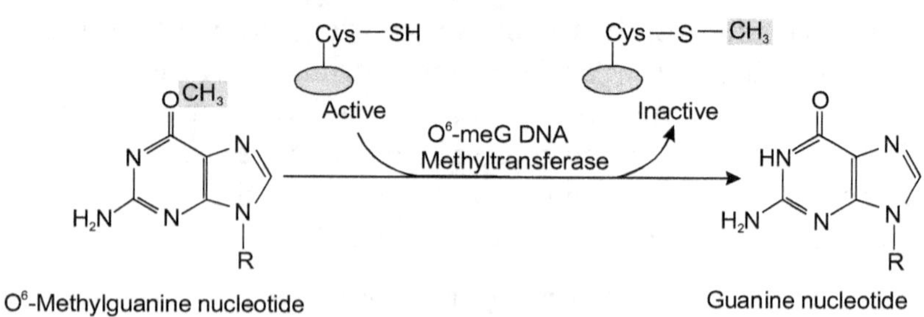

Figure 6.29   Direct repair by DNA photolyase

Figure 6.30   Direct repair of damaged DNA by O⁶-methylguanine-DNA
methyl transferase

Post-Replication Repair

This mechanism of DNA repair (Figure 6.31) was first proposed by Miroslav Radman in an excision-defective strain of *E.coli*. This mechanism responds after damaged DNA has escaped repair and has failed to be completely replicated.

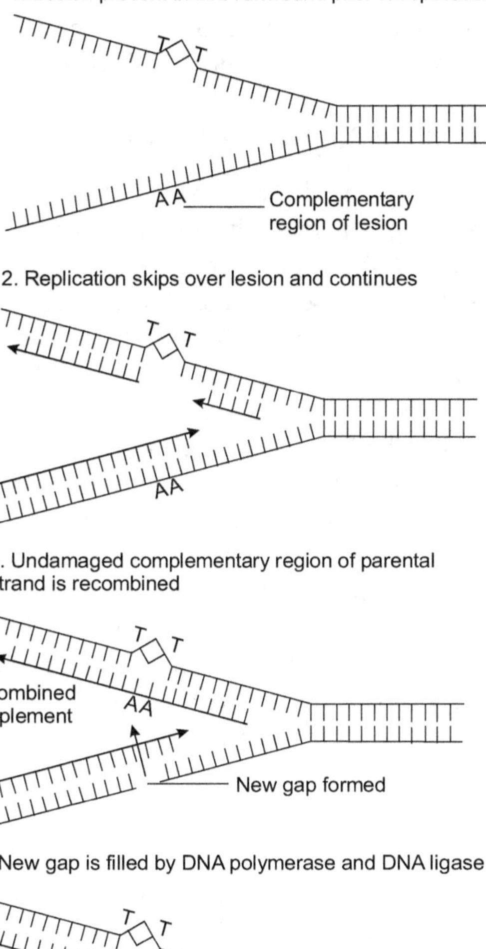

Figure 6.31   Post-replication repair

In a DNA with a lesion like a pyrimidine dimer that undergoes replication, DNA polymerase first stops at the point of lesion and then skips over it, leaving a gap in the newly synthesized strand. In order to counteract this, a protein Rec A directs a recombinational exchange process and fills that gap by inserting a segment initially present in the undamaged complementary strand. This now creates a gap on the "donor" parental strand which is then filled by repair synthesis as replication proceeds. Since a recombinational event is involved, post-replication is also referred to as homologous recombination repair.

Double Strand Break Repair in Mammals

If both the strands of the double-helical DNA are cleaved, like that caused by exposure to ionizing radiation in eukaryotic cells, DNA double strand break repair (DSB repair) pathway is activated. This pathway re-anneals the two DNA segments. This also involves homologous recombinational repair, because damaged DNA is actually recombined and replaced with homologous undamaged DNA. This is required because when both the strands are broken, there is no undamaged parental strand available to use as the source of the complementary DNA sequence during repair. Therefore the genetic information present in the homologous region of either sister homolog is recruited to replace the damaged double-stranded break. The undamaged homologous region is actually recombined into the damaged DNA molecule. This process occurs during the late S/G2 phase of the cell cycle, after DNA replication. This stage only allows the recruitment of the undamaged copy of the sister chromosome to be utilized. It is seen that in yeast, a complex of five proteins called RAD52 complex is utilized, which is able to restore the integrity of the broken double helix.

Another pathway called non-homologous recombinational repair or end joining also is available for this repair. This mechanism does not recruit a homologous region of DNA during repair. This system is activated in G1, prior to DNA replication.

The SOS Response and Mutagenic Repair

The *E. coli* SOS response is an inducible response to DNA damage that results in increased capacity for DNA repair, inhibition of cell division and altered metabolism. The repair is mediated by the activation of about 30 SOS genes. These are normally expressed at low levels due to transcriptional regression by a repressor called Lex A repressor which binds to the operator sequences of SOS bases upstream of each gene. The repressor is capable of inhibiting its own synthesis.

When DNA is extensively damaged, single-stranded regions are exposed which interact with a protein called Rec A to produce an activated complex, Rec A*. This complex causes cleavage of Lex A repressor. The cleaved Lex A protein can no longer bind to DNA and the SOS genes are derepressed. When damaged DNA is no longer present in the cell (DNA damage is repaired), Rec A becomes inactivated and will no longer cleave Lex A. Lex A now accumulates in the cell from the existing Lex A mRNA pool and Lex A gene and other SOS genes are shut down (Figure 6.32). This is an example of negative feedback loop.

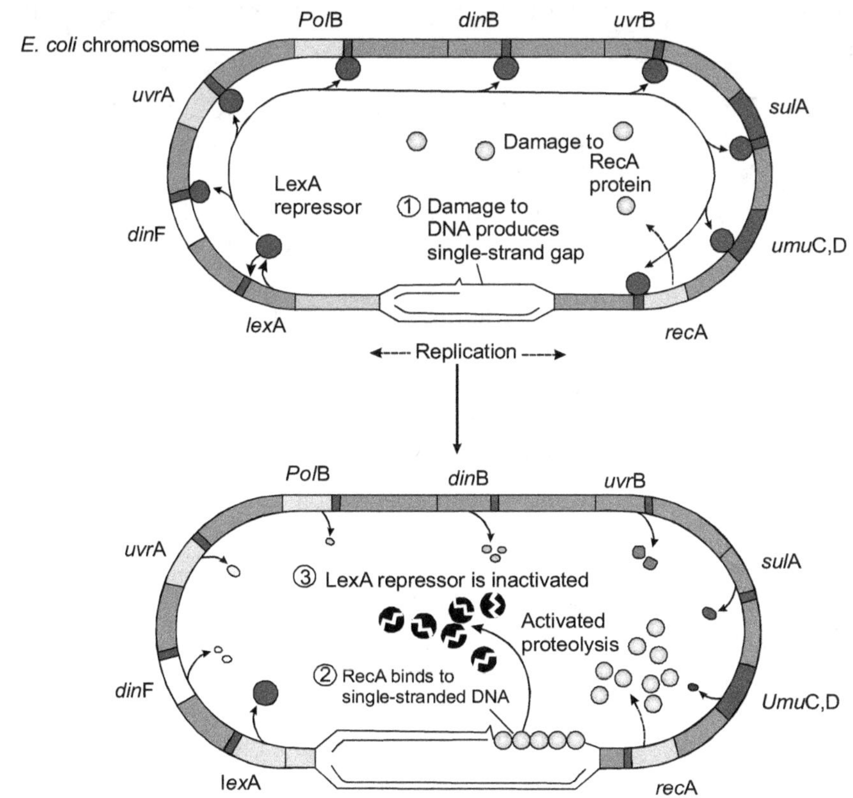

Figure **6.32**   SOS response in *E. coli*

The Lex A repressor is not simply dissociated from DNA during the induction of the SOS response. Lex A is inactivated when it catalyses its own cleavage at a specific ala-gly peptide bond, producing two approximately equal fragments. At physiological pH, this autocleavage reaction requires the Rec A protein. Rec A is not a protease. Its interaction with Lex A facilitates the self-cleavage reaction. This function of Rec A may be referred to as co-protease activity.

Lex A has operator sites near each gene specified in the figure. The normal cell has about 1000 Rec A monomers. During DNA damage (e.g. by UV rays), when single-stranded gaps are got in DNA, Rec A binds to these ssDNA and its co-protease activity is activated. This causes cleavage of Lex A repressor. When this occurs, the SOS given including rec A are induced.

The genes and the proteins encoded by them which are involved in DNA repair as part of the SOS response in *E. coli* are given in Table 6.5.

Table 6.5   Genes and proteins involved in SOS response

Genes	Proteins
polB (din A)	Encodes polymerization subunit of DNA polymerase II, required for replication restart in recombinational DNA repair
uvrA, uvr B	Encodes ABC excinuclease subunits Uvr A and Uvr B
umuC, umuD	Encodes DNA polymerase V, a special polymerase that can replicate past many of the DNA lesions that would normally block replication
sulA	Encodes protein that inhibits cell division, thereby allowing time for DNA repair
recA	Encodes Rec A protein required for error-prone repair and recombinational repair
dinB	Encodes DNA polymerase IV, involved in error-prone repair

Error-Prone Repair

When there is large level of DNA damage like exposure to strong UV rays, another repair path called error-prone repair occurs. When this pathway is active, DNA repair becomes less accurate and a high mutation rate results. The enzymes formed in SOS response are involved in error-prone repair. The Umu D protein is cleaved to a shorter form called Umu D′ and this forms a complex with the Umu C protein to create a special type of polymerase called DNA polymerase II also referred to as polymerase V. This enzyme can replicate past many of the DNA lesions that would normally block replication. At the site of such a lesion, proper base pairing will not take place. Therefore this translesion replication is considered as error-prone. Though the ultimate result is production of mutated DNA, these may be in less frequency compared to complete block of replication.

Another enzyme, DNA polymerase IV, is also induced during the SOS response. It is a product of the *dinB* gene. This enzyme also is involved in error-prone repair.

Spicy Questions and Answers

1. **What would have been the results of Meselson and Stahl's experiment if DNA replication was by "dispersive model"?**

    According to the "dispersive model", it was suggested that the parental DNA strands are cleaved into pieces of random size and then joined with pieces of newly replicated DNA to yield daughter duplexes. In that case, all the next generation DNA molecules would have had the same density and would have appeared as a single band in the hybrid region. They would not have settled either in the light or the heavy region.

2. **What would have been the results of Meselson and Stahl's experiment if DNA replication was by "conservative model"?**

    According to "conservative model" it was suggested that the parental DNA strands do not undergo any cleavage or strand separation, but a new DNA molecule is obtained exactly similar to the parent molecule, i.e., the parent molecule is conserved. In that case, two bands would have been got on caesium chloride centrifugation, one in the light region (that of the new DNA with no heavy isotope of nitrogen) and one in the heavy region (that of the parent molecule with heavy isotope of nitrogen) and there would not have been a band in the hybrid region. The presence of, a clear band in the hybrid region confirmed the "semiconservative model" of DNA replication.

3. A culture of *E.coli* is grown in a medium containing ^{15}N (^{15}N-heavy isotope of nitrogen). They are switched to a medium containing ^{14}N for three generations (one DNA molecule in the 0th generation, becomes two molecules in the first generation, four molecules in the second generation and eight in the third generation). What is the molar ratio of hybrid DNA ($^{15}N–^{14}N$) to light DNA ($^{14}N–^{14}N$) at this point of eightfold increase in the DNA population?

According to the semiconservative model of DNA replication, among the eight molecules, two will be hybrid and six will be light. So the molar ratio of hybrid DNA to light DNA will be $2/6 = 0.33$.

4. You are provided with a DNA molecule in which $G = 24.1\%$, $C = 18.5\%$ and $T = 32.7\%$ in one strand (i) What is the base composition of the other strand? (ii) What is the base composition of the total DNA synthesized from the given DNA?

(i) The base composition of the other strand will be $T = 24.7\%$, $C = 24.1\%$, $G = 18.5\%$, $A = 32.7\%$ (ii) The base composition of the total DNA synthesized from the given DNA will be the same as the base composition of the original parent DNA molecule. In the problem, the data is given for one strand and that of the other strand is calculated and presented in the answer (i). Therefore the base composition of both the strands will be

$A = (24.7 + 32.7)/2 = 28.7\%$

$G = (24.1 + 18.5)/2 = 21.3\%$

$C = (18.5 + 24.1)/2 = 21.3\%$

$T = (32.7 + 24.7)/2 = 28.7\%$

5. If a double-stranded circular chromosome contains 4,800,000 base pairs and if the rate of replication is 1000 base pairs per second, how long will it take to replicate the entire chromosome?

(Hint: replication is bi-directional)

Time taken for replicating the entire chromosome is 4,800,000 divided by $2 \times 1000 \times 60 = 40$ minutes.

6. In the above case, if the time of replication was observed to be less than 40 minutes, how would that have been possible?

The second cycle of replication would have started even before the completion of the first cycle.

7. **A mixture consists of all the ingredients required for DNA replication. All the four dNTPs are labelled with ^{32}P in α phosphate group. After some time, the mixture is treated with trichloroacetic acid which can precipitate the DNA that is formed. dNTPs are not precipitated. (i) If any one of the four dNTPs is not added initially in the mixture, will the precipitate show radioactivity? (ii) Among the four dNTPs, if dATP alone was radiolabelled, will the precipitate show radioactivity? (iii) if all the four dNTPs are radiolabelled with ^{32}P in the β or γ phosphates instead of α phosphate group, will the precipitate show radioactivity?**

   i. If one of the dNTPs is missing, there will not be radioactivity in the precipitate because DNA can be synthesized, dictated by the template stored and the unavailability of even one dNTP will not allow polymerization by DNA polymerase.

   ii. Yes, the precipitate will show radioactivity because it will be incorporated opposite to T in the template strand.

   iii. No, during the addition of one nucleotide to another, the β and γ phosphates are eliminated and only α phosphate group in the nucleotide is involved in phosphodiester bond formation.

8. **If the DNA of humans has 3×10^9 base pairs and if an Okazaki fragment, on an average, is 2000 base pairs long, how many RNA primers would have been synthesized by primase enzyme?**

   For 2×10^3 bp, one RNA primer is synthesized. Therefore for 3×10^9 base pairs, 1.5×10^6 RNA primers will be synthesized.

9. **During DNA replication, both growing chains are synthesized simultaneously. But DNA polymerases can synthesize in the 5′ to 3′ direction only. How is this possible when the two chains of DNA are antiparallel?**

   Because of the synthesis of leading strand by continuous synthesis and the other, i.e., lagging strand, by discontinuous synthesis.

10. **What are the two mechanisms that maintain the high degree of fidelity in DNA replication?**

    i. The template strand dictates the synthesis of the new strand.

    ii. Any incorrect base inserted is repaired immediately by the proofreading mechanism of DNA polymerase.

11. **Indicate the order of the following events that lead to DNA synthesis, starting from G1 phase of cell cycle. (a) mitosis (b) binding of DNA polymerase to origin of replication (c) G1 phase of cell cycle (d) leading and lagging strand synthesis (e) termination of DNA synthesis (f) G2 phase of cell cycle.**

    (c), (b), (d), (e), (f), (a).

12. **If the human genome has 3 billion base pairs and if the rate of polymerization is 3000 base pairs per minute (in each direction at the replication fork) in each origin of replication and if the complete genome is replicated in 6 hours, find out the number of origins of replications. If they are equally spaced, what is the distance between two successive origins (in terms of nucleotide pairs)?**

    Total no. of base pairs = 3×10^9

    For 2×3000 base pairs, time taken = 1 minute (since replication occurs bidirectionally). Therefore for 3×10^9 base pairs, time taken $= 1/2 \times 10^6$ minutes.

    If the time taken for replication is 6×60 minutes, the number of origins $1/2 \times 10^6$ divided by $6 \times 60 = 10^5/72 = 1389$

    No. of nucleotide pairs between two adjacent origins = $3 \times 10^9/1389 = 2.16 \times 10^6$ base pairs.

13. **Why are thymine dimers corrected more rapidly when cells are in the light than when they are in the dark?**

    By a photoreactivation system, thymine dimers are repaired. DNA photolyase is the enzyme involved which is activated by visible light.

Review Questions

1. When the *E. coli* chromosome replicates, about how many Okazaki fragments would have been formed? How are these numerous fragments assembled in the correct order in the new DNA?

2. List the enzymes and proteins required for DNA synthesis. Explain their mode of action.

3. What do you mean by

    i.   Continuous DNA synthesis and discontinuous synthesis

    ii. Leading strand and lagging strand?

4. What are the differences between eukaryotic and prokaryotic DNA polymerases?

5. Define

   i. Kornberg enzyme

   ii. Klenow fragment

   iii. Catenanes

   iv. Concatemers

6. Write an essay on the enzymology of DNA replication.

7. What are the basic requirements for DNA replication?

8. What are replisomes and primosomes?

9. How was the semiconservative mode of DNA replication established?

10. What are the strategies that regulate the synthesis of DNA?

11. Comment on the origin of replication and the termination sites of replication in both eukaryotes and prokaryotes.

12. Discuss the chemistry of DNA polymerase and DNA topoisomerase.

13. What is the difference between the action of topoisomerase I and II?

14. Explain the role of DNA helicase and primase in DNA synthesis.

15. Explain the different models of DNA synthesis.

16. What are telomerases? What is their role in DNA replication? What will happen if telomerase fails to do its function?

17. Write short notes on replication in bacteriophages. Give examples.

18. Discuss the various groups of inhibitors of DNA replication.

19. What is the difference between symmetric and asymmetric model of DNA replication?

20. Draw the chemical reaction mechanism for the formation of a phosphodiester linkage during DNA synthesis. Write a note on the enzyme that promotes the reaction.

21. Draw the chemical reaction mechanism of DNA ligase.

22. Give the schematic representation of the fork model of DNA replication.

23. What role does a primer play in DNA replication? How is it formed?

24. What is meant by sigma replication? How is it degraded?

25. What are the types of DNA damages? How are they repaired?

26. Write short notes on

    i.   Mismatch repair mechanism

    ii. Base excision repair

    iii. Nucleotide excision repair

    iv. Direct repair

    v.   Post-replication repair

27. Discuss the action of DNA photolyase.

28. Comment on SOS response.

29. What are thymine dimers? How are they formed and repaired?

30. Explain the editing function and processivity of DNA polymerase.

7

RIBONUCLEIC ACID AND TRANSCRIPTION

Ribonucleic acid molecules are long, unbranched polymers made up of ribonucleotide units. Each ribonucleotide contains a ribose, phosphate and a purine or pyrimidine base. Each ribonucleotide is connected to the two neighbouring ribonucleotides through $5' \rightarrow 3'$ phosphodiester bond. The purine/pyrimidine bases present in RNA are adenine, guanine, uracil and cytosine. RNA has the ribose moiety with $2'$-hydroxyl group unlike DNA. This hydroxyl group makes RNA very sensitive to alkaline hydrolysis causing cleavage at the $5' \rightarrow 3'$ phosphodiester bonds to produce $5'$-hydroxyl groups. DNA that lacks $2'$-hydroxyl group is resistant to alkaline hydrolysis.

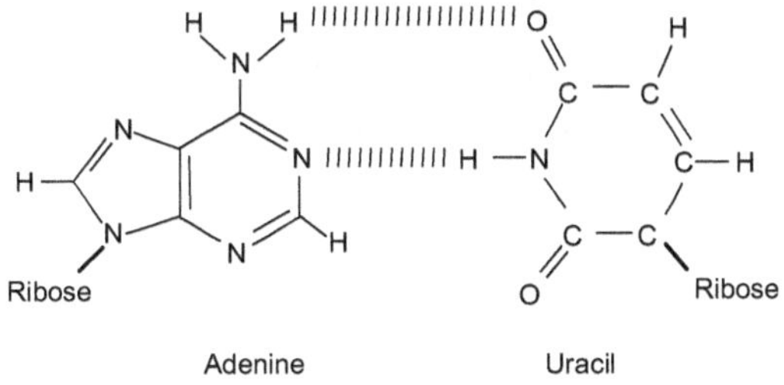

Figure 7.1   Adenine–Uracil base pair

Uracil present in RNA has the same base-pairing properties as thymine, i.e., it can form base pair with adenine (Figure 7.1). However, since RNA molecules are single-stranded there is no complementary strand base-paring as in DNA.

A typical cell contains about ten times as much RNA as DNA. There are four primary types of RNA—ribosomal RNA, transfer RNA, messenger RNA and a class of small eukaryotic RNA molecules found almost exclusively as ribonucleoprotein particles.

Messenger RNA (mRNA) is considered as an informational molecule because it carries the information of amino acid sequences (to be present in the protein product) in the form of triplet codons. Ribosomal RNA (rRNA) is a structural molecule because it constitutes an important component of the ribosomes, the cell organelle that is involved in protein biosynthesis. Transfer RNA (tRNA) is considered both as an informational and structural molecule, because it carries the anticodons for the amino acid sense codons and on charging carries the amino acids to be incorporated in the protein chain during translation and also because it acts as an adaptor molecule promoting anticodon–codon interactions. All the types of RNA are transcribed from DNA base sequences. Significant differences exist between the structures and modes of synthesis of the RNA molecules of prokaryotes and eukaryotes, though the basic mechanisms of their functions are the same.

TRANSCRIPTION

The genetic information is transferred from DNA to RNA and then from RNA to protein molecules. RNA molecules are synthesized by using one strand of DNA as a template. The polymerization reaction is catalysed by the enzyme RNA polymerase. The monomers involved as substrate are UTP, CTP, GTP and ATP (Figure 7.2). The process of transcription involves binding of RNA polymerase to DNA template, initiation, elongation and termination.

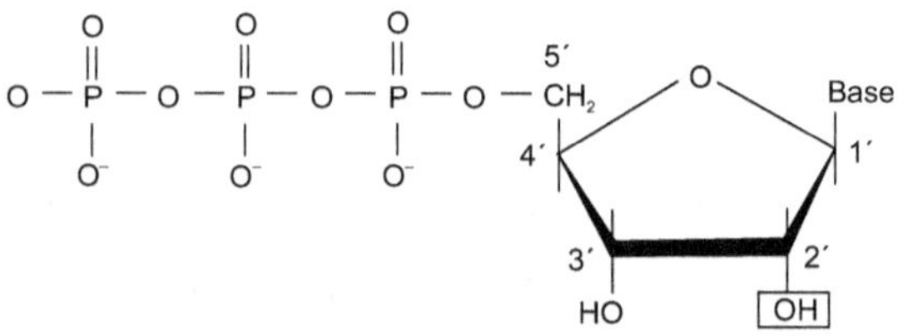

Figure 7.2   A ribonucleoside 5'-triphosphate

BASIC FEATURES OF RNA SYNTHESIS

The precursors in RNA synthesis are the four NTPs. Instead of TTP as in DNA, UTP is one of the four precursors required along with CTP, GTP and ATP. The sugar moiety in the nucleotide are ribose sugar with 2'-OH and 3'-OH group. As in the synthesis of DNA, 3'-OH group of one nucleotide and 5'-phosphate group of the incoming nucleotide are involved in phosphodiester bond formation, with the liberation of pyrophosphate. The sequence of bases in an RNA molecule is determined by the base sequence in the DNA template strand. Each base added to the growing end of the RNA chain is chosen by its ability to base-pair with the bases in the DNA strand. If the sequence of bases in the DNA template is 3'ATCGTA 5', then the order of bases added in the newly synthesized RNA will be 5'TAGCAT 3'.

The 2'-OH group shown in box in Figure 7.2 is replaced by a H in a deoxynucleotide.

An RNA strand is copied only from one strand of a segment of DNA molecule (Strand A in Figure 7.3).

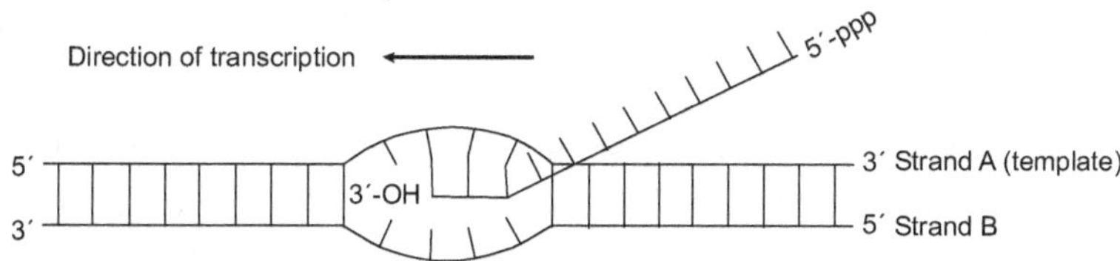

Figure 7.3   Transcription of DNA template

Because the incoming nucleotide is involved in bond formation with the 3'-OH group of the already existing nucleotide, RNA synthesis proceeds in the direction 5' → 3' as DNA synthesis. In contrast to DNA synthesis, RNA polymerases are able to initiate chain growth without the requirement for any primer, i.e., a 3'-OH group is not necessary to initiate addition of nucleotides against the DNA template. Though the monomers incorporated in RNA chain are NMPs, NTPs are the precursors and the first base added retains the triple phosphate in the 5'end. The product of RNA synthesis is referred to as primary transcript which then undergoes some processing to give the different types of mature RNA molecules. Thus, the primary transcript is always terminated with a triple phosphate in the 5'end.

The overall polymerization reaction may be written as

$$n\text{NTP} + \text{XTP} \xrightarrow[\text{RNA polymerase}]{\text{Mg}^{2+}, \text{DNA template}} 3'(\text{NMP})_n.\text{XTP5}' + n\text{PPi}$$

XTP represents the first nucleotide at the 5′-terminus of the RNA chain. NMPs are the mononucleotides in the RNA chain. Each time one NMP is added to the RNA chain, one pyrophosphate is released. Mg^{2+} ion is required for all nucleic acid polymerization reactions.

There are several major differences between prokaryotic and eukaryotic transcription.

RNA polymerase is the key enzyme involved in transcription. Prokaryotes have only one type of RNA polymerase that synthesizes all the types of RNA. But eukaryotes have three different types of the enzyme, RNA polymerase I, II and III, each responsible for the synthesis of particular classes of RNA. RNA polymerase I makes only rRNA, RNA polymerase II synthesizes all mRNA and RNA polymerase III makes tRNA and 5S rRNA. All these types of RNA polymerases are found in the nucleus. Each is a high molecular weight protein, consisting of two large subunits and up to ten small subunits depending on the organism.

E. coli RNA polymerase consists of six polypeptide subunits: two alpha (α) subunits, one beta (β) subunit, one beta prime (β') subunit, one omega (ω) subunit and one sigma (σ)subunit (Figure 7.4).

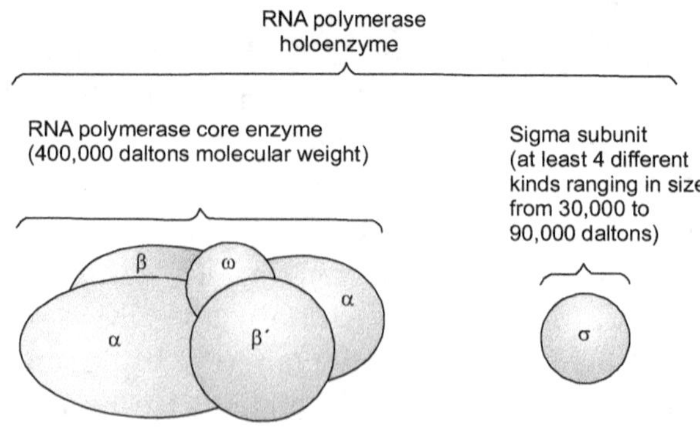

Figure 7.4　Structure of RNA polymerase

Five of these subunits α, α, β, β' and ω are tightly bound together and constitute the **core enzyme**. β' subunit contains two Zn^{2+} ions that participate in the enzyme's catalytic function. The active enzyme also requires the presence of Mg^{2+}. The core enzyme can synthesize RNA from DNA, but it cannot identify the initiation site of the gene. The sigma subunit is essential for initiating transcription at the beginning of a gene and binds to the core enzyme to form the **holoenzyme**. Bacterial cells have several different sigma subunits (Table 7.1) and each is responsible for initiating RNA synthesis for specific groups of genes.

Table 7.1   Bacterial sigma subunits

Subunit (Superscript gives MW × 10³)	Genes whose transcription is initiated
σ^{70}	Housekeeping genes
σ^{54}	Nitrogen metabolism enzyme genes
σ^{38}	Stress response protein genes
σ^{32}	Heat-shock genes
σ^{28}	Genes for flagellar synthesis and chemotaxis

STEPS IN THE SYNTHESIS OF RNA

The synthesis of RNA from DNA template includes four steps: binding of RNA polymerase to the DNA template, initiation of RNA synthesis, elongation and termination.

Binding of RNA Polymerase to the DNA Template

Binding occurs in particular regions called **promoters**. They are sequences of nucleotides in the DNA where several interactions occur. RNA polymerase must recognize a specific DNA sequence, attach in a proper conformation, locally open the DNA strands in order to gain access to the bases to be copied and then initiate synthesis. These events are guided by the base sequence of the DNA and the polymerase σ subunit, without which the promoter is not recognized. In case of some promoters, auxiliary proteins are also required for binding.

The sigma subunit in association with the core enzyme recognizes and binds to the promoter. The promoter site defines the beginning of a gene for transcription. How tightly and how frequently an RNA polymerase molecule binds to the DNA is determined by the promoter sequence. All promoters recognized by a single sigma subunit have similar sequences and the nucleotides that occur most frequently in the promoter are known as a **consensus sequence** (Figure 7.5). The promoter consensus sequence is composed of two short sequences separated by about 20 base pairs. These two groups of sequences are labelled –35 and –10 sequences because they occur 35 and 10 bases before the first base of the DNA that will be transcribed into the first base of the RNA. DNA bases to the left (negative numbers) of the transcription start site are said to be **upstream** and bases to the right (positive numbers) are said to be **downstream**.

The first base of DNA that is transcribed into RNA first base is referred to as +1 base. This marks the start of a gene. The DNA strand that is transcribed is called the **template strand**. The other DNA strand with the $5' \rightarrow 3'$ orientation similar to the RNA transcript is called the **coding strand**. In *E. coli*, the polymerization reaction occurs at a rate of about 40 nucleotides per second at 37°C. Thus a gene of 1000 bp is transcribed in about 30 seconds.

A gene is generally defined as a physical entity like the DNA sequence that includes a start site (promoter), the coding information and a termination sequence. Another term for a gene is a **cistron** which is the information required to make one polypeptide. But in bacteria and in eukaryotic cell organelles namely the mitochondria and chloroplast, more than one polypeptide is often encoded in a single message. Such mRNA are called **polycistronic mRNA** (Figure 7.6) because more than one cistron is transcribed into a single mRNA.

Three important events occur: (i) template binding at a polymerase recognition site (ii) binding to an initiation site and (iii) establishment of an open promoter complex.

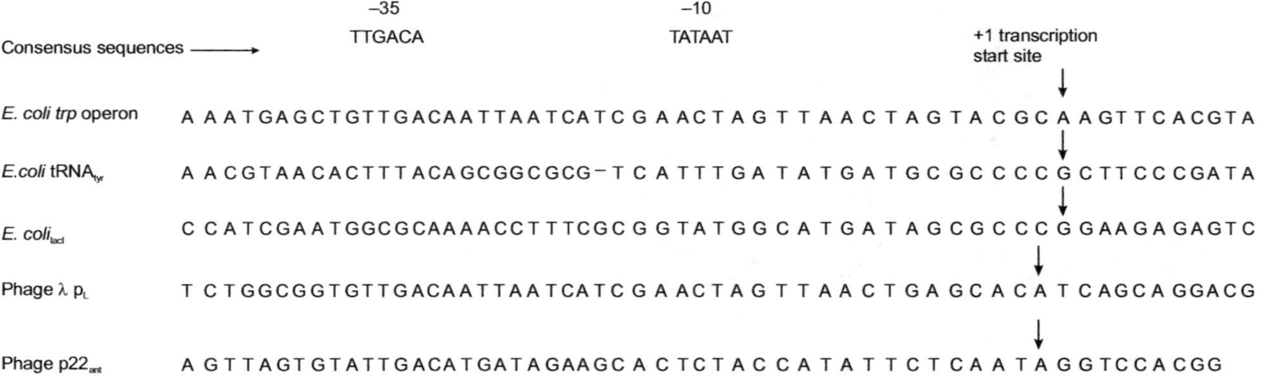

Figure 7.5   Promoter sequences

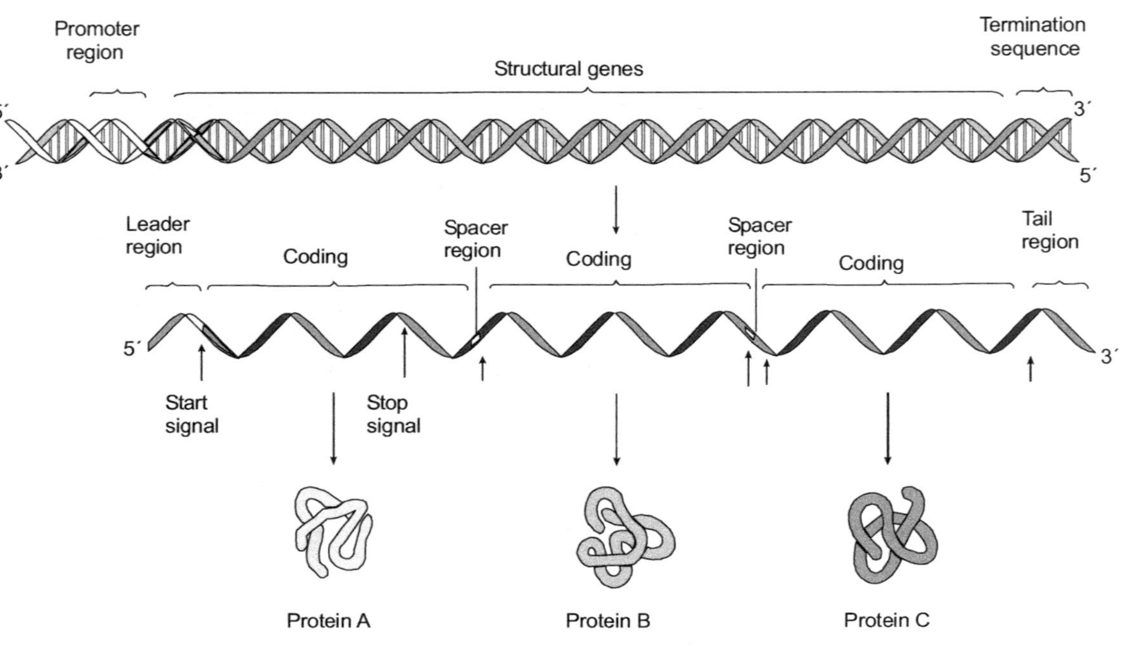

Figure 7.6   A polycistronic mRNA

Promoter regions in prokaryotes   In a region 5–10 bases to the left of the first base copied into mRNA is the right end of a sequence called the **Pribnow box**. A basic sequence derived from a large set of observed similar sequences is the consensus sequence. It is obtained by comparing a large number of sequences from a particular region. Examination of more than 100 *E. coli* promoters showed that the frequency of occurrence of the bases is the following, where the subscript indicates the frequency.

$$T_{70}A_{95}T_{45}A_{60}A_{50}T_{96}$$

The Pribnow box orients the RNA polymerase in a way that synthesis proceeds from left to right. It is the region at which the double helix opens to form the open promoter complex (Figure 7.7).

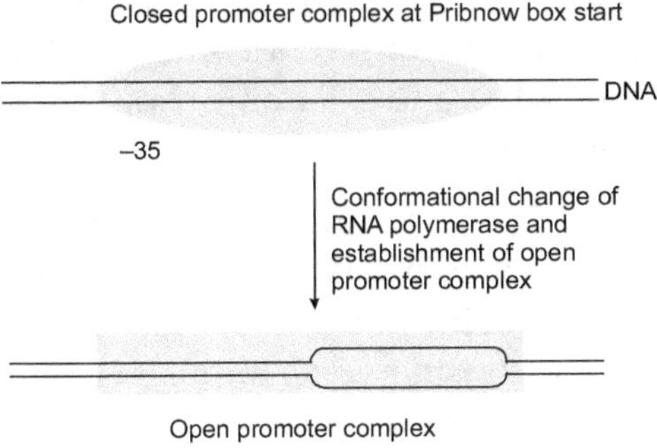

Figure 7.7   Stages of binding of RNA polymerase holoenzyme to a prokaryotic promoter

In the open promoter complex, the first base transcribed is chosen as a reference point and numbered +1. The direction of transcription is called downstream. All upstream bases, which are not transcribed are given negative numbers starting from the reference point (Figure 7.8). The Pribnow box is enclosed between –21 and –4, depending on the particular promoter.

In case of many promoters, there is a second important region to the left of the Pribnow box, whose sequences in different promoters have common features. This six-base sequence is called the **–35 sequence** and has a consensus TTGACA. This is the initial site of binding of RNA polymerase. It is assumed that σ factor, which is quite a large protein, makes initial contact with either or both the –35 sequence and the

Pribnow box and then mediates a conformational change of the core enzyme that enables various regions of the giant enzyme to contact both sequences.

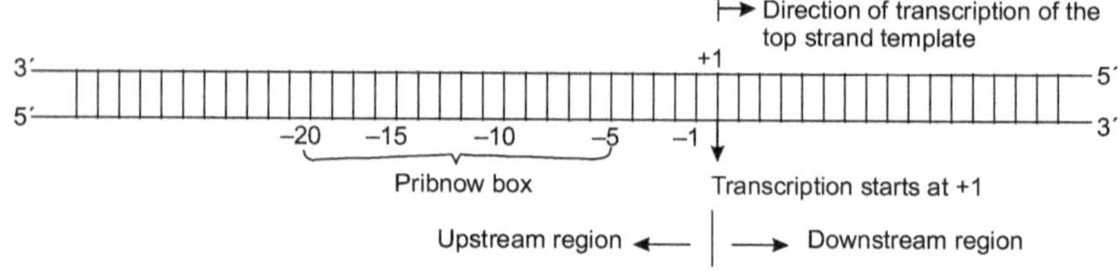

Figure 7.8   Upstream and downstream regions

Once the holoenzyme binds to the Pribnow box, there is the formation of the open promoter complex. This highly stable complex is the active intermediate in chain initiation. The DNA helix in an open promoter complex is locally unwound, starting about 10 bp from the left end of the Pribnow box and extending about 20 bp past the position of the first transcribed base.

It is thought that RNA polymerase itself induces this unwinding and undergoes a conformational change itself in so doing. This melting is necessary for pairing of incoming ribonucleotides. The Pribnow box and –35 sequence are considered as high level or strong promoters. There are also weak promoters in which recognition and/or binding by RNA polymerase is less strong. The number of RNA molecules synthesized from genes with weak promoters is much less than from a strong promoter. Therefore fewer mRNA molecules are made per unit time by genes with weak promoters. Promoter strength is one factor that determines the number of copies of each protein molecule present in the cell.

Some bacterial promoters require an activator protein for effective initiation. For example, λ phage has two promoters—pI and pre. These promoters can be active only in the presence of an auxiliary protein, the λ-encoded cII protein. This protein acts at a region upstream from the Pribnow box. In this upstream region, there is no –35 sequence. But the site of action of this protein is where this sequence would be. This region consists of two identical tetranucleotides flanking a hexanucleotide. Since one turn of dsDNA has 10 bp, the two tetranucleotides are on the same side of the helix and the central hexanucleotide is on the other side. The cII protein binds to the DNA in the major groove containing each tetranucleotide. The hexanucleotide sequence is the binding site for RNA polymerase. RNA polymerase can bind to this site only when cII protein is already bound to the tetranucleotide. Even though RNA polymerase and cII bind to opposite sides of the DNA, because of the great size of the enzyme, contact is made between RNA polymerase and cII protein and these contacts provide the necessary stability to allow transcription to be initiated.

Another example of a promoter that requires an auxiliary protein is the *E. coli lac* promoter. This promoter includes a –35 sequence but does not bind RNA polymerase unless the cyclic AMP (cAMP) receptor protein (CRP) is also bound. In the *lac* promoter there are two CRP–cAMP binding sites, one in the –70 to –50 segment referred to as site I and another in the –50 to –40 segment referred to as site II. Site I contains an inverted repeat sequence that is characteristic of most strong binding sites. Site II is a very weak binding site. But when CRP–cAMP complex is bound to site I, the ability of this complex to bind to site II is highly enhanced. Once site II is occupied, RNA polymerase can bind tightly to the promoter sequence, presumably by binding to CRP–cAMP complex (Figure 7.9).

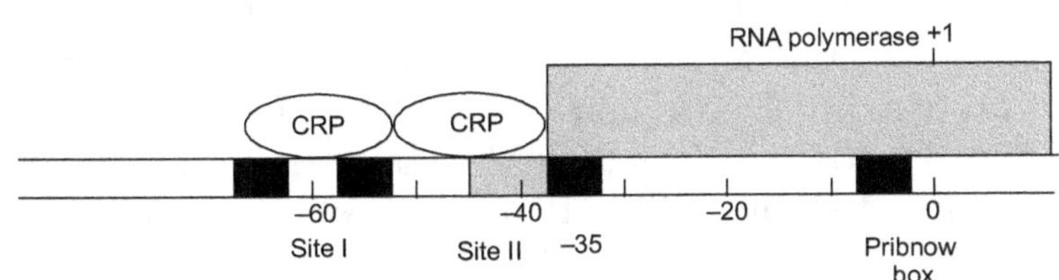

Figure 7.9   *E. coli lac* promoter

In the absence of CRP, RNA polymerase holoenzyme binds to the Pribnow box, but a closed promoter complex is formed. In this case, the opening of the DNA strand to act as a template does not occur.

Promoter regions in eukaryotes   The structure of eukaryotic promoters is generally more complex than the prokaryotic ones. These promoters in the upstream regions, which are hundreds of base pairs from the transcription start site, are lengthy sequences that control the rate of transcription. Specific proteins called transcription factors bind to particular DNA sequences in case of many promoters. The complexity of eukaryotic promoters is due to the fact that eukaryotic DNA is in the form of chromatin, which is not accessible to RNA polymerase.

There are different promoters of RNA polymerase II that synthesizes mRNA.

There is a nearly universal sequence, about 25 bp upstream from the transcription start site, whose consensus is TATAAAT. It is known as the **TATA** box or **Goldberg–Hogness** box which is similar to the prokaryotic Pribnow box. The TATA box is usually flanked by high G + C sequences. Many promoters have another common sequence in the –75 region with the consensus GG(T/C)CAATCT in which T and C are equally frequent at the third position from the left. This sequence is called **CAAT** box. A third element recognized in a few promoters is the **GC** box which has the consensus GGGCGG.

In eukaryotes, at least three *cis*-acting elements bring about efficient initiation of transcription of polymerase III, *"cis"* meaning "next to" or "on the same side". In molecular genetics, *cis* element means adjacent parts of the same DNA molecule. Therefore, examples of such *cis*-acting elements in eukaryotic transcription are the Goldberg–Hogness or TATA box, CAAT box and enhancers.

Also some *trans*-acting factors facilitate transcription. They act "from across" the promoter regions. Examples are the **transcription factors**, special proteins involved in transcription like TFIIA, TFIIB, etc. One of these, TFIID binds directly to the TATA-box sequence and is called the TATA-binding protein (TBP). TBP consists of about 10 polypeptide subunits. Once it binds to DNA, at least seven other transcription factors bind sequentially to TFIID forming a pre-initiation complex, which is then bound by RNA polymerase II.

The transcription factors in eukaryotes appear to play the role of sigma factor in the prokaryotic enzyme. The transcription factors play an important role in the regulation of eukaryotic gene expression.

The TATA box and the other sequences are the binding sites for transcription factors, but not directly for RNA polymerase. The RNA polymerase II does not directly interact with the promoter region and is strictly dependent on the presence of bound proteins— the transcription factors. This is the main difference between the promoters of eukaryotes and prokaryotes. The TATA box determines the base that is first transcribed.

The promoters of RNA polymerase III differ from those of RNA polymerase II in that, the pol III promoters are downstream from the transcription start site and within the transcribed DNA. For example, in the 5S rRNA gene of *Xenopus laevis*, the promoter is between +45 and +95 nucleotides downstream from the start point. It means that RNA polymerase II reaches forward to find the start point (+1), whereas RNA polymerase III reaches backward. In *Xenopus*, initiation of transcription of 5S rRNA requires three transcription factors. One of these is TFIIIA or a 40-kDa protein, which binds to the DNA in the region +45 to +96. It has at least three binding sites: two for the internal promoter and one for RNA polymerase III. This protein is specific for the transcription of 5S rRNA gene. Once the 40-kDa protein is bound to the promoters, many cycles of transcription occur without dissociation of the complex. The complex even survives cell division. It is suggested that the 40-kDa protein binds tightly to the non-coding DNA strand and is passed in each transcription cycle by RNA polymerase III that is bound only to the coding strand during chain elongation.

Another important region in the DNA whose existence and position correlates with the transcriptional activity of a particular gene is the **hypersensitive site**. They are specific regions highly susceptible to the action of DNase. It is seen that in chromatin isolated from chicken reticulocytes that synthesize globin, the globin gene's specific sites are preferentially broken by the action of DNase, but in the oviduct that synthesizes ovalbumin, the chromatin contains hypersensitive sites only in the ovalbumin gene region and not in the globin gene region. Also, hypersensitive sites are not present in the ovalbumin gene region of reticulocyte chromatin.

One more important binding site for transcription factors is the **upstream activation site**, usually present 50–300 nucleotide pairs from the promoter. Even if their position is changed by a few 100 nucleotide pairs, there is no decrease in their effect on gene expression.

Enhancers are sequences of 20–30-bp size associated with transcription of cellular genes. The enhancers have no promoter activity themselves. It is seen that deletion of enhancers reduces transcription of the associated gene about 100-fold. An enhancer's activity, even if moved upstream or downstream of the promoter, is not lost. As the distance between the enhancer and the promoter increases, the enhancing effect is decreased. By genetic engineering, an enhancer can be cut from its normal site and reinserted in reverse orientation with respect to the promoter without loss in its enhancing activity. When two or more promoters are downstream from an enhancer, the magnitude of enhancement is greater for the nearest promoter. The efficiency of enhancement varies with the promoters. The enhancers are suggested either to direct the formation of a particular DNA conformation or to alter the chromatin structure, thereby favouring promoter function. They are also suggested to provide bidirectionary entry for the transcription factor or RNA polymerase.

RNA Chain Initiation

Once the open promoter complex is formed, RNA polymerase is ready to initiate synthesis. RNA polymerase has two nucleotide binding sites called the initiation site and the elongation site. The initiation site binds purine triphosphates, ATP and GTP. Usually ATP is the first nucleotide in the chain. Thus, the first DNA base that is transcribed is thymine (+1 base). The initiating nucleoside triphosphate binds to the enzyme in the open promoter complex and forms a hydrogen bond with the complementary DNA base. The elongation site, otherwise referred to as the catalytic site, is then filled with a nucleoside triphosphate, which can form hydrogen bond with the second base in the DNA template (+2 base). The two nucleotides added are then joined together. The first base is released from the initiation site and initiation is complete. RNA polymerase and the template strand move relative to each other. So, the two binding sites are shifted by exactly one nucleotide. The newly formed dinucleotide remains hydrogen-bonded to the DNA.

Chain Elongation

After seven to eight nucleotides are added to the growing chain, RNA polymerase undergoes a conformational change and loses the σ subunit. This marks the transition from the initiation phase to the elongation phase. Thus, elongation is promoted by the core enzyme. The core enzyme moves along the DNA, binding a nucleoside triphosphate that can pair with the next DNA base and opening the DNA helix as it moves. The DNA helix recloses as synthesis proceeds. The newly synthesized RNA is released from its hydrogen bonds with the DNA as the helix re-forms. Roughly 12 RNA bases are paired to the DNA in the open region.

Elongation does not occur at a constant rate. RNA synthesis slows down when particular regions of DNA are passed, then continues at normal rate, slows down again, accelerates again and so on. This reduction in rate of synthesis of RNA is called a **pause**. This is suggested to be due to the formation of hairpin loop structures in the newly synthesized RNA, due to the presence of complementary bases within a single RNA chain. This is possible if the RNA strand is transcribed from a palindrome sequence of DNA template. Another reason for the pause is suggested to be due to the phosphodiester cleavage of the synthesized RNA by the same enzyme RNA polymerase, which does the polymerization reaction. The degradation activity of the enzyme may be in the reverse direction, similar to the editing function of DNA polymerase.

Termination and Release of the Newly Synthesized RNA

Termination of RNA synthesis occurs at specific base sequences within the DNA molecule. These sequences are of two types—simple terminators and those that require auxiliary termination factors. About 100 factor-independent termination sequences have been determined in bacteria. There are three important regions.

This effect is easily demonstrated *in vitro*. If Rho is added to a reaction mixture after RNA polymerase has passed the Rho-dependent terminator, RNA synthesis continues until the Rho-independent terminator is reached.

1. There is an inverted repeat containing a central non-repeating segment, like

    $$5'\,A\,B\,C\,D\,E - X\,Y\,Z - E'\,D'\,C'\,B'\,A'\,3'$$

    $$3'\,A'\,B'\,C'\,D'\,E' - X'\,Y'\,Z' - E\,D\,C\,B\,A\,5'$$

    This sequence is capable of intrastrand base-pairing, forming a stem and loop in the RNA transcript and possibly a cruciform structure in the DNA strands.

```
              X Y Z
              E    E′
              D    D′
              C    C′
              B    B′
    5′— A    A′— 3′
    3′— A′   A — 5′
              B′   B
              C′   C
              D′   D
              E′   E
              X′Y′Z′
```

The stem-and-loop structure renders resistance for the newly synthesized RNA to degradation by RNase II, an intracellular ribonuclease that is inactive against double-stranded RNA.

2. Near the loop end of the stem, there is a high G+C sequence. RNA polymerase usually slows down when synthesizing the corresponding RNA segment.

3. This is followed by a sequence of A-T pairs, with A in the template strand, yielding in the RNA, a sequence of 6–8 Us (Figure 7.10).

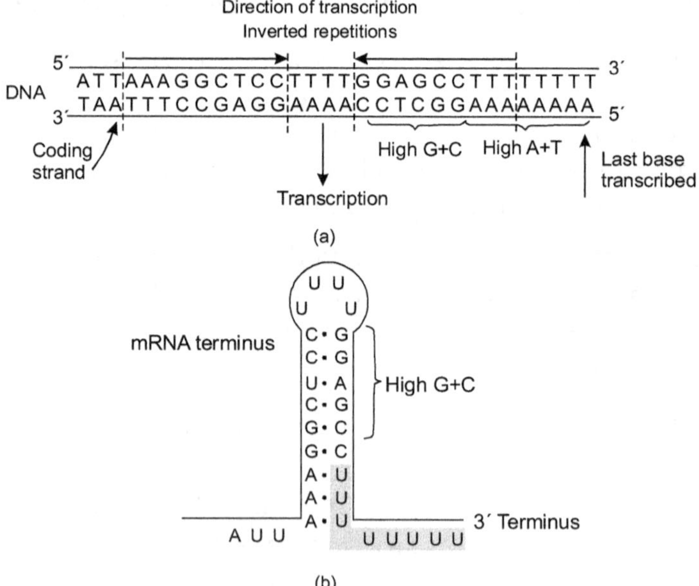

Figure 7.10   Base sequence of (a) the DNA of the *E. coli* trp operon at which transcription termination occurs and of (b) the 3′-terminus of the mRNA molecule.

The inverted repeat sequences are indicated by reversed arrows.

Transcription does not terminate at a unique site. Some RNAs end with five Us and some with six Us.

A second class of terminator region requires a termination protein called Rho. Rho-dependent termination sequences have a form different from the Rho-independent sequences (Figure 7.11). The region in which termination occurs is preceded by a long stem and loop that is followed by a long tract of 70–80 nucleotides free of any double-stranded segment. This segment has short GC-rich regions upstream from the termination site. A poly(U) segment is not present. Rho protein binds tightly to RNA. When bound to segments rich in C, it acquires a powerful ATP-cleaving activity. This is essential for termination of transcription. Rho is found to interact with the β subunit of RNA polymerase. As polymerization proceeds and C-rich segments of RNA are made, the ATPase activity of Rho increases until no further polymerization is possible, because NTPs cannot reach RNA polymerase without degradation. Sometimes Rho is required to prevent chain growth and sometimes Rho is required to release the RNA.

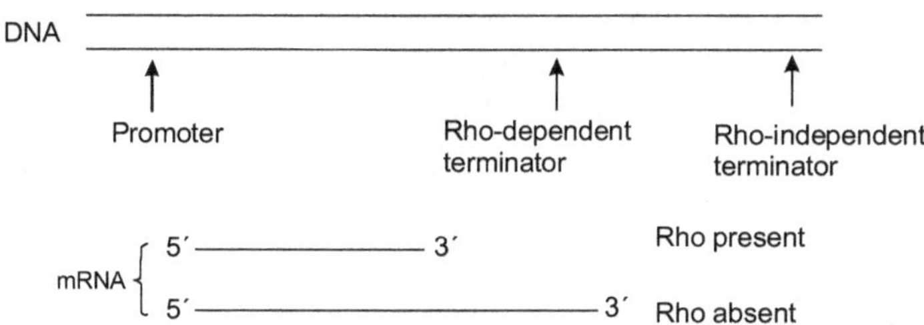

Figure 7.11   The effect of Rho on termination of RNA synthesis

The final step in termination is dissociation of both the core enzyme and the RNA from DNA template. When RNA polymerase core enzyme can no longer advance, the reverse reaction, that is, degradation of phosphodiester bond predominates. The RNA is degraded right through the DNA–RNA hybrid region. Since without the σ subunit, the core enzyme cannot restart transcription, the core enzyme also leaves the DNA. Once again it interacts with a free σ subunit to reform the holoenzyme.

Figure 7.12 describes the four events of transcription.

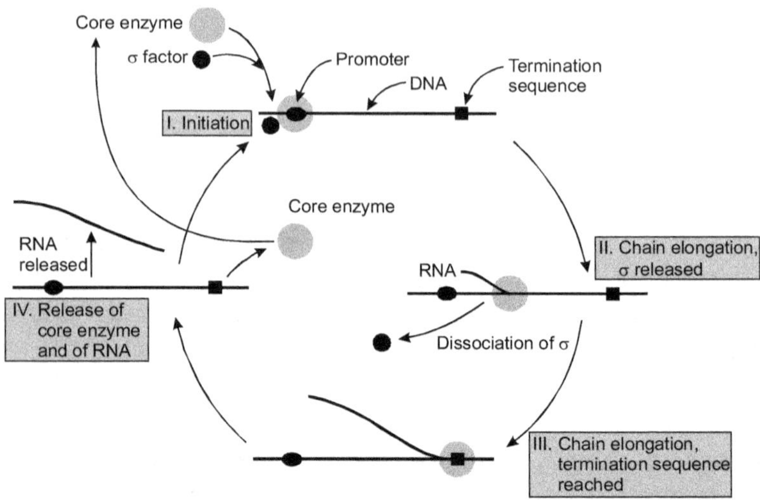

Figure 7.12   Events in transcription

TYPES OF RNA MOLECULES

There are three major types of RNA—messenger RNA, an informational molecule, ribosomal RNA, a structural molecule, and transfer RNA, both an informational and a structural molecule. In eukaryotes, there are another type of RNA molecules located in small ribonucleoprotein particles. All these RNA molecules are transcribed from DNA base sequences.

Messenger RNA (mRNA)

The base sequence of a DNA molecule determines the amino acid sequence of the polypeptide chains. Transcription of the template strand of the DNA gives base sequences of RNA molecule (mRNA), which is used by the protein-synthesizing machinery of the cell to give the proteins of specific amino acid sequence. The nucleotide sequence of the mRNA is read in groups of three bases (codons) from a start codon to a stop codon, with each codon corresponding either to one amino acid or to a stop signal.

In prokaryotes, an mRNA molecule may encode several different polypeptide chains when it is referred to as polycistronic mRNA molecule.

The segment of RNA corresponding to a DNA cistron is often called a **reading frame**, because it is read by the protein synthesizing system. Also, the polycistronic mRNA gives proteins of a single metabolic pathway. For example, in *E. coli*, three proteins (enzymes) are required to metabolize lactose which are obtained from a single mRNA molecule and the ten enzymes required for the synthesis of histidine are encoded in another mRNA molecule. In eukaryotes, each protein is encoded by an mRNA molecule which are referred to as monocistronic mRNA.

The size of mRNA molecules vary within a wide range. If a protein has 50 amino acid residues, 150 nucleotides (3 for one amino acid) are required in a monocistronic mRNA molecule. Polycistronic mRNA molecules may have 3000–7000 nucleotides.

In addition to the reading frames and start and stop sequences required for translation, there are other regions also in the mRNA which are significant. Though protein biosynthesis starts only from the start codon, there is a non-translated mRNA region before the coding region, which is referred to as a **leader**. Sometimes, the leader sequence contains a regulatory region. Untranslated sequences may be found both at the 5'-phosphate and 3'-hydroxyl termini. There may also be intercistronic sequences called **spacers**.

Although an mRNA is transcribed from only one strand of DNA at a given place along a gene, not all mRNA molecules are synthesized from the same DNA strand. In a long stretch of DNA, an mRNA may be seen growing in both directions. Rightward-moving mRNA is transcribed from the r strand. Since the growth of mRNA proceeds from the 5' end to the 3' end, and since a DNA–RNA hybrid is an antiparallel structure, the r strand is drawn with the 3'-OH terminus at the left (Figures 7.13 and 7.14).

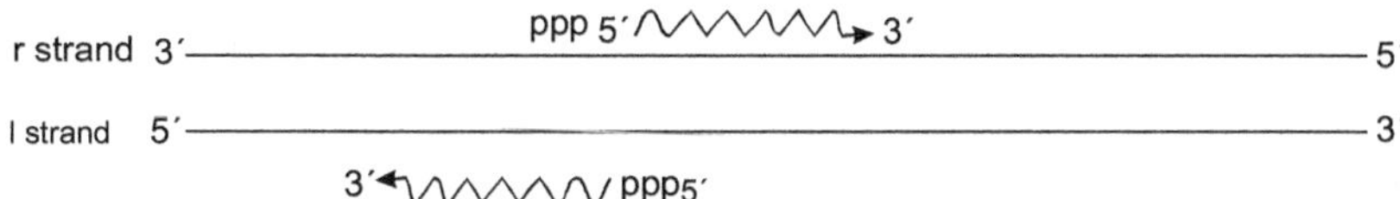

Figure 7.13   l and r strands of a DNA molecule acting as templates for RNA synthesis

BOX 7.1   TYPES OF RNA EDITING

RNA editing refers to the change of nucleotide sequence of a pre-mRNA prior to translation. There are two types of RNA editing—**substitution editing** in which the identities of individual nucleotide bases are altered, and **insertion/deletion editing**, where nucleotides are added to or subtracted from the total number of bases. Substitution editing is used in some nuclear-derived eukaryotic RNAs. It is prevalent in mitochondrial and chloroplast RNAs transcribed in plants. *Physarum polycephalum*, a slime mould, uses both substitution and insertion/ deletion editing for its mitochondrial mRNAs.

Trypanosoma, a parasite that causes African sleeping sickness, uses insertion/deletion editing in mitochondrial RNAs. The number of uridines added to an individual transcript can make up more than 60 per cent of the coding sequence, usually forming the initiation codon and placing the rest of the sequence into the proper reading frame. Insertion/deletion editing in trypanosomes is directed by gRNA (guide RNA) templates, which are also transcribed from the mitochondrial genome. These small RNAs are complementary to the edited region of the final, edited mRNAs. They base-pair with the pre-edited mRNAs to direct the editing machinery to make the correct changes.

Substitutional editing occurs in mammalian nuclear encoded mRNA transcripts. Apolipoprotein B (apo B) exists in both long and short forms. But a single gene encodes both proteins. In human intestinal cells, apo B mRNA is edited by a single C to U change, which converts a CAA glutamine codon into a stop codon and terminates the polypeptide at half its genomically encoded length. The editing is performed by a protein complex that binds to a "mooring sequence" on the mRNA transcript downstream of the editing site. In a second system, the synthesis of subunits constituting the glutamate receptor channels (Glu R) in mammalian brain tissue is also affected by RNA editing. Here, adenosine (A) to inosine (I) editing occurs in pre-mRNA prior to translation, where I is read as guanosine (G) during translation. A family of three adenosine deaminase acting on RNA (ADAR) enzymes are responsible for the editing of various sites within the glutamate channel subunits.

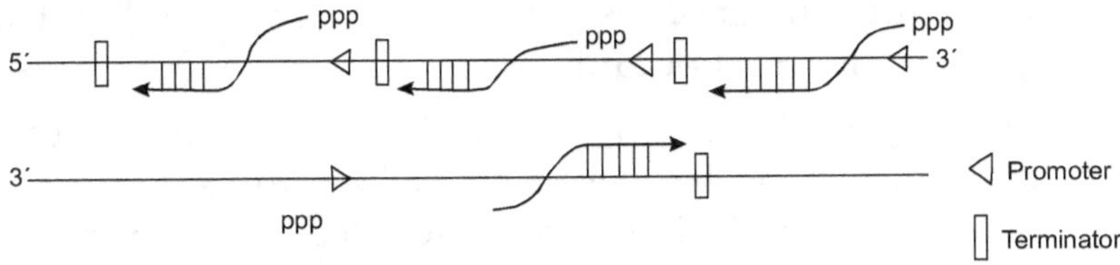

Figure 7.14   Schematic drawing showing that complementary DNA strands can be transcribed, but in different regions in the DNA

Compared to the other types of RNA in prokaryotes, the mRNA's lifetime is short. For bacteria, the half-life of a typical mRNA molecule is a few minutes. Though this feature may seem to be wasteful, it has an important regulatory function. If a particular protein is no longer required by the cell, synthesis of the corresponding mRNA is stopped.

Transfer RNA (tRNA)

Transfer RNA molecules are small adaptor molecules. There are many different tRNA molecules in a particular cell. Each tRNA molecule has several regions, two of which are important. One region has a sequence of three bases that can hydrogen-bond with the bases in the codon present in mRNA. This sequence is called the **anticodon** and this is present in a stem-loop structure referred to as **anticodon arm**. A second site is the amino acid attachment site. The amino acid that binds to this site corresponds to the particular codon in mRNA that forms base pairs with the anticodon of the tRNA.

The tRNA molecules have 73 to 93 nucleotides. Several segments of the molecule can form double-stranded regions that gives the molecule a cloverleaf structure in which open loops are connected to one another by double-stranded stems.

The tRNA nucleotides are numbered from the 5'-phosphate terminus. The 5'-phosphate terminus is always base-paired, which contributes to the stability of tRNA. The 3'-OH terminus is always a four-base single-stranded region having the base sequence XCCA-3'-OH in which X can be any base. This is called the CCA or **acceptor end**. Adenine in the CCA sequence is the site of attachment of the amino acid.

There are many modified bases in tRNA like dihydrouridine (DHU), ribosyl thymine (rT), pseudouridine (ψ) and inosine (I).

There are three large single-stranded loops. The lowermost anticodon arm or loop has seven bases. The anticodon occupies positions 34, 35 and 36. This is always preceded by two pyrimidines followed by a modified purine. Thus the anticodon loop has the following general sequence:

5'-Py-U -XYZ- Pu (modified)-Variable base-
$\underbrace{}_{\text{Anticodon}}$

The loop containing bases 14 to 21 is called the DHU loop. The size is not constant in all tRNA molecules. Some may have up to three extra bases. The loop containing bases 54 through 60 has the sequence T ψ C and is called T ψ C loop.

Overall, there are four double-stranded regions called **stems or arms**. An additional loop, containing bases 44 through 47 is also present called **extra arm** (Figure 7.15).

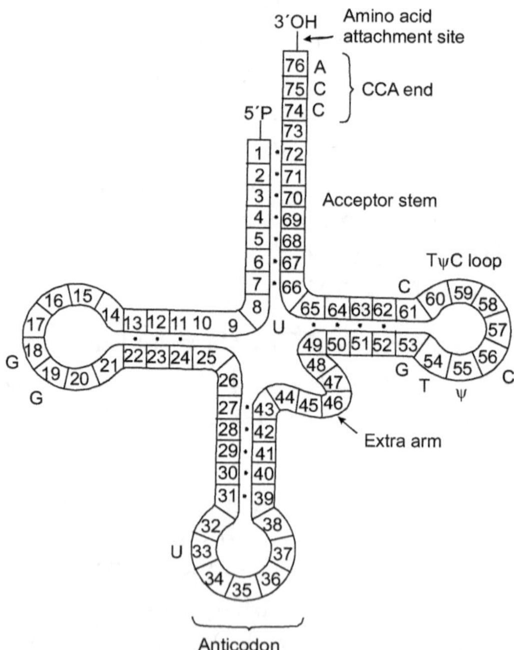

Figure 7.15   The currently accepted "standard" tRNA cloverleaf with its bases numbered. A few bases present in almost all tRNA molecules are indicated.

X-ray crystallographic analysis shows that the three-dimensional structure of tRNA molecule is L-shaped with folds and two helical double-stranded branches each about 60 Å long and mutually perpendicular. One branch consists of the acceptor stem and T ψ C stem and the other consists of the stems of the DHU and anticodon loops (Figure 7.16).

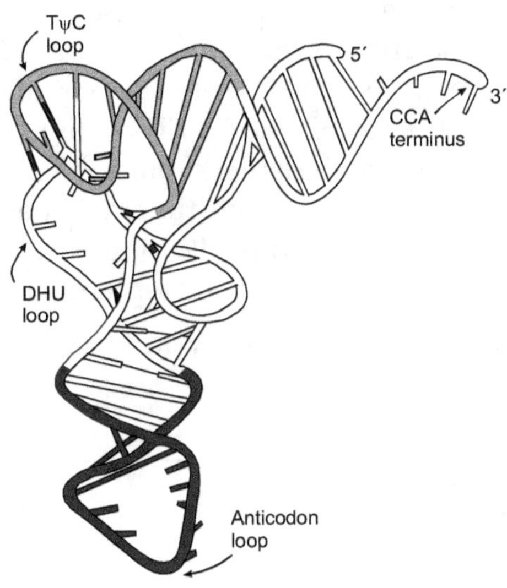

Figure 7.16   Three-dimensional structure of yeast tRNA^{Phe}

The CCA acceptor terminus is at one end of the L and the anticodon loop at the other end, about 70 Å away. The T ψ C loop that interacts with ribosomal RNA is at the corner of the L structure.

In eukaryotes, tRNA molecules are excised from large transcripts called pre-tRNA, that has one or more tRNA sequences. Yeast tRNA gene has 10–30 base introns which is spliced to get mature tRNA.

Ribosomal RNA (rRNA)

Ribosomal RNAs are structural units of the ribosome. In *E. coli*, there are three rRNAs—5S, 16S and 23S rRNA with 120, 1542 and 2904 nucleotides. The S designations refer to the sizes of the rRNAs measured in Svedberg units, which denotes the rate of sedimentation. In eukaryotes, there are four types of rRNAs—5S, 5.8S, 18S and 28S rRNA with 120, 160, 1774 and 4717 nucleotides respectively.

Molecular hybridization studies have established the degree of redundancy of the genes coding for the rRNA components. The *E. coli* genome contains seven copies of a single sequence that encodes all the three components, 5S, 16S and 23S rRNA.

In eukaryotes, many more copies of a sequence encoding the 18S and 28S components are present. In *Drosophila*, approximately 120 copies per haploid genome are each transcribed into a molecule of about 34S which is then processed into 5.8S, 18S and 28S rRNA species. The rRNA genes are called rDNA. In prokaryotes the larger subunit of the ribosome consists of a 23S and 5S rRNA. In eukaryotes, 5S, 5.8S and 28S rRNA are present in the larger subunit. The smaller prokaryotic subunit contains the 16S rRNA, while the smaller eukaryotic subunit consists of the 18S rRNA. Ribosomal RNA molecules perform all the important catalytic functions associated with translation.

The components of the prokaryotic and eukaryotic ribosomes are presented in Figure 7.17.

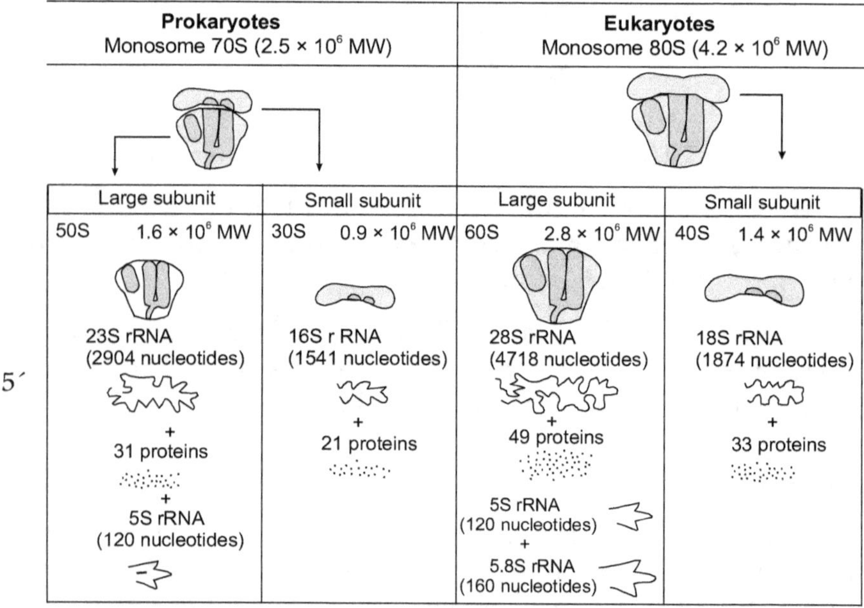

Figure 7.17   The components in the prokaryotic and eukaryotic ribosomes

Other Types of RNA Molecules

Eukaryotic cells contain another type of small RNA molecules of 100–300 nucleotides. Those present in the nucleus are called **small nuclear RNA (snRNA)** and in the cytoplasmic species are called **small cytoplasmic RNA (scRNA)**. These types of RNA molecules are associated with specific proteins, forming small ribonucleoproteins— **small nuclear RNP (snRNP)** and **small cytoplasmic RNP (scRNP)**. The different snRNPs are designated U1, U2… and each contains a unique RNA molecule (U1 RNA, U2 RNA, etc.), some of which participate in RNA processing.

POST-TRANSCRIPTIONAL MODIFICATIONS OR PROCESSING OF RNA MOLECULES

The synthesis of the three major types of RNA molecules is initiated at a promoter and completed at the terminator. However, all are initially referred to as primary transcripts. They undergo some modifications to give the mature RNA molecules. The following three properties of these molecules indicate that they are not the immediate products of transcription, except prokaryotic mRNA molecules:

1. The molecules are terminated by a 5′-monophosphate rather than the expected triphosphate found at the ends of all primary transcripts.

2. Both rRNA and tRNA are much smaller than the primary transcript.

3. All tRNA molecules contain bases other than A, G, C and U and these "unusual" bases are not present in the original transcript.

Thus post-transcriptional modifications or RNA processing are the molecular changes made in the primary transcript to yield mature RNA molecules.

Processing of tRNA Molecules

E. coli tRNA^{tyr} molecule with 85 nucleotides was well-characterized and the nucleotide sequence was studied. In *E. coli* there are two copies of the tRNA^{tyr} gene. Two identical adjacent copies of the DNA are present from which this tRNA is transcribed. Each gene consists of about 350 (not 85, the number of nucleotides in the tRNA molecule) nucleotide pairs. They are separated by a 200-nucleotide pair "spacer". The two genes are transcribed as a single RNA molecule that is cut off after transcription is complete. In a 350-bp gene, transcription starts 41 bp upstream from the 5'end of the tRNA base sequence and stops 224 bp downstream from the 3' terminus of the tRNA. The transcript formed is processed in three stages (Figure 7.18).

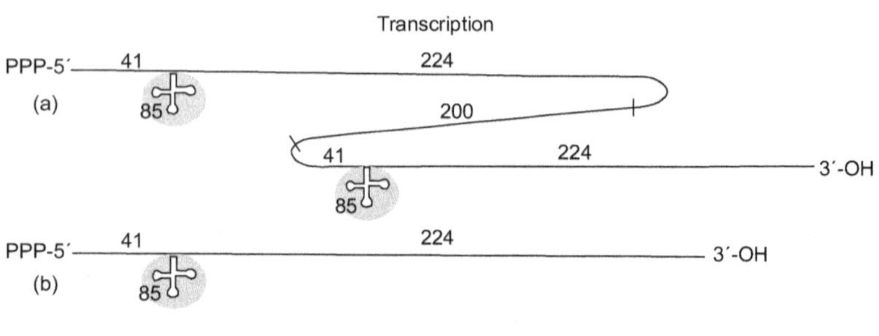

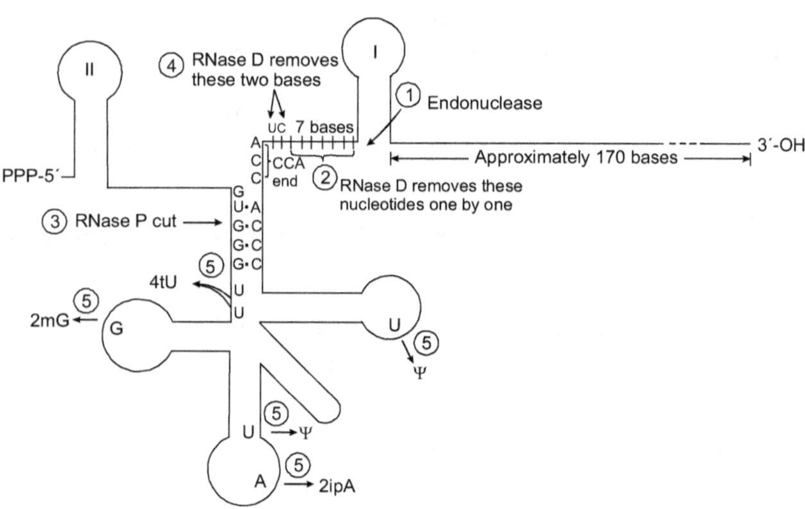

Figure 7.18   Post-transcriptional modification of tRNA

1. *Formation of 3'-OH terminus*   An endonuclease (step1) recognizes a hairpin loop(I) and an exonuclease recognizes CCA base sequence. After endonuclease digestion at site 1, seven bases upstream are removed by an exonuclease called RNase D (step 2). This enzyme initially stops two bases short of the CCA terminus and later removes these two bases after the processing of the 5′ end. This leaves a molecule called pre-tRNA.

$$—p5' - (41\ bases) - tRNA - (2\ bases) - 3'\text{-}OH$$

2. *Formation of 5′-P terminus*   An enzyme called RNase P generates the 5′-P terminus in all *E. coli* tRNA molecules. RNase P removes the excess RNA from the 5′ end of a precursor molecule, including a hairpin loop (II), by an endonucleolytic cleavage (step 3). This generates the correct 5′-end. Once the 5′-P terminus is formed, RNase D removes the two 3′-terminal nucleotides (step 4), leaving a tRNA molecule having the correct length.

RNase P is an unusual enzyme. It contains 86% RNA and 14% protein by weight. Also, the RNA possesses the catalytic activity and the protein ensures the correct folding of the RNA to maximize the catalytic activity. The discovery of these catalytic RNAs called ribozymes indicates that the long-standing dogma that all enzymes must be proteins is invalid.

3. *Production of the modified bases*   The final tRNA processing event is to produce the altered bases in the tRNA (step 5). Selected enzymes produce the necessary charges. In $tRNA_1^{tyr}$, two uridines are converted to pseudouridine (ψ), two uridines to two 4-thiouridine (4tU), one guanosine to 2′-O-methyl guanosine (2mG) and one adenosine to isopentenyl adenosine (2ipA).

The modification of bases in tRNA are as indicated in Figure 7.19.

All tRNA molecules are terminated by CCA 3′OH. If this sequence is absent in some precursor tRNA molecules like in yeast, CCA sequence is added by the enzyme tRNA nucleotidyl transferase.

Multiple copies of a particular tRNA molecule are commonly found in a single transcription unit. For example, there are four copies of one of the $tRNA^{leu}$ in its precursor molecule. The occurrence of different tRNA molecules in a single transcript is also frequent. For example, one $tRNA^{ser}$ and one $tRNA^{thr}$ are present in a single unit in *E. coli*. Some tRNA molecules are even obtained from the transcript that contains rRNA.

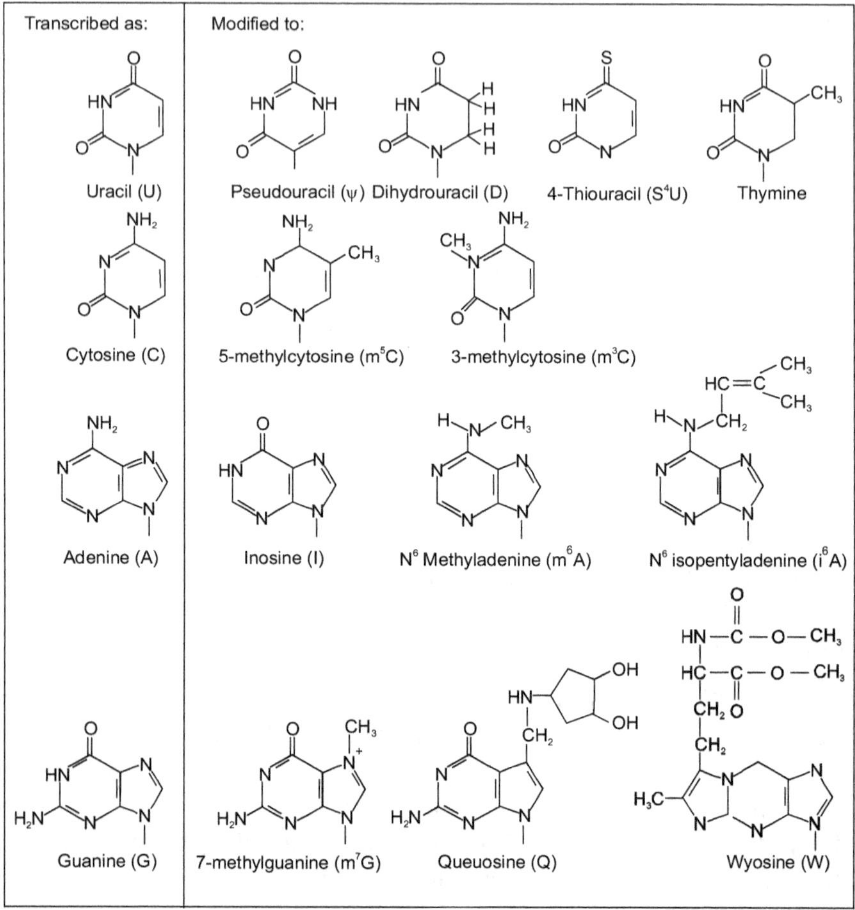

Figure 7.19   Examples of base modifications in the nucleotides of tRNA. In the left-hand column are the bases as they are originally transcribed. In the right-hand column are some common modifications.

BOX 7.2   TRANSCRIPTION OF φX174

Bacteriophage φX174 carries a single strand of DNA known as the "plus" strand. Upon infection in *E. coli*, the plus strand directs the synthesis of the complementary "minus" strand, with which it combines to form a circular duplex DNA called replicative form. The minus strand is the antisense strand which is transcribed, that is, it acts as the template for RNA synthesis. The plus strand is the sense strand, so called because it has the same sequence as the transcribed RNA.

Processing of Ribosomal RNA

Bacterial ribosomes contain three kinds of rRNA—5S rRNA, 16S rRNA and 23S rRNA with 120, 1541 and 2904 nucleotides respectively. These molecules plus several tRNA molecules are cleaved from a continuous transcript having more than 5000 nucleotides of known sequence. Seven primary transcripts have been isolated from *E. coli* containing rRNA which differ by the identity of the tRNA molecules and the location of the tRNA with respect to rRNA sequence. For example, there is a transcript that has four different tRNA molecules and is represented as follows:

16S rRNA-tRNA[ile]-tRNA[ala]-23S rRNA-5S rRNA-tRNA[asp]-tRNA[trp]. This can be represented as 16S rRNA-spacer-23S rRNA-5S rRNA-spacer where tRNAs are present in the spacer regions (Figure 7.20).

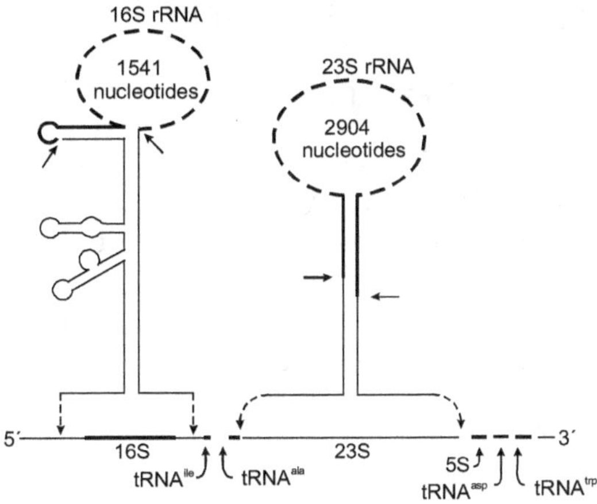

Figure 7.20 A schematic diagram of *E. coli* rRNA transcripts from which 5S, 16S, and 23S rRNA molecules are excised. The regions containing the 16S rRNA and 23S rRNA molecules are shown in expanded form above the line designating the transcript.

An rRNA transcript is usually cut while it is being synthesized, by several enzymes acting in sequence. RNase III cleaves double-stranded RNA in double-stranded stem regions, by making two single-strand breaks in complementary sequences but not opposite to one another. A 5′-phosphate and a 3′-OH group are generated but they are not the termini of the rRNA. Several enzymes are required to complete the processing.

From a single transcript, the different rRNAs are produced in a constant ratio. Since one molecule each of 5S, 16S and 23S rRNAs are present in a ribosome, their production also is maintained in a 1 : 1 : 1 ratio to avoid wastage of the molecules.

The organization of the genes for the four types of eukaryotic rRNAs 5S, 5.8S, 18S and 28S rRNA molecules differ from that observed in prokaryotes. The gene for 5S rRNA is present in a separate chromosome. The genes for the other three types of rRNA are present in another chromosome.

Several hundred copies of a DNA sequence that encode the 18S, 5.8S and 28S rRNA molecules are present in the nucleolus of a typical animal cell. The three ribosomal genes comprise a 40S transcription unit from which the three rRNA segments are excised. rRNA sequences of higher eukaryotes do not contain introns.

Processing of mRNA Molecules

A newly synthesized RNA molecule is referred to as the primary transcript. A structural gene in the DNA gives a specific protein product on transcription and translation. The transcribed RNA after processing yields a mature mRNA molecule. Post-transcriptional modification of mRNA occurs only in eukaryotes. There are three important events in the processing of mRNA molecules:

1. Formation of 5′ cap

2. Formation of 3′ tail and

3. RNA splicing

1. *Formation of 5′ cap*   A residue of 7-methyl guanosine is linked to the 5′-terminal residue of the mRNA through an unusual 5′,5′-triphosphate linkage. The 5′ cap is formed by the condensation of a molecule of GTP with the triphosphate at the 5′ end of the transcript. The guanine is then methylated at N-7 and additional methyl groups are added at the 2′-OH of the first and second nucleotides adjacent to the cap. The methyl groups are derived from S-adenosylmethionine.

The first step is

$$\text{Gppp} + \text{pppApNpNp} \longrightarrow \text{GppppApNpNp} + \text{ppi} + \text{pi}$$

The reaction is brought about by the enzyme guanylyl transferase. This is followed by the action of guanine 7-methyl transferase that methylates guanine at N^7 to form 7-O-methyl guanine. Next step is the methylation of the initial G or A to form either 2′-O-methyl guanosine (2′-O MeG) or 2′-O-methyl adenosine (2′-O MeA).

The resulting structural unit is called a cap:

(7 Me G)-5′ppp 5′-(2′-0-Me(G/A)-3′-p-5′-nucleoside-3′p

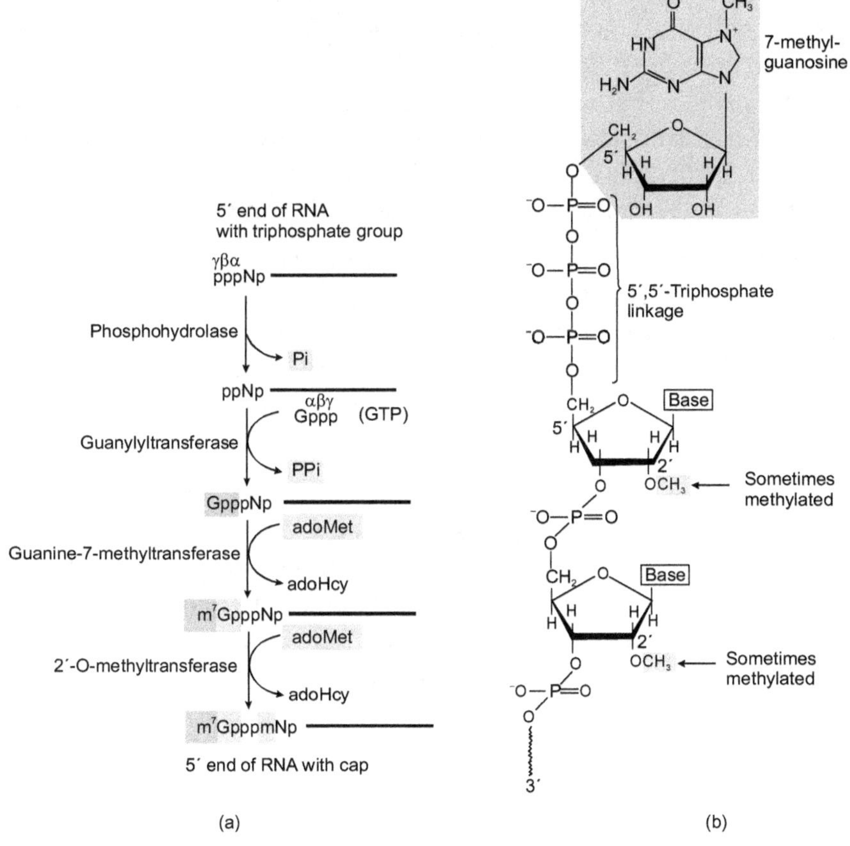

Figure 7.21   The 5′ cap of mRNA. (a) 7-Methylguanosine is joined to the 5′ end of almost all eukaryotic mRNAs in an unusual 5′, 5′-triphosphate linkage (b) Structure of caps at the 5′ ends of eukaryotic mRNA molecules. adoMet—S-adenosylmethionine, adoHcy—S-denosylhomocysteine

If there is no methylation of the 2'-OH position, the cap structure is designated Cap 0, in which 7-MeG is added to the 5' terminus and no further methylation occurs. This structure is not found in animal cells. In viral mRNA and most animal cell mRNA, methylation of ribose also occurs on the second nucleotide, giving the structure called Cap 1 (Figure 7.21).

In some cases, methylation also occurs in the third nucleotide, giving the Cap 2 structure.

Capping occurs shortly after initiation of mRNA synthesis and precedes all other modification. The biological significance of capping seems to be in the efficient protein biosynthesis. There are specific cap binding proteins in the cytoplasm which attach to the cap and stimulate formation of translation complex between mRNA and the ribosome and certain proteins required to initiate translation. Capping may function to protect the mRNA from degradation by nucleases.

2. *Formation of 3' tail*   At the 3' end, most eukaryotic mRNAs have a string of 80 to 250 adenylate residues called the poly(A) tail. The poly(A) tail is added in a multistep process. The transcript is extended beyond the site where poly(A) tail is to be added and then it is cleaved at the poly(A) addition site by an endonuclease component of a large enzyme complex.

The mRNA site where cleavage occurs is marked by two sequence elements: the highly conserved sequence 5' AAUAAA 3', 10 to 30 nucleotides on the 5' side upstream of the cleavage site and a less well-defined sequence rich in G and U residues 20 to 40 nucleotides downstream of the cleavage site (Figure 7.22) . Cleavage generates free 3'-OH group that defines the end of the mRNA to which adenylate residues are immediately added by polyadenylate polymerase, which catalyses the reaction.

$$RNA + nATP \rightarrow RNA\text{–}(AMP)_n + nPPi$$

where n = 80 to 250. This enzyme does not require a template, but requires the cleaved mRNA as a primer.

Not all mRNA molecules are polyadenylated. For example, in animal cells histone mRNA lacks poly(A) and the AAUAAA sequence. Termination of synthesis of this mRNA occurs at a unique position that remains at the 3' end. In yeast, termination and polyadenylation are coupled processes. However, some transcripts may lack the AAUAAA sequence and polyadenylation may still occur. The function of the poly(A) terminus is to increase the stability of mRNA.

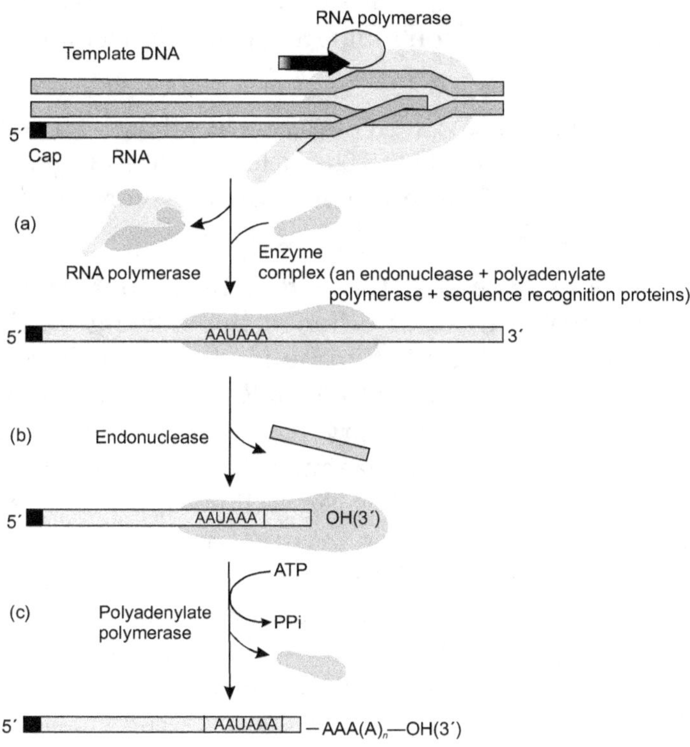

Figure 7.22   Poly(A) tail formation in mRNA

3. RNA splicing   A characteristic of most of the primary transcripts of higher eukaryotes is the presence of untranslated intervening sequences called introns that interrupt the coding sequence called exons. They are excised from the primary RNA transcript. The amount of RNA removed may range from 50–90% of the primary transcript. The excision of the introns and the formation of the final mRNA by the joining of the exons is called **RNA splicing**. The 5′ segment, i.e., the cap and the 3′-tail of a primary transcript is not discarded and therefore is always present in the final mRNA molecule. The number of exons is one more than the number of introns. The number of introns per primary transcript may vary considerably.

Number of introns in some eukaryotic genes are two in α-globin and Ig L-chain gene, four in Ig H-chain gene, seven in ovalbumin gene and 52 in α-collagen gene. The histone genes and interferon genes in higher organisms do not contain introns. In bacteria, a polypeptide chain is generally encoded by a DNA sequence that is co-linear with the amino acid sequence, continuing along the DNA template without interruption.

In eukaryotic mRNAs, most exons are less than 1000 nucleotides long, with many in the 100 to 200 nucleotide size range, encoding stretches of 30 to 60 amino acids. Introns vary in size from 50 to 20,000 nucleotides.

There are four classes of introns. Group I and group II introns have some common features but differ in the splicing mechanisms. Group I introns are found in some nuclear, mitochondrial and chloroplast genes coding for rRNAs, mRNAs and tRNAs. Group II introns are found in the primary transcripts of mitochondrial or chloroplast mRNAs in fungi, algae and plants. Neither class of introns requires a high-energy cofactor like ATP for splicing. The splicing mechanisms in both groups involve two transesterification reaction steps. A ribose 2'- or 3'-OH group makes a nucleophilic attack on a phosphorus and in each step, a new phosphodiester bond is formed at the expense of the old, maintaining the balance of energy. These reactions are similar to the DNA breaking and rejoining reactions promoted by topoisomerases and site-specific recombinases.

Group I splicing reaction requires a guanine nucleoside or nucleotide cofactor. The cofactor is not used as a source of energy. Instead, the 3'-OH group of guanosine is used as a nucleophile in the first step of the splicing pathway. The guanosine 3'-OH group forms a normal 3' → 5'-phosphodiester bond with the 5' end of the intron. The 3'-OH group of the exon that is displaced in this step then acts as a nucleophile in a similar reaction at the 3'-end of the intron. Thus the introns are excised and exons are ligated (Figure 7.23).

In group II introns, the pattern of intron removal is similar, with the difference that the 2'-OH of a specific adenosine in the intron acts as a nucleophile attacking the 5' splice site to form a lariat structure as an intermediate (Figure 7.24).

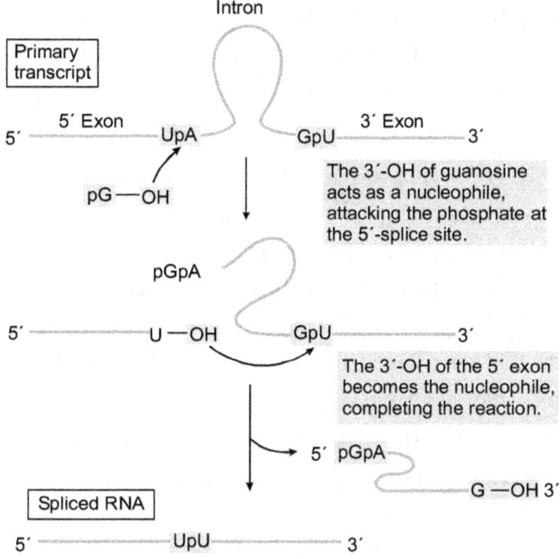

Figure 7.23   Splicing mechanism of group I introns

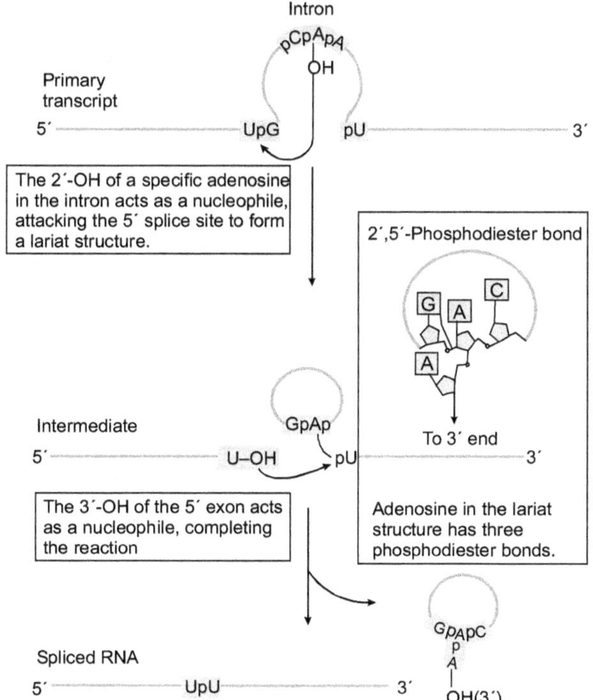

Figure 7.24   Splicing mechanism of group II introns

The splicing of group I and group II introns occurs by self splicing—the RNA molecules themselves performing the catalytic function. Intron classes that are not self-splicing are not designated with a group number.

The group III introns are the largest class of introns found in nuclear mRNA primary transcripts. They undergo splicing by the same lariat-forming mechanism as group II introns. But, splicing of group III introns requires the action of snRNPs (small nuclear ribonucleoproteins (called "snurps"). Each snRNP contains small nuclear RNAs (snRNAs) of eukaryotes, 100 to 200 nucleotides long. Five snRNAs (U1, U2, U4, U5 and U6) are involved in splicing reactions which are present in abundance in eukaryotic nuclei. The RNAs and proteins in snRNPs are highly conserved in eukaryotes from yeast to human.

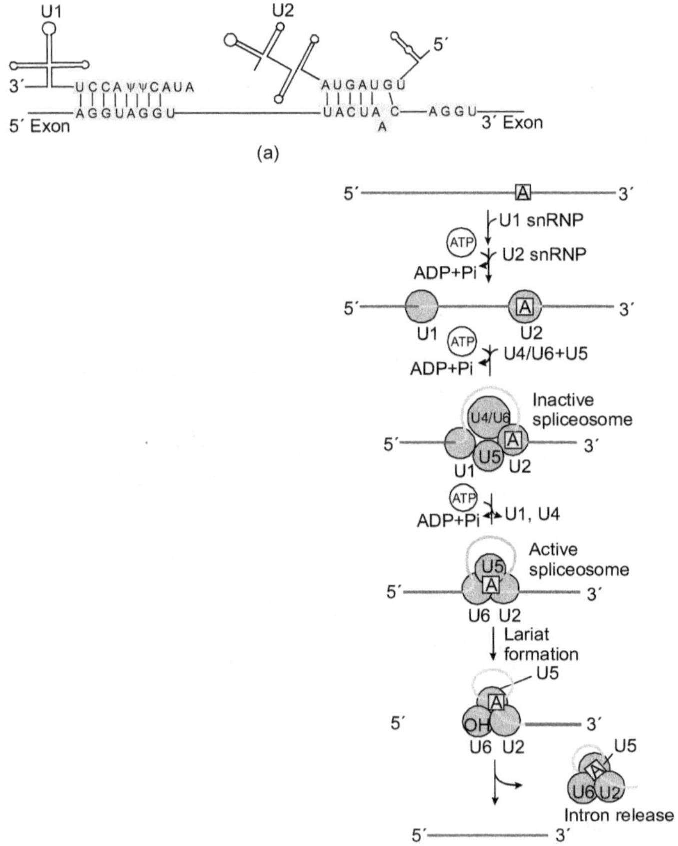

Figure 7.25   Splicing mechanism of group III introns in mRNA primary transcripts

The U1 snRNA has a sequence of nucleotides that is complementary to a sequence near the 5′-splice site of nuclear mRNA introns. The U1 snRNP binds to this region in the primary transcript. Addition of the U2, U4, U5 and U6 snRNPs leads to the formation of a complex called **spliceosome**. The actual splicing reaction occurs within this complex (Figure 7.25).

The group IV introns are found in certain tRNAs. Splicing of this group of introns requires ATP and an endonuclease. The splicing endonuclease cleaves the phosphodiester bonds at both the ends of the intron and the two exons are joined by a mechanism similar to the reaction catalysed by ligase.

The introns are not an exclusive feature of pre-mRNAs. They are also found in tRNA genes, rRNA genes, mitochondrial and chloroplast genes and in a few bacteriophage genes. In nuclear genes that encode mRNAs, several conserved sequences define an intron. Conserved sequences are found at both ends of the intron. The 5′ end has the dinucleotide sequence GU and the 3′ end has the dinucleotide sequence AG. These nucleotides are found in all introns from nuclear mRNA genes. This observation is referred to as **GT-AG** rule, using the sense-strand DNA sequence and is useful in the identification of introns in DNA sequences.

Adjacent to these terminal nucleotides are conserved sequences that vary slightly from intron to intron. The 5′-consensus sequence is GUAAGU and the 3′-consensus sequence is 6Py N C A G (where 6Py refers to six pyrimidines and N, any one of the four nucleotides). An additional consensus sequence Py N Py Py Pu A Py (Py—pyridimine and Pu—purine) is found about 18–40 nucleotides upstream from the 3′ end of the intron.

Therefore, the consensus sequences of a pre-mRNA intron can be represented as

$$\downarrow \qquad\qquad\qquad\qquad\qquad\qquad\qquad\qquad\qquad \downarrow$$

5′ – – – –GUAAGU – – – –PyNPyPuAPy – – – –6PyNCAG – – – –3′

where the arrows represent the splice sites.

INHIBITORS OF TRANSCRIPTION

Two related antibiotics, rifamycin B which is produced by *Streptomyces mediterranei* and its semi-synthetic derivative rifampicin, specifically inhibit prokaryotic transcription by RNA polymerase.

This has made them medically useful bacteriocidal agents against gram-positive bacteria and are used in the treatment of tuberculosis. The β-subunit of RNA polymerase has the rifamycin-binding site. Rifamycin inhibits neither the binding of RNA polymerase to the promoter nor the formation of the first phosphodiester bond, but it prevents further chain elongation. The inactivated RNA polymerase remains bound to the promoter, blocking its initiation by uninhibited enzyme. Once RNA chain initiation has occurred, rifamycins have no effect in the subsequent elongation process. Rifamycins are useful research tools because they permit the transcription process to be dissected into its initiation and elongation phases.

The antibiotic cordycepin (3'-deoxyadenosine) is an adenosine analog that lacks the 3'-OH group and it inhibits the bacterial RNA synthesis. If present in an RNA synthesis system, it will be incorporated in the growing RNA opposite to T in the DNA, but prevents further elongation step.

Intercalating agents like actinomycin D inhibit both RNA and DNA polymerases. Actinomycin D is a useful antineoplastic agent produced by *Streptomyces* which binds tightly to duplex DNA. Therefore it strongly inhibits both transcription and DNA replication, by interfering with the activity of DNA polymerase and RNA polymerase. X-ray studies indicate that the DNA assumes a B-like conformation in which the actinomycin's phenoxazone ring system is intercalated between the DNA's central GC base pairs. Consequently, the DNA helix is unwound by 23° at the intercalation site and the central GC base pairs are separated by 7.0 Å. The DNA helix is severely distorted from the normal B-DNA conformation such that its minor groove is wide and shallow resembling A-DNA. Two chemically identical cyclic depsi peptides which resemble actinomycin D have both peptide bonds and ester linkages, which assume different conformations. They extend in opposite directions from the intercalation site along the minor groove of the DNA. The complex is stabilized through the formation of base–peptide and phenoxazone–sugar–phosphate backbone hydrogen bonds as well as by hydrophobic interactions in a way that explains the preference of actinomycin D to bind to DNA with its phenoxazone ring intercalated between the base pairs of a 5'–G C–3' sequence.

Other intercalating agents, including ethidium and proflavin also inhibit nucleic acid synthesis by a similar mechanism. Many of these agents are valuable antibiotics and/or antineoplastic substances.

Streptolydigin is an inhibitor of prokaryotic transcription. It binds to the β-subunit of RNA polymerase like rifamycin and preferentially blocks chain elongation whereas rifamycin blocks initiation.

The poisonous mushroom *Amanita phalloides* (death cap) which is responsible for a majority of fatal mushroom poisonings, contains several types of toxic substances including a series of unusual bicyclic octapeptides known as amatoxins. α-Amanitin is an amatoxin that forms a tight 1 : 1 complex with RNA polymerase II and a looser one with RNA polymerase III and specifically block their elongation steps. α-Amanitin is therefore used as an important tool for mechanistic studies of these enzymes. RNA polymerase I as well as mitochondrial, chloroplast and prokaryotic RNA polymerases are insensitive to α-amanitin. About 5 to 6 mg of amatoxin present in about 40 g of fresh mushroom is highly toxic and can even be fatal to humans. But they act very slowly. Death occurs due to liver dysfunction, but no earlier than several days after mushroom ingestion.

Spicy Questions and Answers

1. **Genes that encode rRNA and tRNA do not encode proteins. Why are they called genes if they do not encode proteins?**

   Any DNA sequence that encodes an RNA molecule can be referred to as a gene. A gene need not necessarily encode a protein.

2. **Describe the functions of other types of RNA that do not fall into the three major classes of RNA.**

   The three major classes of RNA are mRNA, tRNA and rRNA. Apart from these RNAs, there are other types of RNA that include the following:

   1. primer RNAs that participate in DNA synthesis

   2. telomerase RNA template that participates in the telomeric DNA synthesis

   3. snRNAs, i.e., small nuclear RNAs that participate in the removal of introns as components of spliceosomes

4.   small RNAs that direct secretory proteins to the receptors on the cell membranes

5.   ribozymes—RNA molecules that catalyse biochemical reactions.

3. **RNA polymerase II may encounter situations where it stalls, halting transcription before termination occurs. How can this situation be remedied?**

The transcription factor TFIIS makes a stalled RNA polymerase II to back up. By this movement, RNA polymerase removes a few nucleotides from the 3′ end of the RNA and attempts to start the elongation process once again.

4. **Why are coupled transcription and translation not possible in eukaryotes?**

Transcription and translation are coupled in prokaryotes. In eukaryotes they are compartmentalized. In prokaryotes, the mRNA is ready to be translated as soon as it is transcribed. Actually, translation begins even before transcription has terminated. As soon as the growing mRNA chain separates from DNA, ribosomes attach to it and begin translation on the 5′ end of the molecule, following right behind the RNA polymerase while it is still transcribing the mRNA. In eukaryotes, RNA is transcribed in the nucleus and the mRNA, tRNA and rRNA, all required for translation, are then exported from the nucleus into the cytoplasm. Since transcription and translation are coupled events in prokaryotes, there is no chance for the modification of mRNA. On the other hand, eukaryotic mRNA undergoes post-transcriptional modification in the nucleus before it enters into the cytoplasm.

5. **Why is it important for ribosomes to follow close behind prokaryotic RNA polymerase during elongation?**

To prevent the premature termination of transcription, the ribosomes block the sites where transcription is likely to terminate in the reading frame.

6. **The (A + T)/(G + C) ratios of *E. coli* and *Mycobacterium tuberculosis* differ significantly. In which classes of genes will there be a large difference in the ratio among the two species—in the genes that code for tRNA, rRNA or mRNA?**

Genes that code for mRNAs. The genes that encode tRNAs and rRNAs are highly conserved among prokaryotes because they encode molecules that are similar in structure.

7. In *E. coli*, elongation does not proceed at a constant rate. Describe how it proceeds and what factors may influence the rate of transcription. Why is it essential for the cells to control the rate of elongation in prokaryotes?

    E. coli RNA polymerase elongates the RNA in periodic bursts of RNA synthesis, each followed by a brief phase. Certain proteins are seen to affect the rate of elongation. Although the core enzyme of *E. coli* RNA polymerase is capable of elongation in the absence of any other proteins *in vitro*, certain proteins present in the cell control the rate of elongation *in vivo* like Nus A protein, which binds to the core enzyme in the place of the σ factor. Nus A slows down elongation when RNA polymerase encounters certain nucleotide sequences, ensuring that transcription proceeds at the proper rate. Nus A keeps the rate of transcription similar to the rate of translation so that the ribosomes are able to follow the RNA molecule right behind the RNA polymerase.

8. The following represent deoxyribonucleotide sequences derived from the template strand of DNA:

    Sequence 1: CTTTTTTGCCAT

    Sequence 2: ACATCAATAACT

    Sequence 3: TACAAGGGTTTCT

    (a) **For each strand, determine the mRNA sequence that would be derived from transcription**

    (b) **Determine the amino acid sequence that is encoded by these mRNAs**

    (c) **For sequence 1, what is the sequence of the partner strand?**

    (a) mRNA sequence of

    1. 5′ GAAAAAACGGUA 3′

    2. 5′ UGUAGUUAUUGA 3′

    3. 5′ AUGUUCCCAAAGA 3′

    (If the sequences mentioned in the problem are in 3′ → 5′ direction)

    (b) amino acid sequences encoded by these mRNAs are

    1. glu-lys-thr-val

    2. cys-ser-tyr-stop

    3. met-phe-pro-ser

    (c) Sequence of partner strand for (1) is 5′ GAAAAAACGGTA 3′

9. **Messenger RNA molecules are very difficult to isolate in prokaryotes because they are quickly degraded in the cell. Can you suggest a reason for this? Eukaryotic mRNAs are more stable and exist longer in the cell. Is this an advantage or disadvantage for a pancreatic cell making large quantities of insulin?**

Because mRNA molecules are mostly single-stranded structures with very little folding because of complementary base pairing within the string, they are quite labile. But eukaryotic mRNAs are modified to have cap in the 5′ end and poly(A) tail in 3′ end. These features protect the mRNAs from degradation. This is one of the features of regulation of eukaryotic gene expression which is based on the stability of mRNA molecules. If mRNAs have longer life, they can be used to produce more copies of the required protein like insulin.

10. **What is the relationship between sedimentation coefficient of RNA molecules and the number of nucleotides present in them?**

In *E. coli*, there are three rRNAs called 5S, 16S and 23S rRNAs. These designations refer to the sizes of the rRNAs measured in Svedberg units, which denote how rapidly a molecule sediments under centrifugation. Larger particles have more mass and generally sediment more rapidly than smaller particles giving them a higher sedimentation coefficient value. Density of the molecule is also a factor to be considered, because greater the density, the faster the sedimentation and therefore the higher the Svedberg value. In case of RNA molecules, Svedberg units correspond to the size of RNA molecules in nucleotides, but the relationship is not linear. For example, in *E. coli*, the 16S rRNA has 1542 nucleotides, whereas the 23S rRNA is larger with 2904 nucleotides. Svedberg values cannot be added. For example, mammalian ribosomes have two subunits with Svedberg values of 40S and 60S but two units associate to form 80S ribosome.

11. **For each of the following statements, indicate which of the three eukaryotic RNA polymerases is being described:**

(a) **RNA polymerase, that transcribes the greatest diversity of genes in terms of nucleotide sequences**

(b) **RNA polymerase that transcribes the smallest genes**

(c) **RNA polymerase that transcribes most of the RNA in the cell**

(d) **RNA polymerase that uses promoters on the 5′ end of the gene**

(e) **RNA polymerase located primarily in the nucleolar organizer region**

(f) **RNA polymerase that transcribes pre-mRNAs**

(g) **RNA polymerase that transcribes 5S rRNA**

(a) RNA polymerase II that transcribes the genes giving pre-mRNA or hnRNA which after post-transcriptional modification gives mRNA.

(b) RNA polymerase III that transcribes tRNA genes and genes that code for other small rRNAs.

(c) rRNA molecules account for a large portion, up to 80–90% by mass, of all RNA transcribed in the cell. They are transcribed by RNA polymerase I.

(d) RNA polymerase I that transcribes rRNA molecules and RNA polymerase II that transcribes mRNA have the promoters on the 5´ end of the gene, that is in the upstream region.

(e) The genes for the precursor rRNA occupy a specific region of the nucleus called the nucleolus. The portion of cellular DNA that contains the tandemly repeated precursor rRNA genes is called the nucleolus organizer region (NOR) and is contained within the nucleolus. RNA polymerase I transcribes only in the nucleolus.

RNA polymerases II and III transcribe in the part of the nucleus outside the nucleolus in an area called nucleoplasm.

(f) Unprocessed mRNA otherwise called pre-mRNA or heterogeneous nuclear RNA (hnRNA) is transcribed by RNA polymerase II.

(g) In a few eukaryotic species, RNA polymerase I transcribes the 5S rRNA as part of the large precursor rRNA and in most eukaryotes, RNA polymerase III transcribes the 5S rRNA from a separate gene. RNA polymerase III also transcribes all the tRNA genes.

Review Questions

1. List the differences between prokaryotes and eukaryotes regarding transcription and RNA processing.

2. Describe the characteristics of the three major types of RNA and their functions in the cell.

3. Describe the role of sigma factor and distinguish between the holoenzyme and the core enzyme of RNA polymerase in prokaryotes.

4. How is transcription initiated and terminated in *E. coli* ?

5. Define the process of transcription. Where does this process fit into the central dogma of molecular genetics?

6. What are the characteristic features of eukaryotic promoters?

7. Describe the post-transcriptional modifications of the tRNA molecules.

8. What are the major processing events taking place in the formation of a mature eukaryotic mRNA molecule?

9. What are spliceosomes? Explain their functions.

10. Write an essay on RNA synthesis under the following headings:

    i.   binding of RNA polymerase to DNA template

    ii. initiation

    iii. elongation

    iv. termination

11. Describe the roles of the following proteins in transcription: rho, Nus A, TFIIA, TFIIB, TFIID, TFIIE, TFIIF and TFIIS.

12. What are the different types of rRNA molecules in prokaryotes and eukaryotes? How are they processed?

13. Comment on intron removal and exon splicing.

8

THE GENETIC CODE

In the late 1950s, before it became clear that mRNA is involved in transferring the genetic information from DNA to proteins, it was thought that DNA itself might directly encode proteins. At that time, ribosomes had already been identified and also it was found out that an unstable template intermediate, an RNA molecule, was involved in the transfer of the genetic information from DNA to proteins. However, the RNA molecules associated with the ribosomes were found to be stable. In the year 1961 Francois Jacob and Jacques Monad postulated the existence of messenger RNA (mRNA). Once mRNA was discovered, it was clear that even though genetic information is stored in DNA, the code that is translated into proteins resides in RNA. The central question then was how only the four letters—nucleotides—could specify 20 words—the amino acids. Thus, the central dogma of molecular genetics was evolved for the flow of the genetic information (Figure 8.1).

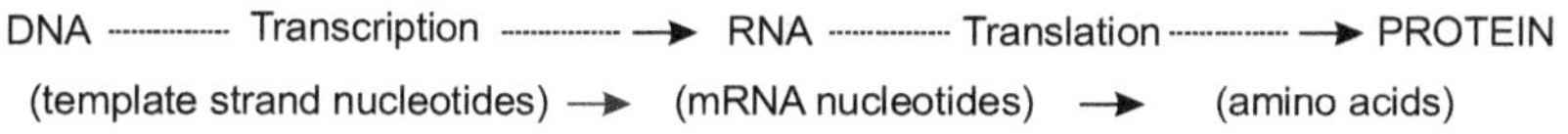

Figure 8.1 Flow of genetic information from DNA to protein through RNA

FEATURES OF THE GENETIC CODE

The Triplet Nature of the Code

In the early 1960s, Sydney Brenner argued on theoretical ground that the code must be a triplet. One nucleotide cannot code for

FIRST (5') LETTER

	SECOND LETTER				THIRD (3') LETTER
	U	C	A	G	
U	UUU, UUC } Phe; UUA, UUG } Leu	UCU, UCC, UCA, UCG } Ser	UAU, UAC } Tyr; UAA Ochre (terminator); UAG Amber (terminator)	UGU, UGC } Cys; UGA Opal (terminator); UGG Trp	U, C, A, G
C	CUU, CUC, CUA, CUG } Leu	CCU, CCC, CCA, CCG } Pro	CAU, CAC } His; CAA, CAG } Gln	CGU, CGC, CGA, CGG } Arg	U, C, A, G
A	AUU, AUC, AUA } Ileu; AUG Met (initiator)	ACU, ACC, ACA, ACG } Thr	AAU, AAC } Asn; AAA, AAG } Lys	AGU, AGC } Ser; AGA, AGG } Arg	U, C, A, G
G	GUU, GUC, GUA } Val; GUG (initiator)	GCU, GCC, GCA, GCG } Ala	GAU, GAC } Asp; GAA, GAG } Glu	GGU, GGC, GGA, GGG } Gly	U, C, A, G

Figure 8.2 The genetic code table

one amino acid, because there are only four nucleotides. A combination of two nucleotides will provide only 16 unique code words ($4^2 = 16$) which are not sufficient to code for the 20 amino acids. A triplet code may provide $4^3 = 64$ codons, more than enough to code for the 20 amino acids (Figure 8.2). This is much simpler than a four-lettered code (4^4), which would specify 256 codons, too much in surplus for the 20 standard amino acids.

Degeneracy of the Genetic Code

Crick and his colleagues drew other inferences from their experiments. In the genetic code, 64 in number, most codons must function in the specification of amino acids. They reasoned that if each amino acid had only one codon, then only 20 out of the 64 possible codons could be used for coding amino acids. In this case, most frameshift mutations should have affected one of the remaining 44 "non-coding" codons in the reading frame and hence a nearby frameshift of the opposite polarity mutations should not have suppressed the original mutation. Therefore, the code was deduced to be degenerate which means that more than one codon can specify a particular amino acid. The genetic code is found to be unambiguous, meaning that each triplet specifies only a single amino acid.

Non-overlapping Nature of the Code

Sydney Brenner also argued that the code was non-overlapping. He considered theoretical nucleotide sequences encoding a protein consisting of three amino acids. In the nucleotide sequence G T A C A, for example, parts of the central triplet, TAC are shared by the outer triplets, GTA and ACA (Figure 8.3). Brenner argued that if this is the case, then only certain amino acids should be found adjacent to the one encoded by the central triplet.

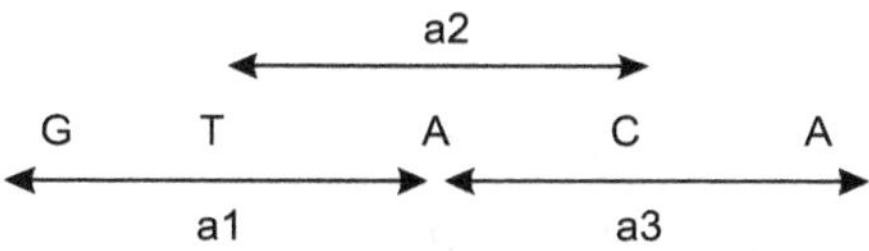

Figure 8.3   Overlapping codons

A given nucleotide sequence can be read in any of its three reading frames. But most DNA sequences encode a protein product in only one reading frame. Therefore usually genes do not overlap, but there are a few exceptions. In several viruses, the same DNA sequence codes for two different proteins by employing two different reading frames. The DNA of bacteriophage ϕX174, consists of 5386 nucleotide residues. It is not long enough to code for the ten different proteins produced by its genome. A comparison of the nucleotide sequence of ϕX174 chromosome with the amino acid sequence of the proteins it encodes revealed several overlapping gene sequences. Several of the genes are nested within other genes. In several cases, the initiation codon of one gene overlaps the termination codon of the other gene.

Similar observations are made in phage λ and cancer-causing simian virus 40 (SV 40) and in RNA phages such as Q β and Q 17.

For any given central amino acid, only 16 combinations (2^4) of three amino acid sequences (a tripeptide) are theoretically possible. Brenner concluded that if the code were overlapping, tripeptide sequences within proteins should be somewhat limited. Such restrictions were not observed in the tripeptide sequences of the proteins. For any central amino acid, many more than 16 different tripeptides were observed.

Another major argument against an overlapping code involved the effect of a single nucleotide change. If the code is overlapping, two adjacent amino acids would be affected by a point mutation. But mutations in the gene coding for the protein coat of tobacco mosaic virus (TMV), human hemoglobin and the bacterial enzyme tryptophan synthetase revealed only single amino acid change. This confirmed the non-overlapping nature of the genetic code.

The third argument against an overlapping code was proposed by Francis Crick in 1957. He argued that any affinity between nucleotides and an amino acid required the role of an adaptor molecule that could covalently bind to the amino acid, yet be capable of hydrogen-bonding to a nucleotide sequence. This molecule was later found to be the tRNA and during protein biosynthesis it was seen that the ribosome accommodates two tRNA molecules at a time, which again confirmed the non-overlapping nature of the genetic code.

The Genetic Code is Commaless and has no Punctuation

Between 1958 and 1960, information related to the genetic code continued to accumulate. Based on genetic evidence, Crick hypothesized that the code is commaless, that is, he believed that there is no internal punctuation along the reading frame. It was established that once translation commences, the codons on the mRNA are read one after the other with no breaks between them.

The Start and Stop Codons

The codon AUG is the most commonly used start codon. Sometimes, GUG is used. UAA, UAG and UGA were identified as stop codons.

Wobble Hypothesis

Most of the amino acids are specified by two, three or four different codons which is referred to as degeneracy of the genetic code. When we observe the pattern of degeneracy, it is seen that the first two letters of the code for a particular amino acid are the same, with only the third differing. Crick observed a pattern in the degeneracy at the third position. In 1966, he postulated the wobble hypothesis.

When it was found out that 61 of the 64 codons are sense codons coding for the 20 standard amino acids, it was expected that each of these 61 codons are recognized by 61 tRNA molecules with the specific anticodon. The fidelity of protein synthesis is mainly determined by the correct pairing of the mRNA codons in $5' \rightarrow 3'$ direction and the tRNA anticodons in $3' \rightarrow 5'$ direction, so that the amino acids attached to these tRNAs are involved in peptide bond formation and the nucleotides in the mRNA are truly translated into amino acids of the protein.

The one-codon–one-anticodon/one-tRNA idea was found wrong when the first tRNA, yeast tRNA[ala] was sequenced. Alanine has four codons GCU, GCC, GCA and GCG. This tRNA was found to recognize three of these codons namely GCU, GCC and GCA. Study of the sequence of this tRNA molecule, especially in the anticodon region that comprises of the 34th, 35th and 36th nucleotides from the 5′ end of the tRNA (that form the anticodon), showed that there is a non-hydrogen-bonded region in the tRNA molecule with the base sequence 5′– GC3′. In this, G and C form base-pairing with C and G of alanine codons. The 5′ side of the anticodon was found to have an unusual nucleoside inosine (a nucleoside of ribosyl hypoxanthine). Inosine was found to base-pair with U, C and A in the 3′ side of the alanine codons GCU, GCC and GCA. However, it does not form base pair with the fourth codon of alanine GCG. The bond between I in the 5′ side of anticodon and U, C and A in the 3′ side of the codon are referred to as non-Watson–Crick base pairing (Figure 8.4).

The chemical origin of inosine in the 5′ side of the anticodon in the tRNA molecule is interesting. The anticodon will not start with A in the 5′ side. Whenever A appears in the first position (5′ side of the anticodon: 34th nucleotide) in the unmodified tRNA transcript, it is deaminated by an enzyme called anticodon deaminase to hypoxanthine, the base of inosine. Inosine forms base pair with cytidine, adenosine and uridine.

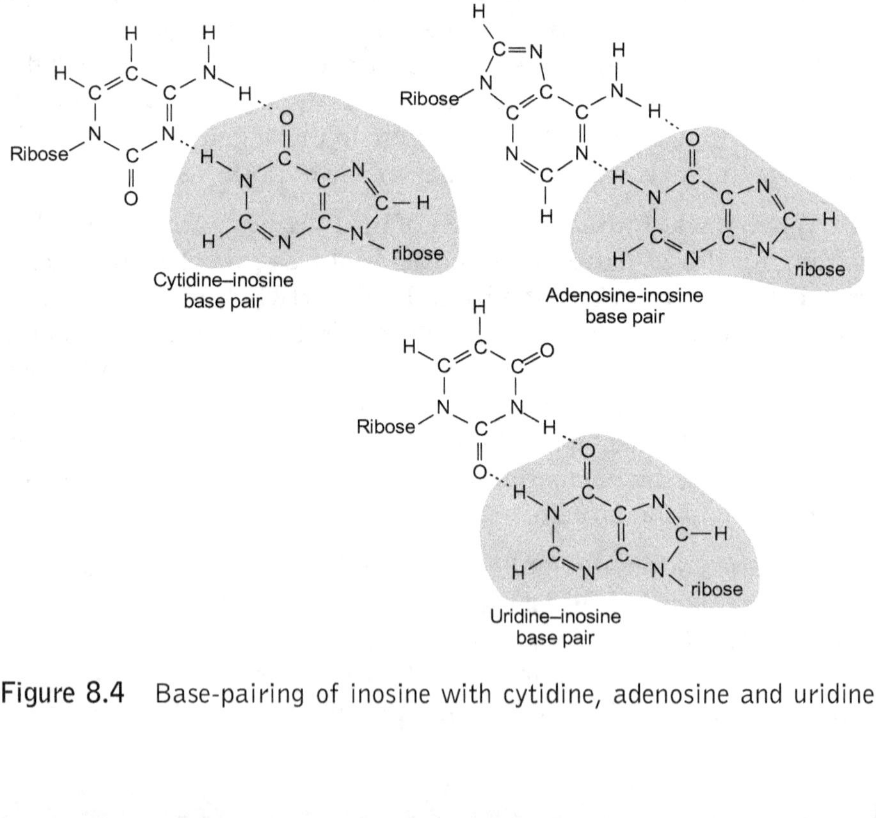

Figure 8.4   Base-pairing of inosine with cytidine, adenosine and uridine

In 1965, Francis Crick made a proposal called Wobble hypothesis, that explains both the response of some tRNA molecules to several codons and the pattern of redundancy of the code. Till that time it was considered that in nucleic acids, base pairing occurs only between G and C, A and T or A and U—the first two in DNA and the last one in RNA. This is true of DNA because the regular helical structure of double-stranded DNA imposes two steric constraints, i.e., two purines cannot form a base pair because there is not enough space for a planar purine–purine pair and two pyrimidines cannot form a base pair because they cannot reach one another. Crick proposed that since the codon–anticodon interaction involves six bases, the paired structure may not be the stranded double helix. He showed that at the third position of the codon, there is non-standard base-pairing as shown in Table 8.1.

Table 8.1 The Wobble pairing

Third position codon base (3´ side)	First position anticodon base (5´ side)
A	U, I
G	C, U
U	G, I
C	G, I

This shows that if there are four codons for an amino acid (e.g. Glycine codons—GGU, GGC, GGA and GGG), the first three codons are recognized by a single tRNA molecule with the anticodon 3′CCI 5′ and the last codon by the anticodon 3′CCC 5′. However, in the first box of the genetic code table, UUU and UUC code for phenylalanine and UUA and UUG code for leucine. In this case AAI cannot be the anticodon, because it will recognize not only phenylalanine codons but also leucine codon UUA. Therefore AAG is the anticodon in a tRNA (tRNA[phe-ala]) that will recognize UUU and UUC, accounting for less specificity in base-pairing in the 3′ side of codon and 5′ side of anticodon (as indicated in Table 8.1). The two leucine codons, UUA and UUG, are recognized by the anticodon AAU in tRNA[leu.] The next box (vertical column) again has four more leucine codons—CUU, CUC, CUA and CUG—which are recognized by leucine anticodons 3'GAI 5' that recognizes the first three codons and 3'GAC 5' that recognizes the last codon. Totally three anticodons (three tRNA [leu)] are enough to recognize all the six codons of leucine.

The codons AUG (methionine) and UGG (tryptophan) are non-redundant and therefore they need one anticodon each UAC and ACC to recognize their codons respectively. There is no tRNA having an anticodon complementary to UGA, UAA and UAG—the stop codons.

Figure 8.5 gives the anticodons (or the different tRNA molecules) required to recognize all the 61 sense codons.

Thus in total, 31 tRNA molecules are found to be sufficient to recognize all the 61 sense codons is as follows:

5´ end   3´ end

AAG-tRNA^{phe}	AGI-tRNA^{ser}	AUG-tRNA^{tyr}	ACG-tRNA^{cys}
AAU-tRNA^{leu}	AGC-tRNA^{ser}		ACC-tRNA^{trp}
GAI-tRNA^{leu}	GGI-tRNA^{pro}	GUG-tRNA^{his}	GCJ-tRNA^{arg}
GAC-tRNA^{leu}	GGC-tRNA^{pro}	GUU-tRNA^{gln}	GCC-tRNA^{arg}
UAI-tRNA^{ile}	UGI-tRNA^{thr}	UUG-tRNA^{asp}	UCG-tRNA^{ser}
UAC-tRNA^{met}	UGC-tRNA^{thr}	UUU-tRNA^{lys}	UCU-tRNA^{arg}
CAI-tRNA^{val}	CGI-tRNA^{ala}	CUG-tRNA^{asp}	CCI-tRNA^{gly}
CAC-tRNA^{val}	CGC-tRNA^{ala}	CUU-tRNA^{glu}	CCC-tRNA^{gly}

Figure 8.5   Anticodon table with the respective tRNA

Another way of calculation of minimum number of tRNA molecules to recognize the 61 sense codons:

1. There are two amino acids (met and trp) each with one codon. The number of tRNAs required to recognize these two codons is two.

2. There are nine amino acids (phe, tyr, his, gln, asn, lys, asp, glu, cys) each with two codons. The number of tRNAs required to recognize these eighteen codons is nine.

3. There is one amino acid (ile) with three codons. The number of tRNA required to recognize these three codons is one.

4. There are five amino acids (val, pro, thr, ala, gly) with four codons each. The number of tRNAs required to recognize these twenty codons is (5 × 2) ten.

5. There are three amino acids (leu, arg, ser) with six codons each. The number of tRNAs required to recognize these eighteen codons is (3 × 3) nine.

Therefore a total minimum number of 31 tRNAs are required with 31 anticodons to recognize the 61 sense codons.

These explanations prove that because of the degeneracy of the genetic code, the anticodon of a given tRNA is able to base-pair with several different codons. The base-pairing involving the third base of the codon is less stringent, allowing wobble at this site. Figures 8.6 and 8.7 show the codon–anticodon interactions involving the wobbling at the third base of the codon (3′ end) and the first base of the anticodon (5′ end).

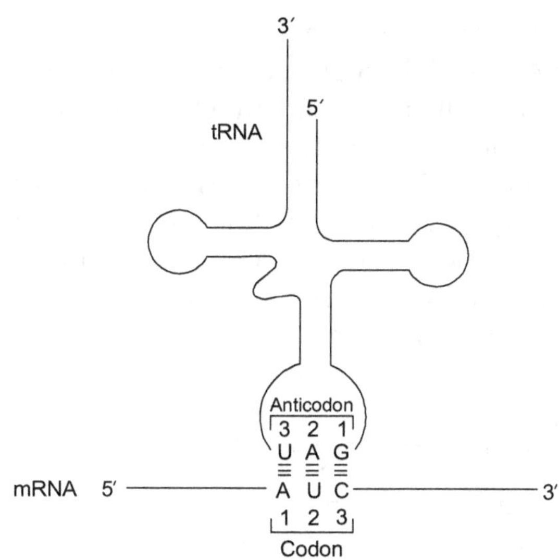

Figure 8.6   Pairing relationship of codon and anticodon–Antiparallel alignment of the two RNAs

	3 2 1	3 2 1	3 2 1
Anticodon	(3′) G –C –I	G–C –I	G–C –I (5′)
Codon	(5′) C –G –A	C –G –U	C –G –C (3′)
	1 2 3	1 2 3	1 2 3

Figure 8.7   Wobble base pairing—inosine forming base pair with U, C and A in the third position of codon

Eight of the 16 boxes of the genetic code table contain one amino acid per box. These eight groups of codons are termed unmixed families of codons. A box is determined only by the first and second positions. In these boxes, if the first two bases are the same, irrespective of the third base, they code for the same amino acid. Mixed families code for two amino acids or for stop signals and one or two amino acids. Six of the mixed family boxes are split in half so that the codons are differentiated by the presence of a purine or a pyrimidine in the third base.

In the unmixed codon family, any mutation in the third position produces another codon for the same amino acid. Wobble in the third position and codon arrangement ensures that less than half of the mutations in the third codon position result in a different amino acid incorporation.

Some mutations resulting in a change of one sense codon to another sense codon may result in change of one amino acid to another with similar properties, still producing a functional protein. All the codons with U as the middle base are for amino acids that are hydrophobic (phe, leu, ile, met, val). Mutations in the first or third positions for any of these codons still codes a hydrophobic amino acid. Both of the two negatively charged amino acids, aspartic acid and glutamic acid, have codons that start with GA. The codons of aromatic amino acids (phe, tyr, trp) start with uracil. Such patterns minimize the negative effects of mutation.

Colinearity of Gene and Polypeptide

The genetic information is stored in linear sequences of nucleotide pairs in DNA which is transcribed and translated and converted to polypeptides with linear sequences of amino acids. It was proved that the amino acid sequences of the polypeptides and the nucleotide pair sequences of the genes coding for these polypeptides are colinear, that is, the first three base pairs of a gene specify the first amino acid of the polypeptide, the next three base pairs specify the second amino acid and so on in a colinear fashion.

The Genetic Code is Nearly Universal

Between 1960 and 1978, it was assumed that the genetic code was universal, common to viruses, bacteria and eukaryotes. The cell-free systems derived from bacteria were found to translate eukaryotic mRNAs. Within eukaryotes, mRNAs from mice and rabbits were injected into amphibian eggs and were found to be efficiently translated. For many eukaryotic genes, like those of hemoglobin molecules, the amino acid sequence of the encoded proteins adhered to the coding dictionary established from bacteria studies.

In 1979, it was reported that DNA from mitochondria of yeast and humans had different coding language. This observation led to the identification of several exceptions to the universal code (Table 8.2).

Table 8.2   Exceptions to the universal code

Triplet	Normal code word	Altered code word	Source
UGA	termination	trp	Human and yeast mitochondria, Mycoplasma
CUA	leu	thr	Yeast mitochondria
AUA	ile	met	Human mitochondria
AGA	arg	termination	Human mitochondria
UAA	termination	gln	*Paramecium tetrahymena*

DECIPHERING OF THE GENETIC CODE

Several landmark experiments done in different laboratories helped in establishing the features of the genetic code.

Mutation Studies to Prove the Triplet Nature of the Codon

The experimental work of Francis Crick, Leslie Barnett, Brenner and R.J. Watts-Tobin in 1961, gave the first solid evidence for the triplet nature of the code. Mutation studies were done in the B cistron of the rII locus of phage T4. B cistron is one of the two functional sites in this locus, which in mutant form, causes rapid lysis and distinctive plaques.

Such mutants are successful in infecting strain B of *E. coli* but cannot reproduce on *E. coli* K12 strain. Crick and his co-workers used acridine dye proflavin to induce mutation. This mutagenic agent intercalates within the double helix of DNA, causing insertion or deletion of one or more nucleotides during replication.

Such an insertion of a single nucleotide causes the reading frame to shift, changing the specific sequence of all subsequent triplets to the right of the insertion. On translation, the amino acid sequence of the encoded protein will be altered due to frameshift mutation. When this mutation occurs in the rII locus, T4 will not reproduce on *E. coli* K12.

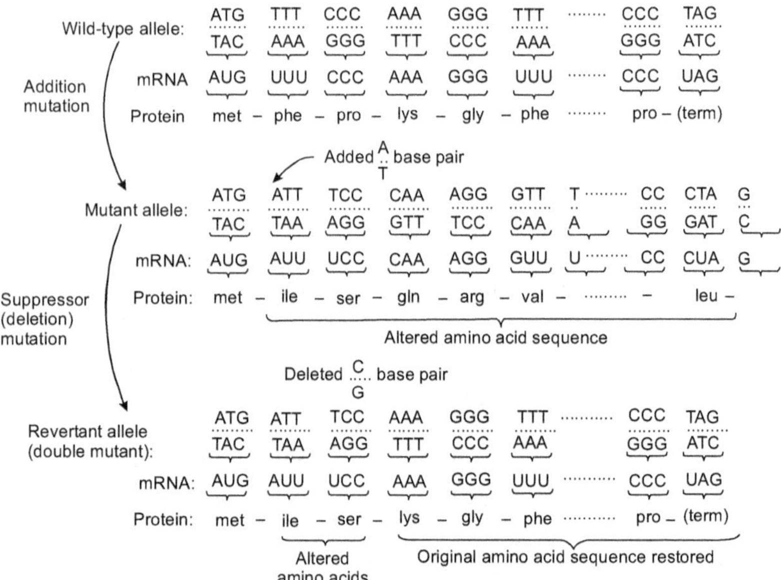

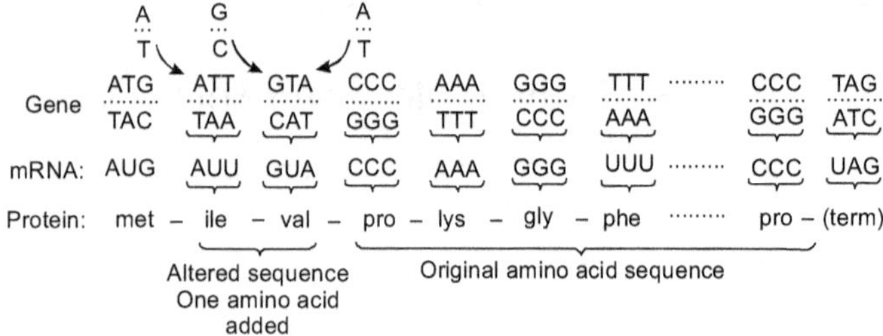

Figure 8.8   Mutation studies to prove the triplet nature of the codon

Crick and his co-workers found out that if phages with these induced mutations were treated again with proflavin, other insertions or deletions were expected to take place. This seemed to result in the formation of a revertant phage, that behaved like wild type phage that could infect *E. coli* K12. If the original mutant contained an insertion

(+) and a second event causes a deletion (–) close to the insertion site, the original reading frame is restored. In this way, they analysed various types of mutants in rII locus and compared the results. They found that when two (+) were together or when two (–) were together, the correct reading frame was not re-established. But when three (+) or three (–) were present together, the original reading frame was re-established (Figure 8.8). These observations strongly supported the triplet nature of the code.

Crick and his co-workers thus proved that the coding ratio is 3 or a multiple of 3, that is, three nucleotides form one codon on the mRNA.

In 1961, Marshall Nirenberg and J. Heinrich Matthaei characterized the first specific coding sequences. This served as a cornerstone for the complete analysis of the genetic code. Two important experimental tools namely an *in vitro* cell-free protein-synthesizing system and an enzyme, polynucleotide phosphorylase that allowed the production of synthetic mRNAs helped in deciphering the genetic code. These synthetic mRNAs served as templates for polypeptide synthesis in the cell-free system.

BOX 8.2   INCORPORATION OF SELENOCYSTEINE DURING PROTEIN BIOSYNTHESIS

Selenocysteine is an unusual amino acid—a derivative of cysteine in which the sulphur atom is replaced by a selenium atom. This is incorporated in some proteins in both prokaryotes and eukaryotes in response to the codon UGA, which is normally a stop codon. In *E. coli,* the products of four genes—*Sel A, Sel B, Sel C* and *Sel D*—are required to incorporate selenocysteine in the protein. *Sel C* gene produces tRNA[ser] which is a suppressor tRNA with anticodon UCA. This tRNA[ser] is charged with serine by seryl-tRNA synthetase. The genes *Sel A* and *Sel B* help in the conversion of seryl tRNA[ser] to selenocysteinyl-tRNA[ser]. The protein product of *Sel B* gene is called SELB which is homologous in sequence to EF-TU an elongation factor used in protein biosynthesis. Therefore it replaces EF-TU during translation by recognizing selenocysteinyl tRNA and UGA stop codon. Also it competes with the termination factors in the recognition of the stop codon and promoting codon–anticodon interaction between UGA and selenocysteinyl-tRNA[ser].

Synthesis of Polypeptides in a Cell-Free System

In the cell-free systems, amino acids can be incorporated into polypeptide chains. This system must contain all the essential factors required for protein synthesis: ribosomes, tRNAs, amino acids, protein synthesizing enzymes, translation factors, high-energy compounds, etc. In order to follow and to have a track of protein synthesis, one or more of the amino acids must be radioactive. Also an mRNA must be added that can serve as the template to be translated.

In 1961 mRNA had yet to be isolated. But the use of polynucleotide phosphorylase allowed the artificial synthesis of RNA templates that could be added to the cell-free system. This enzyme was isolated from bacteria which catalysed the reaction:

$$n(\text{NDP}) \xrightleftharpoons{\text{Polynucleotide phosphorylase}} (\text{NMP})_n + n\text{Pi}$$
$$\text{RNA}$$

The enzyme functions metabolically in bacterial cells to degrade RNA. *In vitro*, if concentration of NDP is increased, the reaction can be forced in the opposite direction to synthesize RNA. In contrast to RNA polymerase, polynucleotide phosphorylase does not require DNA template and requires NDP and not NTP as the substrates. Therefore, addition of each ribonucleotide is random, based on the relative concentration of the four NDP added to the reaction mixture. The probability of insertion of a specific ribonucleotide is proportional to the availability of that molecule, relative to other available ribonucleotides.

Use of Homopolymer RNA Molecules

Nirenberg and Matthai initially synthesized RNA homopolymers each consisting of only one type of ribonucleotide. Therefore the mRNA added to the *in vitro* system was UUUUUUU-----, AAAAAA-----, CCCCCC-----or GGGGGG-----. Using each of these mRNA they determined the amino acids incorporated into the newly synthesized protein. The researchers determined this by labelling one of the 20 amino acids added to the *in vitro* system and conducting a series of experiments, each with a different amino acid made radioactive.

When poly (U) synthetic mRNA was used, it was found to direct the incorporation of only ^{14}C-phenylalanine, and a homopolymer of phenylalanine was obtained. Assuming a triplet code, the first codon was assigned as UUU coding for phenylalanine. In the same way, AAA was found to code for lysine and CCC for proline. Poly G could not serve as a template because the molecule was found to fold back on itself. Therefore, the assignment of amino acid for the code GGG had to await other approaches.

Use of Mixed Co-polymers

Nirenberg, Matthaei and Ochoa and their co-workers then used RNA heteropolymers. They used two or more different NDPs in combination to form the synthetic mRNA. They found that if the relative proportion of each type of NDP is known, then the frequency of any particular triplet codon occurring in the synthetic mRNA could be predicted. If this mRNA is added to the cell-free system and if the percentage of any particular amino acid present in the new protein is ascertained, it was possible to predict the composition of triplets specifying the amino acids.

For example, if A and C (as ADP and CDP) are added in the ratio 1A : 5C, then there is 1/6 possibility for an A and 5/6 possibility for a C to occupy a position in the codon. The chance of getting AAA is $(1/6)^3 = 0.4$ per cent. The chance of getting 2A and 1C is $(1/6)^2 (5/6) = 2.3$ per cent. The codons AAC, ACA and CAA have 2A and 1C. Together, all three 2A : 1C triplets account for 6.9 per cent. In the same way, the chance of getting 2C and 1A is $(5/6)^2 (1/6) = 11.6$ per cent (total 34.6 per cent). CCC occurs with a frequency of $(5/6)^3 = 57.9$ per cent.

After adding ADP and CDP in the ratio 1A : 5C, the percentage of amino acids incorporated into the synthesized protein was calculated. The results are represented in Table 8.3. The data helped in predicting the probable base composition of their respective codons.

Because proline appears 69 per cent which is close to 57.9 + 11.6 per cent, we can deduce that proline is coded by CCC and by one triplet of 2C : 1A. Histidine at 14%, is probably coded by one 2C : 1A (11.6%) and one 1C : 2A (2.3%). Threonine, at 12% is likely to be coded by only one 2C : 1A. Asparagine and glutamine each appear to be coded by one of the 1C : 2A triplets and lysine by 3A.

The results and interpretation of a mixed co-polymer experiment using A and C in the ratio 1 : 5 is presented in Table 8.3.

Table 8.3 Results and interpretation of a mixed co-polymer experiment using 1A : 5C

Percentage of amino acid in the protein		Probable base composition assignments
Lysine	<1%	AAA
Glutamine	2%	1C : 2A
Asparagine	2%	1C : 2A
Threonine	12%	2C : 1A
Histidine	14%	2C : 1A, 1C : 2A
Proline	69%	CCC, 2C : 1A

In a similar manner, using all possible pairs of combination of NDPs in different proportion, the composition of triplet code words corresponding to all 20 amino acids were identified. But the specific sequences of triplets were still unknown.

Triplet Binding Assay

In 1964 Nirenberg and Philip Leder developed the triplet binding assay (Figure 8.9), which led to specific sequence assignments of triplets. The principle of the techniques was that when ribosomes were mixed with RNA sequence as short as three ribonucleotides, they will bind and form a complex. The triplet acts like a codon in mRNA, which is capable of attracting the complementary sequence within tRNA. Such a triplet sequence in tRNA that is complementary to a codon of mRNA was termed anticodon.

RNA triplets of known sequence were synthesized to serve as templates. The tRNA-amino acid that bind to the particular triplet RNA–ribosome complex was to be determined. The amino acid to be tested was made radioactive and the corresponding charged tRNA was produced.

The radioactively charged tRNA, the RNA triplet and ribosomes were incubated together on a nitrocellulose filter. The filter was found to retain the larger ribosomes but not smaller components like charged tRNAs. Radioactivity was detected on the filter only when the RNA triplet–ribosome complex was bound with the radioactively labelled amino acid containing tRNA. This indicated that the triplet nucleotide on the RNA is the codon for that particular amino acid. This experiment was referred to as triplet binding assay that involves complex formation involving the ribosome, RNA triplet and aminoacyl-tRNA in which the amino acid is radioactively labelled.

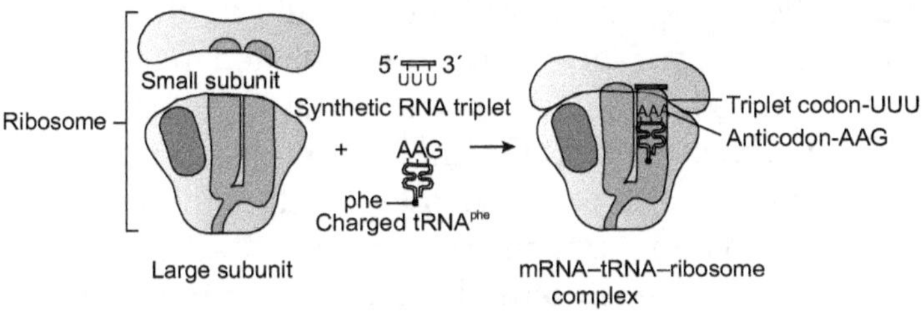

Figure 8.9 Triplet binding assay

This experiment helped in assigning 26 triplets to 9 amino acids as indicated in Table 8.4.

Table 8.4 Amino acid assignments to specific trinucleotides derived from the triplet binding assay

Trinucleotides	Amino acid
UGU UGC	Cysteine
GAA GAG	Glutamic acid
AUU AUG AUA	Isoleucine
UUA UUG CUU	Leucine
CUC CUA CUG	Leucine
AAA AAG	Lysine
AUG	Methionine
UUU UUC	Phenylalanine
CCU CCC	Proline
CCG CCA	Proline
UCU UCC	Serine
UCA UCG	Serine

Later, 50 of the 64 triplets were assigned to amino acids. These findings led to two major conclusions that the genetic code is degenerate, that is, one amino acid may be specified by more than one triplet and that the code is also unambiguous, that is, a single triplet specifies only one amino acid.

Use of Repeating Co-polymers

In the early 1960s Gobind Khorana chemically synthesized long RNA molecules consisting of short sequences repeated many times. First he created shorter sequences (di, tri and tetranucleotides) which were then replicated many times and finally joined enzymatically to form the long polynucleotides. A dinucleotide made in this way is converted to a message with two repeating triplets. A trinucleotide is converted to a message with three potential triplets depending on the point at which initiation occurs and a tetranucleotide creates four repeating triplets (Figure 8.10).

$(UG)_n$ ----------- UGU GUG UGU GUG UGU GUG
            1   2   1   2   1   2

$(AGC)_n$ ----------- AGC AGC AGC AGC AGC AGC AGC
           1   1   1   1   1   1   1

GCA GCA GCA GCA GCA GCA GCA
  2   2   2   2   2   2   2

CAG CAG CAG CAG CAG CAG CAG
  3   3   3   3   3   3   3

$(AGCU)_n$ ----------- AGC UAG CUA GCU AGC UAG CUA
           1   2   3   4   1   2   3

Figure 8.10 Repeating co-polymers and codons

When these synthetic mRNAs were added to a cell-free system, the predicted number of amino acids was incorporated. Table 8.5 shows the amino acids incorporated using repeated synthetic co-polymers of RNA.

A repeating trinucleotide sequence UUCUUCUUC... produces three possible triplets, UUC, UCU and CUU depending on the initiation point. When placed in a cell-free translation system, the polypeptide containing phenylalanine, serine and leucine are produced. But the repeating dinucleotide sequence UCUCUCUCUC... produces the triplets UCU and CUC incorporating leucine and serine in the polypeptide. Comparison of both the results indicates that the triplets UCU and CUC specify leucine and serine but do not specify which triplet specifies which amino acid. Also it could be concluded that either CUU or UUC encode leucine or serine.

Table 8.5 Amino acids incorporated using repeated synthetic co-polymers of RNA

Repeating co-polymer	Codons produced	Amino acids in polypeptides
UG	UGU	Cysteine
	GUG	Valine
AC	ACA	Threonine
	CAC	Histidine
UUC	UUC	Phenylalanine
	UCU	Serine
	CUU	Leucine
AUC	AUC	Isoleucine
	UCA	Serine
	CAU	Histidine
UAUC	UAU	Tyrosine
	CUA	Leucine
	UCU	Serine
	AUC	Isoleucine
GAUA	GAU	None
	AGA	
	UAG	
	AUA	

The results with the repeating tetranucleotide sequence UUAC give more specific information. The possible triplets with this sequence are UUA, UAC, ACU and CUU. We are interested in the triplet CUU. Three amino acids are incorporated, leucine, threonine and tyrosine. Because CUU must specify only serine or leucine and since, only leucine is obtained, it could be concluded that CUU specifies leucine.

Once this is established, we can determine all other specific assignments. Of the two triplet pairs remaining (UUC and UCU from the first case and UCU and CUC from the second case), we have to find the triplet in common that encodes serine which is found to be UCU. By elimination, UUC is found to encode phenylalanine and CUC to encode leucine. Therefore, four specific triplets encoding three different amino acids have been assigned from these experiments. Thus, Khorana reaffirmed triplets that were already deciphered and filled in the leftover gaps in the genetic code table.

Identification of the Stop Codons

When the polymer poly (GUAA) was used as synthetic RNA, large polypeptides were not synthesized. Only two peptides were found namely the tripeptide val-ser-lys and the dipeptide ser-lys. Already it was assigned that valine is encoded by GUA, serine by AGU and lysine by AAG. Poly (GUAA) can be read as

```
1  GUA AGU AAG UAA GUA AGU AAG ···

2  G   UAA GUA AGU AAG UAA GUA AGU ···

3  GU  AAG UAA GUA AGU AAG UAA GUA ···

4  AGU AAG UAA GUA AGU AAG UAA ···
```

The corresponding polypeptides obtained were

1.	val	ser	lys	?
2.	?	val	ser	lys	?
3.	lys	?		
4.	ser	lys	?	

In each case, synthesis stops just before a UAA codon. This indicated that UAA must be a stop codon. Later UAG and UGA were also identified as stop codons. It was found that there is no tRNA that can bind to UAA, which is the reason that it is a stop codon.

As a laboratory joke, the three stop codons were given the names, amber codon (UAG), ochre codon (UAA) and opal codon (UGA).

Identification of the Start Codons

In the *in vitro* system, protein synthesis can start at any base in the polymer used to direct the synthesis. However *in vivo* protein synthesis cannot begin at any base in an RNA molecule. A start codon is required. The codon AUG is the most commonly used start codon. Sometimes, GUG is used.

The various experiments performed to decipher the genetic code revealed and established the dictionary of 61 triplet codon–amino acid assignments. The remaining three triplets are the termination signals, not specifying any amino acid.

Colinearity Between Positions of Gene Mutations and Altered Polypeptide Amino acids

This feature was proved by C. Yanoyfsy and his colleagues who did work on one of the two polypeptides in tryptophan synthetase in *E. coli*. They identified that there was a perfect correlation between the map positions of mutations in the tryptophan synthetase a gene and the positions of the resultant amino acid substitutions in the tryptophan synthetase α -polypeptide (Figure 8.11).

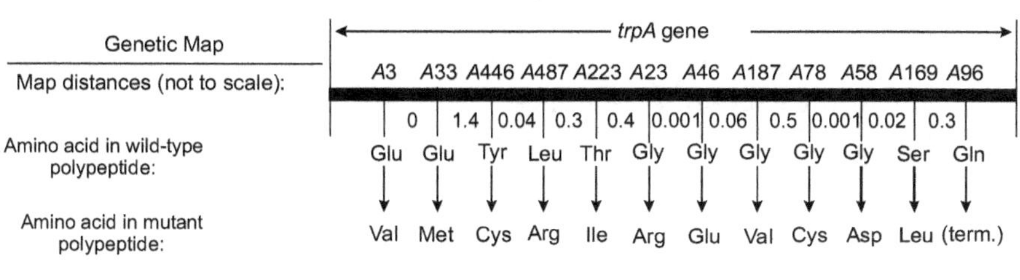

Figure 8.11 Mutation studies in *trpA* gene in *E. coli*

A. Sarabhai and his co-workers showed a similar colinearity between the positions of mutations in the gene of the T4 phage that codes for the major structural protein of the phage head and the positions in the polypeptide affected by the mutations. They studied amber (UAG-chain termination) mutations and showed a direct

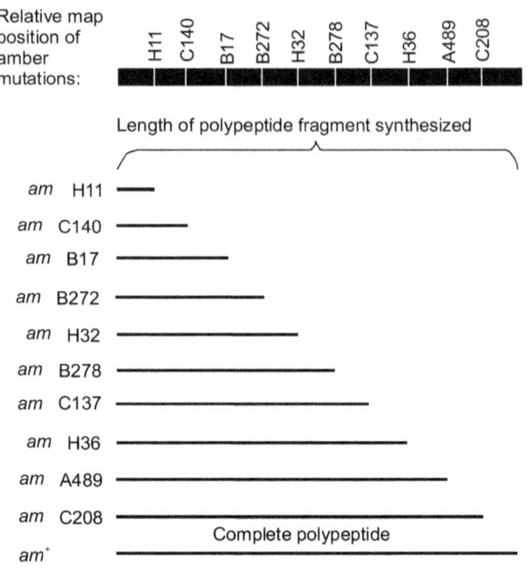

Figure 8.12 Colinearity between the map positions of amber mutations in *T4* gene and the positions of the resulting polypeptide chain terminations

correlation between the length of the polypeptide fragment produced and the position of the mutation within the gene (Figure 8.12).

Spicy Questions and Answers

1. **What base-pair sequence in a segment of a gene in *Drosophila* would code for the dipeptide sequence H$_2$N–met-trp-COOH?**

    Met-Trp

    5′ AUG UGG 3′   codons in mRNA

    3′ TAC ACC 5′   DNA transcribed strand

    5′ ATG TGG 3′   non-transcribed DNA strand

    [Also a stop codon can be indicated after Trp codon and its corresponding DNA segment and be indicated]

2. **Write all the possible mRNA sequences that can code for the simple tripeptide sequence leu-met-tyr.**

    5′ (Any one of the six codons of leucine)–AUG–(Any one of the two tyrosine codons) 3′

    Therefore, a total of 6 × 1 × 2 = 12 mRNA sequences are possible.

3. **Can we predict the base sequence of the mRNA that encoded a particular protein, if the sequence of amino acids in that protein is known?**

   No. Because most of the amino acids have more than one codon, even if the sequence of amino acids are known, we may not know which of these codons are in the mRNA. However, the sequence of nucleotides (codons) which are the only codons for some amino acids like methionine and tryptophan can be predicted with high certainty.

4. **The template strand of a segment of double-stranded DNA has the sequence**

   5′ CTTAACACCCCTGACTTCGCGCCGTCG 3′

   (a) **What is the base sequence of the mRNA that can be transcribed from this strand?**

   (b) **Give the sequence of amino acids encoded by this mRNA.**

   (c) **If the complementary DNA strand is transcribed and translated, will we get the same sequence of amino acids?**

   (a) 3′ GAAUUGUGGGGGACUGAAGCGCGGCAGC 5′

   or

   5′ CGACGGCGCGAAGUCAGGGGGCUGUUAAG 3′

   (b) arg-arg-arg-glu-val-arg-gly-val-lys

   (c) No, the same sequence of amino acids will not be obtained. Because, DNA is transcribed from $5′ \rightarrow 3′$ direction and if the other DNA strand acts as the template for mRNA synthesis, the sequence of nucleotides will not be the same.

5. **How would you make a polyribonucleotide that can serve as an mRNA predominantly coding for many phenylalanine residues and a small number of leucine and serine residues? Can any other amino acid also be encoded by this mRNA?**

   Polyphenylalanine can be obtained using a ribonucleotide sequence $(UUU)_n$. If a small fraction of C is in the mRNA, CUU and UCU codons are possible, accounting for the presence of leucine and serine in the polypeptide along with UUC, the codon of phenylalanine. Nucleotides of known sequence can be incorporated using the enzyme polynucleotide phosphorylase using NDPs as substrates. If only UDPs are used, only poly(U) synthetic RNA is obtained. If a small fraction of CDP is also added, C is also incorporated in the ribonucleotide, accounting for leucine and serine codons.

If the concentration of CDP is slightly increased, sequences of nucleotides UCC (serine codon), CUC (leucine codon) and CCU (proline codons) are also possible accounting for increased frequency of serine, leucine and proline in the polypeptides.

6. **Write all possible anticodons for the glycine codons a) Which position in the anticodon is the primary determinant of its codon specificity? b) Which of these anticodon–codon pairings has a wobbly base pair? c) In which pair do all the three positions exhibit strong Watson and Crick hydrogen bonding?**

   Glycine codons are GGU, GGC, GGA, GGG

   Possible anticodons are 3' CCI 5' and 3' CCC 5'

   (a) 3' side and the middle position of the anticodon

   (b) Anticodon 3' CCI 5' pairing with the codons, 5' GGU 3', 5' GGC 3' and 5' GGA 3'

   (c) Anticodon 3' CCC 5' pairing with the codon 5' GGG 3'

7. **A bacteriophage ϕX174 has a genome of size 5386 bases and includes genes for 10 proteins A–J, with the number of amino acids 455, 120, 86, 152, 91, 427, 175, 328, 38 and 56. What should be the size of the DNA to encode these ten proteins? How can you reconcile the size of the ϕX174 genome with its protein-coding capacity?**

   The number of DNA bases or the size of the DNA required to produce 10 proteins A–J will be $1928 \times 3 = 5784$. (ϕX174 has single-stranded circular DNA. Addition of the amino acids in all the ten proteins and multiplication by three will give the number of nucleotides coding for one amino acid, which is the total number of DNA bases required for the ten proteins). Since the size of ϕX174 DNA is less than this (5386 b), it can be argued that there are regions within one gene, where another gene overlaps.

8. **If the average molecular mass of an amino acid is assumed to be 100 daltons, about how many nucleotides will be present in an mRNA coding sequence specifying a single polypeptide with a molecular mass of 27,000 daltons?**

   The mRNA coding sequence specifies a polypeptide with $27000 \div 100 = 270$ amino acids. Therefore the number of nucleotides present in the mRNA will be $270 \times 3 = 810 + 3$ stop nucleotides $= 813$.

9. **What is coding ratio?**

Coding ratio is the number of nucleotides in an mRNA required to represent an amino acid. Since there are four different nucleotides in RNA and 20 different amino acids in proteins, the minimum acceptable coding ratio is 3 : 1.

10. **If a co-polymer is used as mRNA with a random sequence having equimolar amounts of A and U, what amino acids will be incorporated and in what ratio?**

The possible codons in the mRNA will be AAA, UUU, AAU, AUA, UAA, UUA, UAU, AUU and the amino acid incorporated will be lys, phe ala, asn, ile, leu and tyr. These amino acids are likely to be incorporated with equal frequencies except isoleucine which is incorporated double the times as the other amino acids, since there are two codons specifying this amino acid (AUA and AUU). Also there is a stop codon UAA. When this is encountered, polypeptide synthesis is terminated.

11. **A single tRNA can insert serine in response to three different codons UCC, UCU and UCA. What is the anticodon sequence of this tRNA?**

5'IGA 3'

12. **Polymers of $(GUA)_n$ result in the formation of only two different poly-amino acids rather than three, though there are three reading frames. What is the reason?**

Polymer $(GUA)_n$ can be read in three ways.

	1	GUA GUA GUA ········
Reading frames	2	AGU AGU AGU ········
	3	UAG UAG UAG········

The polypeptide products possible are $(val)_n$ and $(ser)_n$. UAG is the stop codon and no polypeptide is possible with the third reading frame.

Review Questions

1. What are the features of the genetic code?

2. How was the genetic code first decoded? What refinements have since been incorporated in the technique?

3. What is meant by degeneracy of the genetic code?

4. How was colinearity of the nucleotides in the gene and the amino acids in the polypeptide established?

5. How was synthetic mRNA prepared and used in deciphering the codons?

6. Explain wobble hypothesis.

7. Calculate the minimum number of tRNA molecules required to recognize all the sense codons of the standard amino acids.

8. How was the triplet nature of the genetic code proved?

9. Justify the statement that the genetic code is not universal.

10. In what sense and to what extent is the genetic code degenerate and universal?

9

MUTATIONS AND MOLECULAR MECHANISM OF MUTAGENESIS

Mutation is the basis for evolution and the only means by which new genetic variation is created. The word "mutation" is derived from the Latin verb *mutare* which means "to move or change". The word mutation was first used in genetics to describe spontaneous heritable changes in the appearance of an organism. With the discovery of DNA as the genetic material, scientists recognized mutations as alterations in the nucleotide sequence of DNA. A mutation in the DNA of a gene may alter the amino acid sequence encoded by that gene, which may alter or eliminate the function of the gene, which in turn, may affect the phenotype of an individual who inherits the mutation.

AN OVERVIEW OF MUTATION

The term mutation was coined in 1901 by Hugo De Vries to describe the variation he noticed in crosses involving the evening primrose, *Oenothera lamarckiana*, which included gene mutation, i.e., changes in the chemical composition of DNA.

Mutation can be defined as any heritable change in the genetic material. Mutations can occur at any time and in any cell. The genes contain the hereditary characters that are faithfully transmitted from parents to offspring during reproduction. These genes are located on chromosomes. Genes are always DNA which encode their information in definite sequences of nucleotide pairs. DNA molecules are always accurately duplicated and passed on to the next generation by semiconservative mode of DNA replication. However, there may be sudden, heritable changes in the DNA (genetic material) which is referred to as mutation. The term mutation not only refers to the change in the genetic material, but also to the process by which the change occurs. An organism which is normal and which has not undergone any mutation is referred to as wild type and an organism that exhibits an altered phenotype due to change in the genotype because of mutation is referred to as mutant type. If the genotype changes include changes in chromosome number or gross changes in the chromosome structure, it is termed as chromosome aberration. The term mutation is frequently used in the narrow sense to refer only to changes in individual genes.

Mutation is the ultimate source of all genetic variation. It provides the raw material for evolution. Recombination just rearranges the genes into new combinations. Natural selection just preserves the combinations best adapted to the desired environmental conditions. If there is no mutation, all genes will exist in only one form. There will not be different alleles for a gene. Mutation is required for the evolution of organisms and for organisms to adapt to environmental changes. Because mutation of genes results in altered phenotypes, some level of mutation is essential to provide genetic variability. At the same time, if mutation occurs too frequently, it may disrupt the transmission of the genetic information from generation to generation.

The phenotypic effects caused by mutation may range from minor alterations that are detectable only by biochemical methods to drastic changes in important processes which may result in unrestrained cell proliferation causing cancer or at the other extreme, the death of the cell or the entire organism. The effect of a mutation is determined by the type of the cell that is involved and by stage in the life cycle or development of the organism when mutation occurs.

Mutations provide the basis for most of the genetic studies. Mutations provide gene markers which could be followed during transmission of genetic information from parents to offspring. If there are no phenotypic variabilities due to mutation, genetic studies may be impossible. For example, if all pea plants showed same phenotypic characters, Mendel would not have come to any conclusion.

Certain organisms like viruses, bacteria, fungi and fruit flies lend themselves to the induction of mutations that can be easily detected and studied within their short life cycle.

BOX 9.1   REPLICA PLATING TECHNIQUE

There is a basic doubt about the nature of mutation, that is, whether mutation is a purely random event with environmental stress merely preserving pre-existing mutations or whether mutation is directed by environmental stress. For example, the prevalence of a population of bacterial strain resistant to a particular antibiotic has to be identified and verified as to whether such mutant strains are developed among the wild type susceptible strains due to exposure to the particular antibiotic or even before exposure to this stress condition. This was first accomplished by J. Lederberg and E.M. Lederberg using a technique called replica plating.

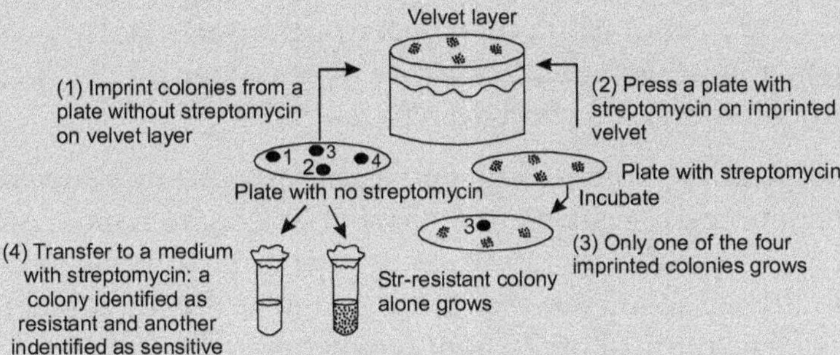

Replica plating technique

The bacterial cells under study are diluted and spread on the surface of nutrient agar medium in Petri plates. After some time, each bacterium will produce a visible colony on the surface of the agar. This is called the master plate which is inverted and pressed on a velvet layer that is present tightly bound to a wooden block. Some bacterial cells from each colony stick to the velvet. Now a Petri plate with nutrient agar medium containing streptomycin (or any other antibiotic) is pressed on the velvet layer. This plate with the antibiotic is called selective plate. Only those bacterial cells which are already streptomycin-resistant will form colonies on the selective plate containing streptomycin (colony No. 3 in the figure, out of the 4 colonies whose positions are indicated in the master plate).

This helps in identifying the bacterial colonies in the master plate as streptomycin-resistant or streptomycin-sensitive. One of these sensitive colonies (Nos. 1, 2 or 4) and the only resistant colony (No. 3) are separately picked up from the master plate and tested by adding to streptomycin medium when it will be confirmed that only colony No. 3 grows in the presence of the antibiotic.

CLASSIFICATION AND TYPES OF MUTATION

Classification Based on the Response to Environmental Conditions

Adaptive and spontaneous mutation   As early as the 1940s it was assumed that bacterial cells exhibited genetic (heritable) variations in response to the environmental conditions. For example, when exposed to bacteriophage T1, certain *Escherichia coli* strains developed resistance to viral infection. Their descendents also turned to be resistant to T1 infection. This was referred to as adaptation hypothesis. In 1943, Salvador Luria and Max Delbruck gave the first experimental evidence to prove that mutations may occur spontaneously within the chromosome independent of any external agent. This is called as spontaneous mutation or random mutation.

The Luria–Delbruck fluctuation test clearly distinguished between the two hypotheses—adaptation and spontaneous mutation. According to the concept of adaptation, every bacterial cell has a constant probability of acquiring resistance when made to contact phage particles. The number of resistant cells produced will depend only on the number of bacteria and phages added to each petri plate and will be independent of all other external conditions. As a result there will be no fluctuation in the number of resistant cells produced from plate to plate and from one experiment to another as predicted by adaptation hypothesis.

On the other hand, according to the concept of spontaneous mutation, if resistance is acquired randomly as a result of mutation, which may occur even before the bacterium comes in contact with the phage, it will finally result in the production of relatively large number of resistant bacterial cells because of subsequent reproduction of mutant bacterial cells. If spontaneous mutation occurs very late after the bacterium comes in contact with the phage, very few resistant cells will be produced. Therefore this hypothesis of spontaneous mutation suggests that the number of resistant cells will fluctuate significantly from experiment to experiment and from one plate to another. In Luria–Delbruck experiment, a great fluctuation was observed thus supporting the hypothesis of spontaneous mutation.

However, two independent investigations published in 1988 by John Cairns and Barry Hall and their colleagues, proved the possibility of adaptive mutation to operate in bacteria in response to environmental pressure. This is also referred to as directed mutations. The results of their work suggested that some bacteria may select mutations to get adapted to the environment. Cairns' experiment involved the use of lac⁻ mutant bacterial strains, which cannot utilize lactose as the carbon source. First, bacterial cells were made to grow in a liquid culture medium containing an adequate carbon source except lactose. Both lac⁻ and lac⁺ cells (lac⁻ spontaneously mutated to lac⁺) were able to grow in this medium and reproduce well. Aliquots of these cells were then plated on petri plates with minimal medium containing lactose as the sole carbon source. The lac⁻ cells present in large number were found to survive but they were not able to proliferate and form colonies. Cells that have mutated to lac⁺ in the original liquid culture formed colonies and were detected. Those lac⁻ cells present in the original liquid culture (which have not mutated to lac⁺ in original liquid culture medium), after transfer to the petri plate containing lactose as sole carbon source, were found to undergo spontaneous mutation to lac⁺ and they were also detected since they formed colonies. Thus Cairns' work proved that adaptive mutation might occur in bacteria in response to environmental stress. Cairns and his co-workers also established that only lac⁻ cells were mutated to lac⁺ in response to lactose in the medium and not any other mutant gene was found to be mutated to wild type.

Cairns' work suggested that bacterial cells have genetic set-up which can respond to external conditions by producing adaptive mutations. This idea seemed to be contrary to the concept that mutations may be spontaneous events which may occur as errors during DNA replication.

In a similar line, Barry Hall and his co-workers reported a similar adaptive response by *E. coli* to growth on a chemical, salicin. This was found to occur through a two-step genetic change, including the removal of an insertion sequence (IS) element from DNA. Based on such evidences, it is suggested that under stressful nutritional conditions like starvation, bacteria may be capable of activating mechanisms which may create a hypermutable state in genes in order to achieve survival.

Classification Based on the Origin of Mutation

Spontaneous mutation Mutation occurs during normal cellular activities, spontaneously without the knowledge of exposure to any known mutagen, mainly during DNA replication due to tautomerization of the bases or error in replication.

Induced mutation Mutation occurs due to exposure to known mutagenic agents or environmental conditions.

Leaving aside the controversial theory of adaptive mutation, in general, mutations can be classified as spontaneous or induced mutation. Spontaneous mutations occur in nature and no known agents are associated with their occurrence. They may also be caused by mutagenic agents that may be present in the environment. Induced mutations are caused by external agents which act as potent mutagens—agents that cause mutation—and increase the mutation rate. The mutagens are of different types, physical, chemical and biological agents.

Spontaneous mutation Mutations are considered as random events. Any cell can be involved in any kind of mutation at any time. However, every gene mutates spontaneously at a characteristic rate.

Watson and Crick described the double helix structure of DNA and proposed the mechanism of semiconservative replication based on specific base-pairing between complementary DNA base pairs. They also proposed a mechanism to explain spontaneous mutations. Watson and Crick suggested that the structure of bases in DNA are not static. Hydrogen atoms can move from one position to another in the purine and pyrimidine bases like movement from an amino group to a ring nitrogen. Such chemical fluctuations are referred to as tautomeric shifts. Though this event is rare, it is of great significance because they alter the base-pairing specificity of the purines and pyrimidines.

The shifts of hydrogen atoms are between the 3rd and 4th position in pyrimidines and between the 1st and 6th positions in the purines. The tautomeric forms of the four DNA bases are given in Figure 9.1. Thymine and guanine are in the more stable keto forms. Because of tautomerization, they may attain a less stable enol form. Similarly, cytosine and adenine are in the more stable amino forms. On tautomerization, they may attain a less stable imino form. The bases may exist in their less stable tautomeric forms only for a very short period of time. But, if a base exists in the rare less stable form at the time when it is being replicated or being incorporated into a nascent DNA chain, a mutation may occur. In the rare enol form, thymine forms base pair with guanine instead of adenine, and guanine in enol form forms base pair with thymine instead of cytosine. Cytosine in rare imino form forms base pair with adenine instead of guanine and adenine in rare imino form forms base pair with cytosine instead of thymine.

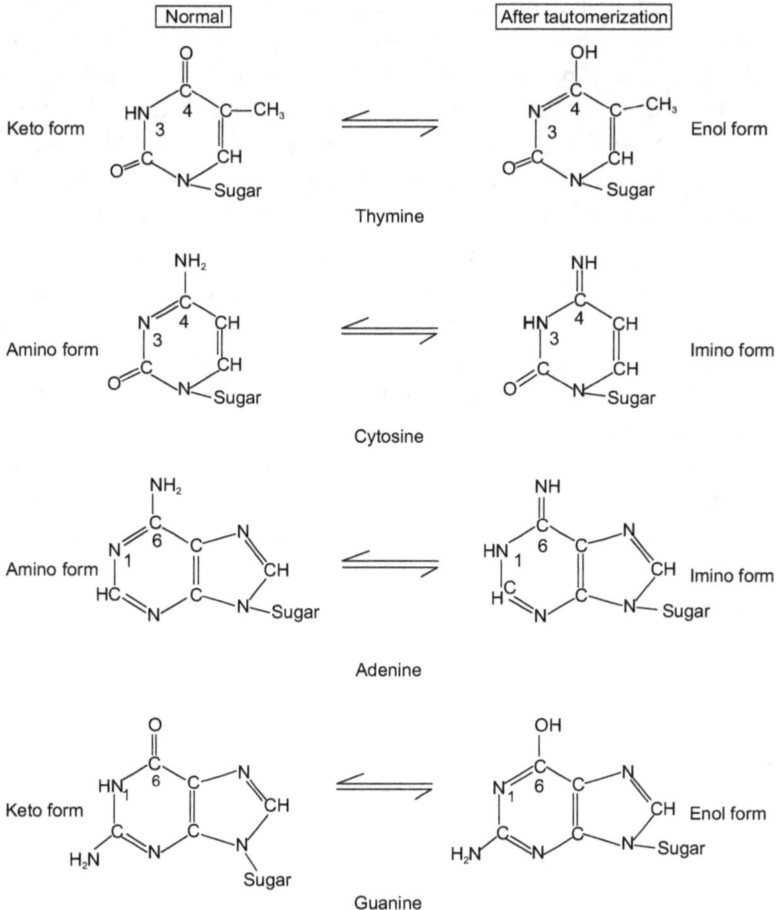

Figure 9.1 Tautomeric forms of the four common bases in DNA

Certain genetic diseases have been identified in humans which are caused by spontaneous mutations in either the egg or the sperm which are involved in fertilization.

Sickle-cell anaemia is a condition identified 1 in 500 Africans which is caused by a single missense mutation (sense of one codon altered, i.e., codon for one amino acid altered to that for another amino acids) at codon 6 of the gene for β-globin chain of hemoglobin. As a result of this mutation which may occur spontaneously, glutamic acid at position 6 in the normal β-globin protein is changed to a valine in the mutant protein. This has a profound effect on hemoglobin, the oxygen-carrier protein of erythrocytes. The mutant hemoglobin molecule loses its capacity to carry oxygen, becomes insoluble in erythrocytes and forms crystalline arrays. Therefore the erythrocytes of affected persons becomes rigid and their flow through capillaries is blocked, causing severe pain and tissue damage. The erythrocytes of heterozygous individuals regarding this gene are found to be resistant to the parasite causing malaria, which is endemic in Africa.

Retinoblastoma is associated with retinal tumours in children that is caused due to spontaneous mutation in somatic cells. The hereditary form of retinoblastoma results from a germ-line mutation in one Rb allele and a second somatically occurring mutation in the other Rb allele. When Rb$^+$ Rb$^-$ retinal cell undergoes somatic mutation, it becomes Rb$^-$ Rb$^-$; as a result, there is proliferation of the cell in an uncontrolled manner, giving rise to retinal tumour. In sporadic retinoblastoma, two separate events of somatic mutations occur, resulting in the formation of Rb$^-$ Rb$^-$. As only one somatic mutation is required for tumour development in children with hereditary retinoblastoma, it occurs at a much higher frequency than the sporadic form.

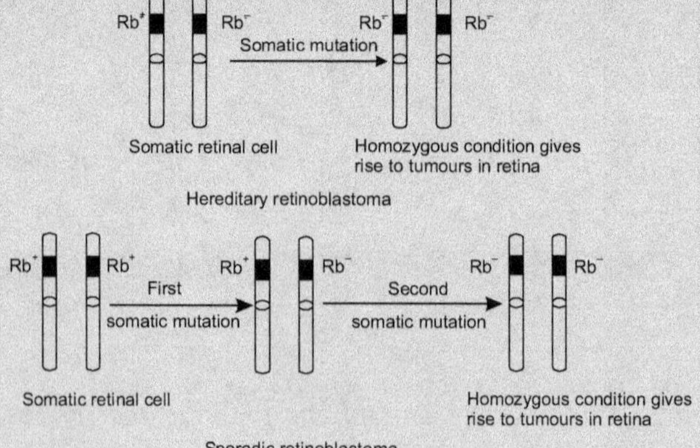

Several nucleotide bases undergo spontaneous loss of their exocyclic amino groups, i.e., they undergo deamination. For example, in a typical cell, deamination of cytosine in DNA to uracil occurs in about one in every 10^7 residues of cytosine per day. This corresponds to about 100 spontaneous events per day in a mammalian cell. However, deamination of adenine and guanine is about 100 times slower. Once uracil is formed, the DNA molecule is recognized as a foreign molecule and uracil is removed by the DNA repair system. Unrepaired uracils in DNA may lead to permanent changes in the nucleotide sequence during replication, because they tend to pair with adenine. Therefore, cytosine deamination would gradually lead to a decrease in GC base pairs and an increase in AT base pairs.

Another important spontaneous reaction observed in deoxyribonucleic acids is the hydrolysis of N-β-glycosyl bond between the base and deoxyribose sugar. This occurs at a higher rate for purines than for pyrimidines. As many as one in 10^5 purines (10,000 per mammalian cell) are lost from DNA every 24 hours.

DNA polymerase is the key enzyme involved in DNA replication which adds the nucleotides from $5' \rightarrow 3'$ direction, dictated by the template parental strand. The same enzyme possesses $3' \rightarrow 5'$ exonuclease activity by which it removes a nucleotide that has been wrongly added during polymerization. This is known as the proofreading action or editing function of the enzyme. If this activity of DNA polymerase is lost, a wrong nucleotide added remains permanently in the new DNA strand, thereby altering the base pair in that position in the next round of replication.

DNA polymerases were found to insert one incorrect nucleotide for every 10^4 to 10^5 correct ones. These mistakes occur because a base is briefly in an unusual tautomeric form allowing it to hydrogen-bond with an incorrect partner.

Induced mutation   Mutations caused by known agents referred to as mutagens are induced mutations. The mutagens are of three broad types.

1. *Physical mutagens*   Physical mutagens include ionizing radiations like α, β, γ or X-rays and non-ionizing radiations like UV light.

2. *Chemical mutagens*   Chemical mutagens are chemical substances that interact with DNA to create base changes like base analogs, base modifiers and intercalating agents.

3. *Biological mutagens*   Certain biological agents like Mu phages, transposable elements and certain RNA viruses bring about mutation in the DNA molecules.

Apart from these mutagens, mutator mutations are types of mutations where certain mutations influence the mutability of other genes.

Mutation caused by ionizing radiations   Electromagnetic radiation of wavelength shorter than visible light, below 0.1μm, has higher energy which is of two types—ionizing radiation namely X-rays, gamma rays and cosmic rays and non-ionizing radiation like ultraviolet rays. Ionizing radiations such as X-rays which possess high energy penetrate living tissues, collide with atoms and cause the release of electrons, resulting in the formation of free radicals that attack molecules like DNA. The low-energy ultraviolet rays dissipate their energy to atoms that they encounter and cause excitation. The increased reactivity of the atoms present in DNA molecules is the basis of the mutagenic effects of ionizing radiation and ultraviolet light.

In 1927, H.J. Muller first demonstrated that mutation could be induced by an external factor. He proved that X-ray treatment induced sex-linked recessive lethal mutations in *Drosophila melanogaster*. He developed a technique called ClB method for a simple and accurate identification of lethal mutations in the X chromosome of *Drosophila*. The method involves the use of female *Drosophila* heterozygous for a normal X chromosome and another X chromosome called ClB chromosome specially constructed for Muller's experiment (Figure 9.2).

The ClB chromosome has three components: 1) "C" refers to cross-over suppressor. A long inversion is made in the chromosome that prevents recombination of genetic markers on the ClB chromosome and alleles on the normal X chromosome. This is to make sure that the markers on the ClB chromosome stay together through meiosis. 2) "l" refers to a recessive lethal in the ClB chromosome. 3) "B" refers to the presence of partially dominant mutation that causes bar eye phenotype, which is a narrow, slit-shaped eye. Because it is partially dominant, females heterozygous for ClB chromosome can be identified readily. Both 'l' and 'B' mutations are located within the inverted segment of the ClB chromosome.

Male flies were irradiated with X-rays and then mated with ClB females. Four types of offsprings were produced.

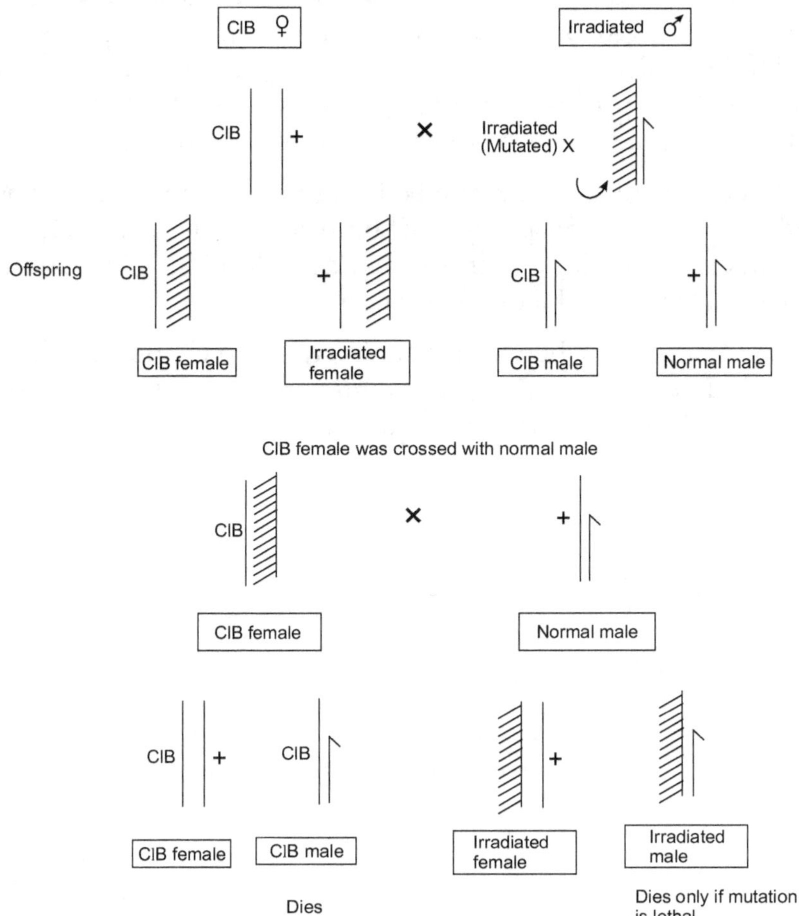

Figure 9.2   ClB technique to detect mutation

Thus in the second mating, all male progenies died. In case irradiated X chromosome had not induced lethal mutation, the progeny of the second type of mating will be female and male in the ratio 2 : 1.

Mutations caused by nonionizing radiations like ultraviolet rays   Ultraviolet rays do not have sufficient energy to induce ionizations.

But certain compounds like purine and pyrimidine bases readily absorb these radiations and enter into a more reactive or excited state. Because of their lower energy content, UV rays penetrate the tissues very slowly and affect the cells on the surface layers in multicellular organisms. But they are potent mutagens for unicellular organisms. DNA molecules absorb UV rays at a wavelength maximum of 254 nm. At this wavelength only, maximum mutagenicity occurs in the DNA. Pyrimidines absorb UV rays more strongly at 254 nm rather than purines. Two products are formed—pyrimidine hydrates and pyrimidine dimers, which are referred to as photoproducts.

On UV irradiation as indicated in Figure 9.3 formation of cytosine hydrate results in mispairing of the bases during replication and formation of thymine dimers causes destabilization of DNA double-helical molecule and blocks replication.

The relationship between mutation rate and UV dosage is highly variable. It depends on the type of mutation that results, the organism that is involved in mutation and also the conditions employed.

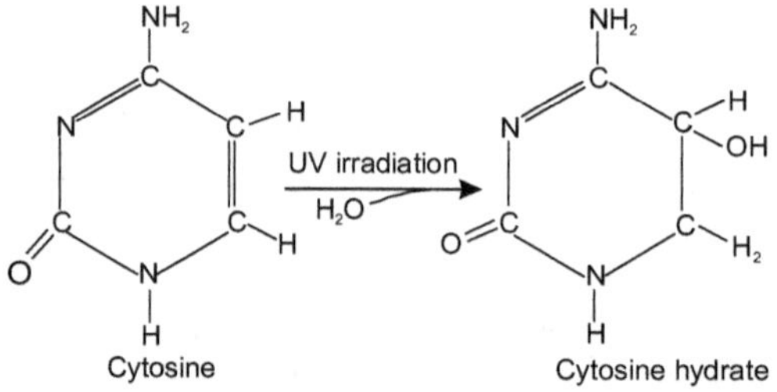

a) Cytosine undergoes hydrolysis to form cytosine hydrate

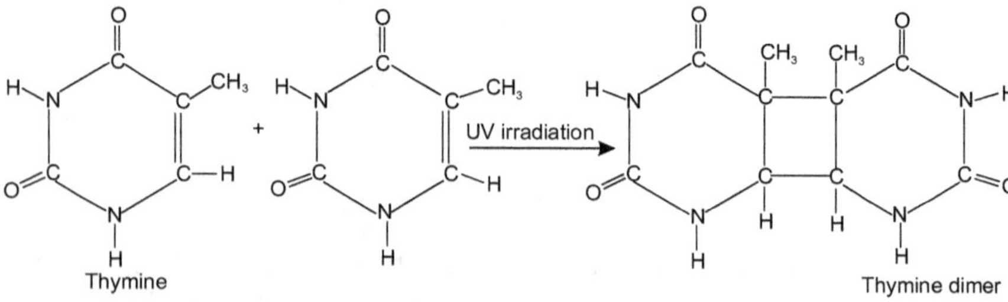

b) Two adjacent thymine bases present in the same DNA strand undergo cross linking to form thymine dimers

Figure 9.3 Pyrimidine photoproducts of UV irradiation

Using this technique, Muller was able to demonstrate an increase in mutation rate up to 150-fold after X-ray treatment.

Another technique was devised for the detection of mutations in *Drosophila*, in this case sex-linked mutations with visible effects on phenotype. Attached X chromosomes were used in this technique. They are two X chromosomes which are joined to a single centromere and they undergo compulsory nondisjunction (failure of separation of homologous chromosomes during anaphase). A special type of female with two attached X chromosomes plus a Y chromosome (XXY) when mated to a male with mutation on the X chromosome of the male, it will be expressed in the male progeny that survives (Figure 9.4).

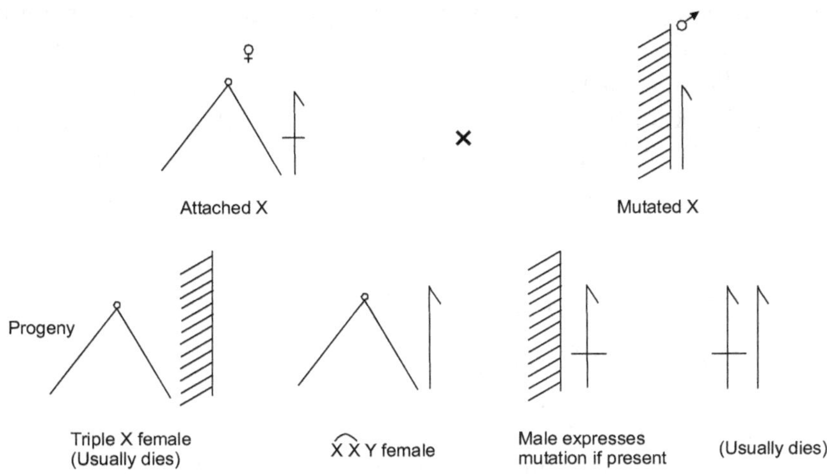

Figure 9.4   Attached X chromosome technique to detect mutation

The male progeny receives the X chromosome from the male parent rather than from the female parent that happens in normal mating. If the male parent is treated with a mutagen like X-rays, the frequency of recessive visible mutations can be easily assessed by screening the male progeny.

Classification Based on the Cell Type

Somatic mutation and germinal mutation   Mutations may occur in any cell and at any state in the cell cycle. If the mutation occurs in a somatic cell (cells other than reproductive cells) it is called

somatic mutation. The somatic cell is capable of producing identical cells, and thus the mutant change will be perpetuated only in somatic cells that descend from the original somatic cell in which mutation has occurred. The delicious apple and the navel orange are mosaics in somatic tissues. The desirable qualities in these fruits appear as a result of spontaneous mutation in single somatic cells.

If mutations occur in germ cells, it is known as **germinal mutation** and the effects are expressed immediately in the progeny. Germinal mutations, like somatic mutation, can occur at any stage of the reproductive cycle. If mutation occurs in a gamete, it is referred to as **gametic mutation** and only a single member of the progeny is likely to have the mutant gene. If dominant mutations occur in germ cells, their effects are expressed immediately in the progeny. If the mutations are recessive, their effects are obscured in diploids. If mutation occurs in gonial cells, several gametes may receive the mutant gene which may enhance its potential for perpetuation. However, the dominance of the mutant allele and the stage in the reproductive cycle when mutation occurs are the major factors that decide the likelihood of the mutant allele to manifest in an organism and a population.

Classification Based on the Direction of Mutation

Forward, backward and suppressor mutations   The mutation that occurs in a wild type gene leading to a form that gives rise to a mutant phenotype is usually referred to as **forward mutation**. Sometimes, the designations "wild type" and "mutant type" are quite arbitrary. They may represent just two different phenotypes. For example, we consider both brown and blue eye colour in humans as wild type. But in a population with almost all individuals having brown eye character, the alleles for blue eyes are considered as mutant.

On many occasions, mutation events are reversible. That is, a subsequent mutation may occur that may restore the original wild type phenotype. This is termed as **back mutation, reverse mutation or reversion**.

Reversion may occur in two ways: (1) the original phenotype may be restored by a true back mutation at the same site in the gene as the original mutation, thereby restoring the original wild type nucleotide sequence (2) the original phenotype may be restored by a second mutation at a different locus in the genome, which may compensate for the first mutation. The second type of mutation is called **suppressor mutation** because it suppresses the effects of the first mutation. Suppressor mutations may occur at a distinct site in the same gene as the original mutation or in different genes, even in different chromosomes.

Suppressor mutations can be distinguished from true back mutations by backcrossing the revertants to the original wild type. If wild type phenotype is restored by a suppressor mutation, the original mutation is still present and this can be separated from the suppressor mutation by recombination (Figure 9.5).

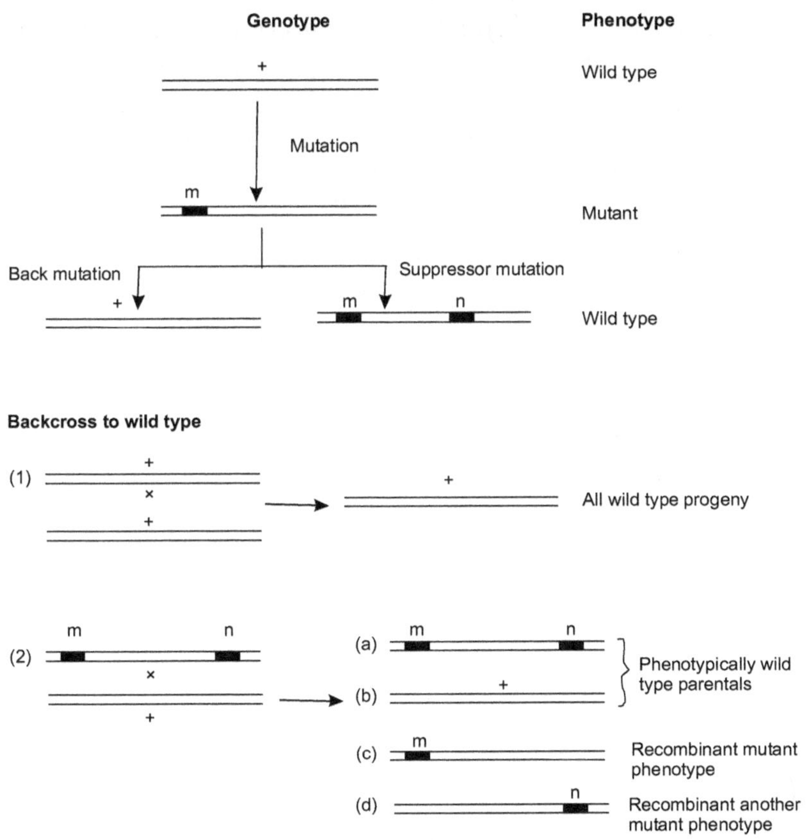

Figure 9.5   True back mutation and suppressor mutation

Suppressor mutations and tRNAs with altered codon recognition   In *E. coli* and yeast, mutations were noticed in tRNA genes which resulted in altered codon recognition by the tRNA anticodons. These mutations were identified as suppressor mutations that suppressed the effect of an already occurred mutation. One example was the suppression of an amber mutation. UAG stop codon is referred to as amber codon. When one of the sense codons within the reading frame of mRNA is changed to amber codon resulting in nonsense mutation, immature termination of polypeptide chain synthesis is expected. This nonsense mutation is suppressed by a mutant tRNA referred to as suppressor tRNA molecule. The mutant tRNA is capable of recognizing the nonsense codons (UAG, UAA or UGA). An amber suppressor tRNA mutation (amber SU3 mutation) occurs in the tRNA^{tyr2} gene—one of the two tRNAs in *E. coli*. The anticodon of the wild type tRNA^{tyr2} was found to be 5′ G′UA 3′ where G′ is a derivative of guanine. The anticodon of the mutant tRNA^{tyr2}

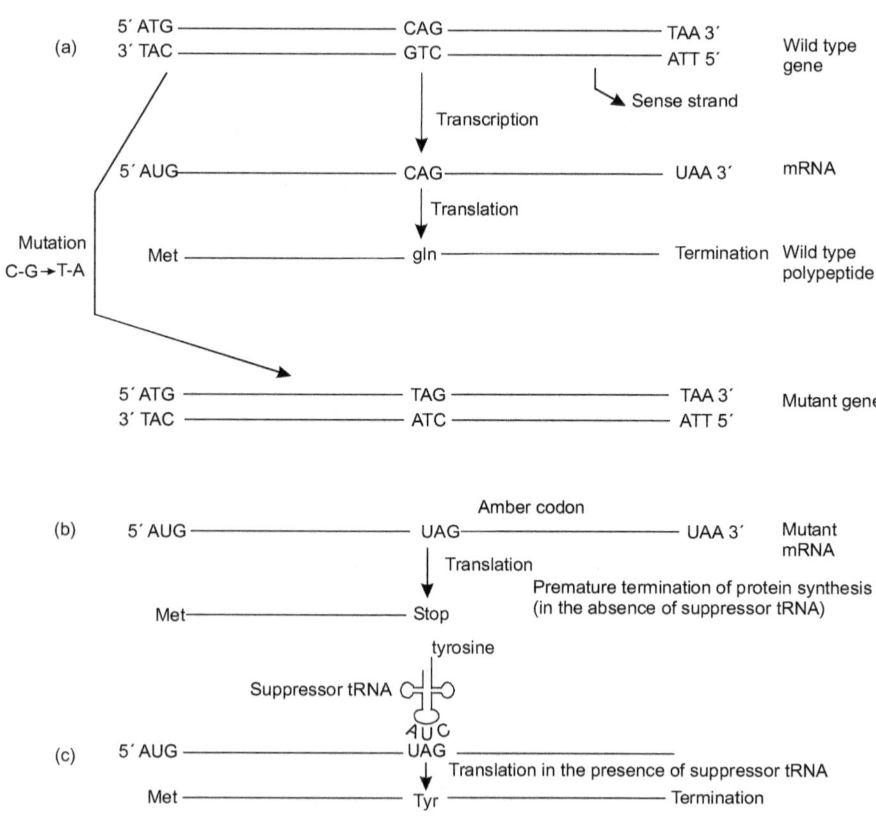

Figure 9.6   Role of suppressor tRNA molecule

(suppressor tRNA) is 5'CUA 3'which is now able to base-pair with the 5'UAG 3'amber codon.

3'AUC 5' anticodon

5'UAG3' codon

Therefore, the suppressor tRNA molecules permit the complete polypeptide to be synthesized from a mutant mRNA bearing a nonsense codon. The polypeptide product will be functional as long as the amino acid inserted by the suppressor tRNA molecule is acceptable at that position.

In Figure 9.6, it has been shown that tyrosine is introduced in the place of glutamine (present in wild type) by suppressor tRNA molecule. The polypeptide so formed containing tyrosine may or may not be functional. However, suppression is observed only if the tyrosine-containing polypeptide is active.

Suppressor mutations thus produce two different mutations from mutated state back to wild type phenotype. These two types are: **intragenic suppressor mutation** in the same gene which was originally mutated, but at a different site, restoring wild type function and **intergenic suppressor mutation** in another gene, restoring wild type function, e.g. that brought about by suppressor tRNA molecules, as indicated in Figures 9.5 and 9.6.

Classification Based on the Trait

Dominant and recessive mutations   Alleles are alternate forms of a gene. Diploids have two alleles—if both are identical, they are said to be homozygous and if both are different, they are said to be heterozygous for the gene. A **recessive mutation** is one in which both alleles must be mutant, in order to show the phenotype of the mutant, i.e., the individual should be homozygous for the mutant allele. In contrast, **dominant mutation** is one in which the individual has one mutant allele and another normal allele showing the phenotype of the mutant (Figure 9.7).

Recessive mutations inactivate the affected gene and there is "loss of function". There may be removal of the gene from the chromosome, disrupted expression of the gene or alteration in the structure of the gene product, thereby altering its function. Dominant mutations result in "gain of function" like increase in the activity of the gene product and new activity for the protein.

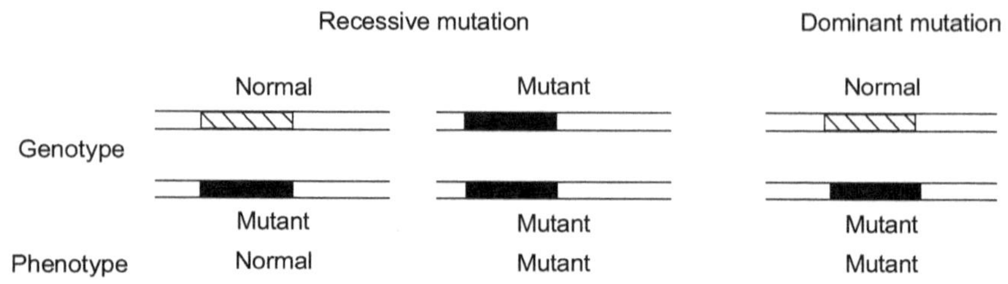

Figure **9.7**   Dominant and recessive mutation

Sometimes dominant mutations may be associated with loss of function. In some cases, two copies of a gene are required for normal function and if one copy is removed it may lead to a mutant phenotype. Such genes are referred to as **haplo-insufficient**. Sometimes, mutation in one allele may lead to a structural change in the protein which may interfere with the function of the wild-type protein encoded by the other allele. They are referred to as **dominant negative mutations**.

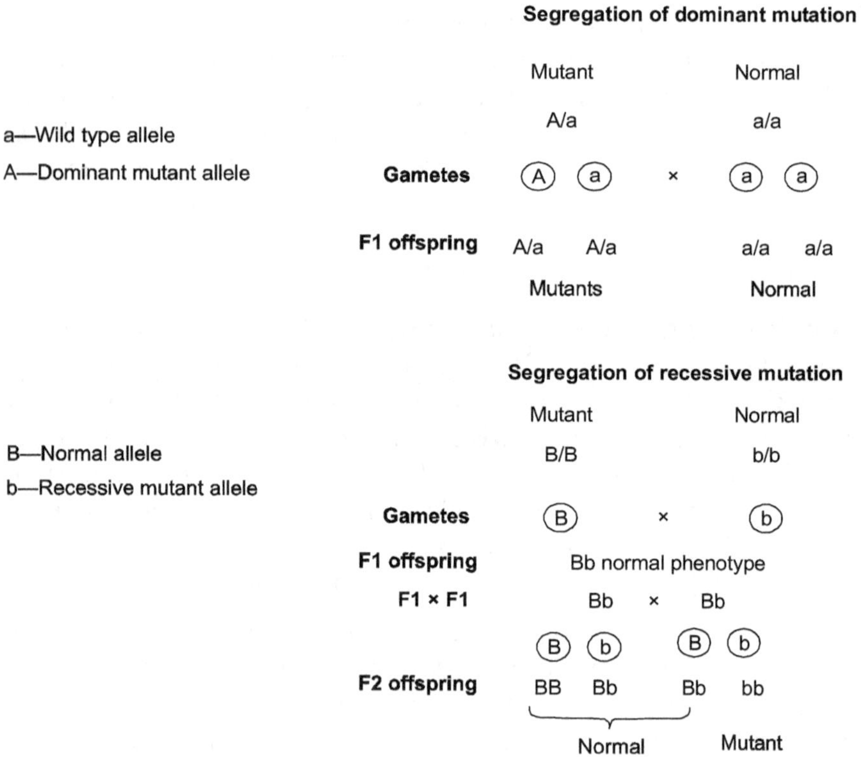

Segregation of dominant mutation

	Mutant	Normal
	A/a	a/a
Gametes	(A) (a) ×	(a) (a)
F1 offspring	A/a A/a	a/a a/a
	Mutants	Normal

a—Wild type allele

A—Dominant mutant allele

Segregation of recessive mutation

	Mutant	Normal
	B/B	b/b
Gametes	(B) ×	(b)
F1 offspring	Bb normal phenotype	
F1 × F1	Bb × Bb	
	(B) (b) (B) (b)	
F2 offspring	BB Bb Bb bb	
	Normal	Mutant

B—Normal allele

b—Recessive mutant allele

Figure **9.8**   Segregation of dominant and recessive mutation

Some alleles are associated with both recessive and dominant phenotype. For example, *Drosophila* heterozygous for the mutant Stubble (Sb) allele has short and stubby body hairs rather than the normal long, slender hairs. The mutant allele is dominant in this case. In contrast, *Drosophila* homozygous for this allele dies. Thus the recessive phenotype is lethal and the dominant phenotype is not lethal.

The inheritance of recessive and dominant mutations differ as shown in Figure 9.8.

Alleles that differ at the same nucleotide site are termed as **homoalleles.** Intragenic recombination between homoalleles cannot result in the production of new alleles.

Alleles that differ due to mutations of nucleotides at different sites are called **heteroalleles.** Intragenic recombination between two heteroalleles can result in the production of new alleles–one will be wild type and the other will be having two different nucleotide mutations (Figure 9.9).

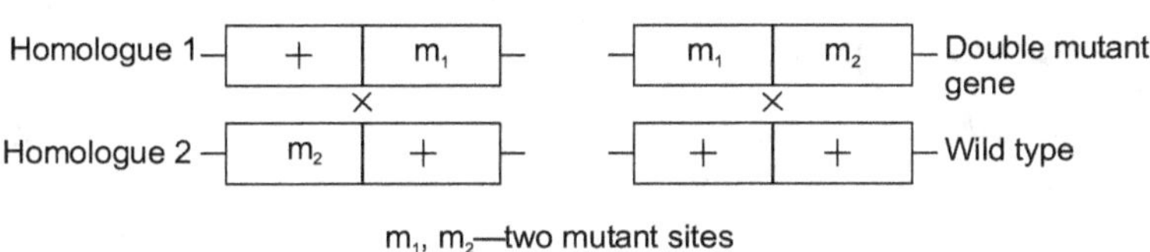

Figure 9.9 Homoalleles and heteroalleles

Two mutants are considered to be functionally allelic, i.e., present in the same gene or cistron, if they complement in the *cis* position and not in the *trans* position. However, if a heterozygote contains two mutations either in the *cis* or in the *trans*-position in different functional units, i.e., present in different genes or cistrons, complementation can produce a normal phenotype. This is because for each mutant functional unit or one DNA molecule, there is a corresponding normal functional unit on the other DNA molecule. Thus, two mutants are functionally non-allelic if they complement in either the *cis* -or the *trans*- position (Figure 9.10).

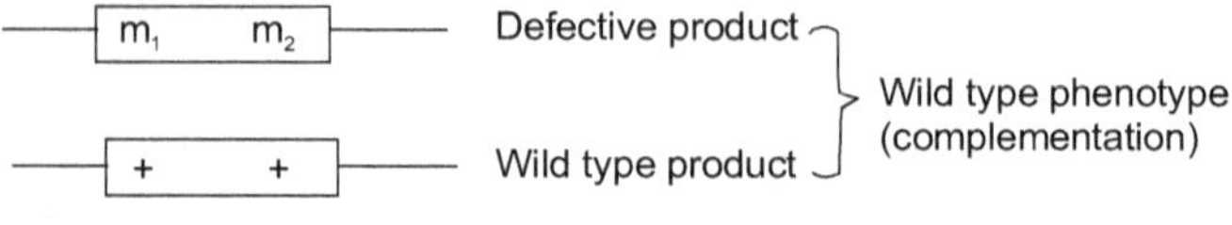

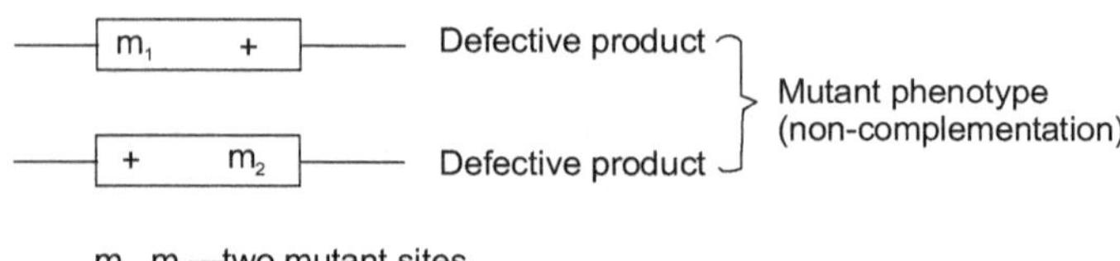

Figure 9.10   Functionally allelic and functionally non-allelic mutants

Mutations can be classified on the basis of several other criteria.

Classification Based on the Size of Mutation

1. ***Point mutation***   A change in a very small segment of DNA, usually one nucleotide pair.

2. ***Gross mutation or multiple mutation***   Changes involving many nucleotide pairs of DNA.

Classification Based on the Types of Nucleotide Changes

1. ***Insertion or addition mutation***   One or more nucleotide pairs inserted in the original sequence in the DNA.

2. ***Deletion mutation***   One or more nucleotide pairs deleted from the original sequence.

   In both insertion and deletion events involving more than one nucleotide pair (in multiple mutation), the changes may be in adjacent positions or in two different positions.

3. ***Substitution mutation***   One or more nucleotides substituted for an already existing nucleotide.

   Substitution mutation is of two types:

1. *Transition mutation* where one purine base is substituted by another purine base or one pyrimidine base is substituted by another pyrimidine base.

2. *Transversion mutation* where one purine base is substituted by one pyrimidine base or one pyrimidine base is substituted by one purine base.

Classification Based on the Effect on Gene Product/Protein/Codons

1. **Silent mutation**   Changes in the codon, usually in the third position, that does not change the amino acid.

2. **Nonsense mutation**   Changes in the codon that changes a sense codon, coding for an amino acid to a nonsense (stop) codon which results in premature termination of protein biosynthesis.

3. **Missense mutation**   Changes in the codon that changes a sense codon coding for one amino acid to another sense codon coding for another amino acid which results in alteration in the amino acid residues in the protein product.

4. **Neutral mutation**   A type of missense mutation, one amino acid replaced by another amino acid, but the new amino acid is very similar to the original one, like one acidic amino acid replaced by another.

5. **Frameshift mutation**   Deletion and addition mutations alter the nucleotide sequences, i.e., the reading frame is shifted so that a number of missense and nonsense mutations are noticed.

Classification Based on the Effect on Gene Function

1. **Loss-of-function mutation**   Any type of mutation that results in a lack of gene function. It is recessive in nature.

2. **Gain-of-function mutation**   Any type of mutation that results in a new or different gene function. This is dominant in nature.

Other Types of Mutations

1. **Nutritional mutation**   Mutation that causes inability to synthesize amino acid or vitamin or utilize carbon sources for energy.

2. ***Biochemical mutation***   Mutation that causes block in biosynthetic or degradative metabolic pathways.

3. ***Behaviour mutation***   Mutation that affects the behaviour pattern of an organism like altering mating behaviour or circadian rhythms of animals.

4. ***Regulatory mutation***   Mutation that affects the regulation of gene expression.

5. ***Conditional mutation***   Mutation that may be present in the genome of an organism, but becomes evident only under certain conditions. Best examples are temperature-sensitive mutations. At certain "permissive" temperatures, a mutant gene product functions normally, but loses its functional capability at a "restrictive" temperature.

MAGNITUDE OF THE PHENOTYPIC EFFECTS OF MUTATION

Change in mutation rate   The alleles mutate at different rates and can be distinguished based on their rate of mutation.

Production of isoalleles   These are alleles that produce identical phenotypes in homozygous or heterozygous conditions, but can be distinguished when they occur in combination with other alleles.

Effect on the viability   They are of three types: **subvitals** where the relative viability is greater than 10% but less than 100% compared to the wild type; **semilethals** which cause more than 90% but less than 100% mortality; **lethals** which kill all individuals before adult stage.

MOLECULAR MECHANISM OF MUTAGENESIS CAUSED BY MUTAGENS

Mutation in DNA strands is caused by many chemicals which are known to have slight to very large mutagenic effects. The first chemical mutagen discovered was mustard gas. C. Auerbach and her associates first analysed the mutagenic effects of mustard gas and related compounds during World War II.

Chemical mutagens are of two types: 1) those that are mutagenic to replicating and non-replicating DNA, e.g. alkylating agents and nitrous acid.) those that are mutagenic only to replicating DNA, e.g. acridine dyes and base analogs. Figure 9.11 gives the structures of selected chemical mutagens.

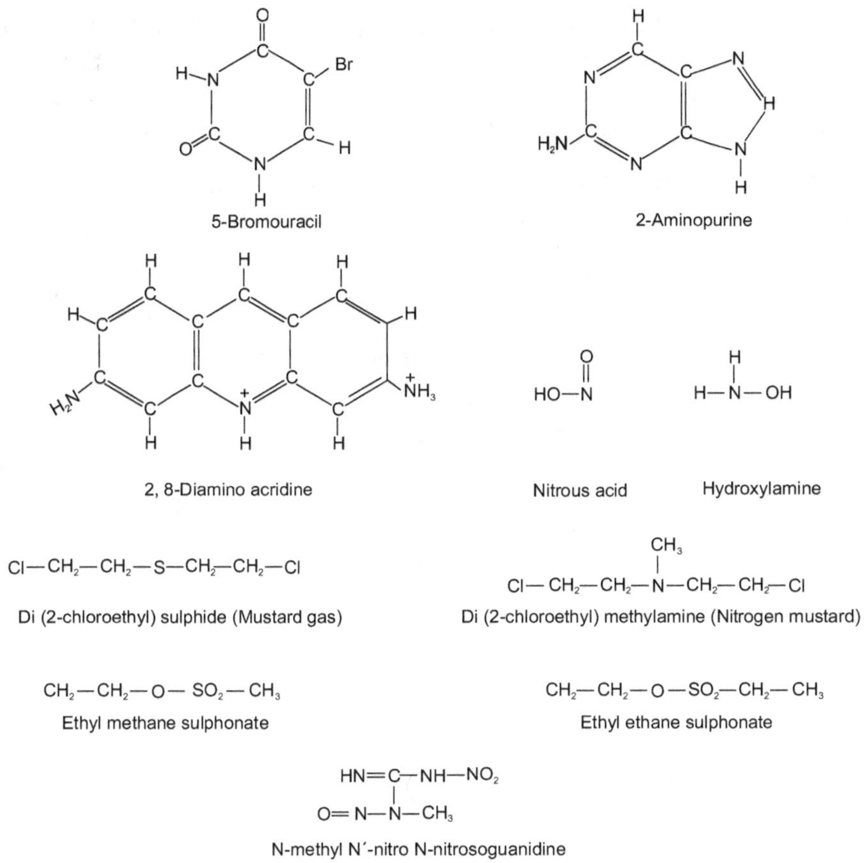

Figure 9.11   Certain chemical mutagens

Base Analogs

These are chemicals which have structural similarities to the purine and pyrimidine bases present in DNA. Therefore, when they are present in the system, they are metabolized and incorporated into DNA during replication. This results in mispairing of the nucleotides leading to mutation. Two common base analogs are 5-bromouracil and 2-aminopurine. 5-bromouracil is a pyrimidine which is an analog of thymine. Bromine in the 5th position of the pyrimidine ring has similar properties as that of the methyl group in the 5th position in thymine. However, bromine alters the charge distribution in the ring and brings about a tautomeric shift. 5-bromouracil in keto form base-pairs with adenine, and in enol form forms pair with guanine. Its keto form is more stable. If 5-bromouracil is present in rare enol form as a nucleoside triphosphate at the time of its incorporation into the nascent strand of DNA, it will be incorporated opposite to guanine present in the template strand.

This results in AT base pair formation after two rounds of replication in the place of GC base pair, because 5-bromouracil in its stable keto form pairs with adenine (Figure 9.12). This is an example of transition mutation.

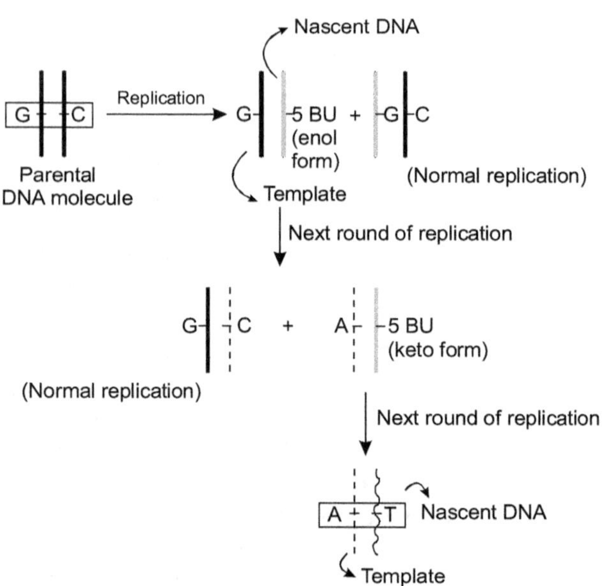

Figure 9.12   Mutation caused by the base analog, 5-bromouracil where G : C base pair is changed to A : T base pair.

On the other hand, if 5-bromouracil is incorporated in its stable keto form opposite to adenine in place of thymine and if it undergoes tautomeric shift to enol form in the subsequent replication, it is likely to bring about AT $\rightarrow$ GC transition (Figure 9.13).

The mutagenic action of 5-bromouracil can also be explained in another way. Nucleoside triphosphates (NTPs) are required for DNA

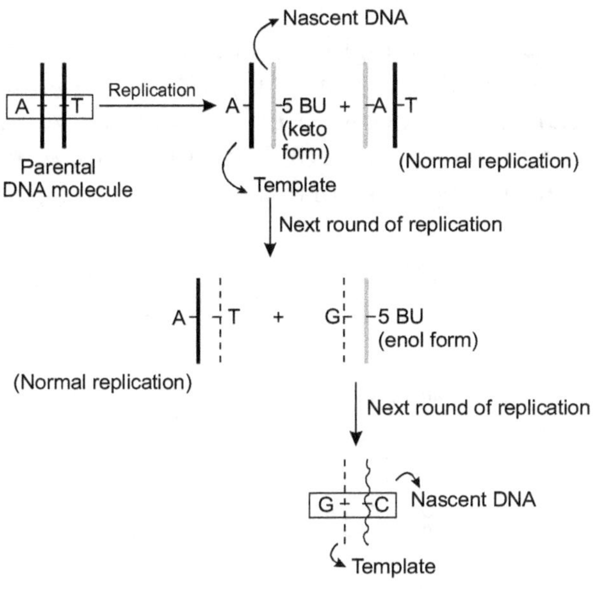

Figure 9.13   Mutation caused by the base analog, 5-bromouracil where A-T base pair is changed to G-C base pair

synthesis as dNTPs. The concentrations of these NTPs in the cells are regulated by the concentration of thymidine triphosphate (TTP) which will help in getting sufficient levels of NTPs and in turn dNTPs for DNA synthesis. Excess TTP is found to inhibit the synthesis of dCTP. 5-bromouracil is an analog of thymine. Therefore 5-bromouracil nucleoside triphosphate also inhibits the synthesis of dCTP. However TTP synthesis continues in the cell. This results in increased ratio of TTP to dCTP which leads to misincorporation of T opposite to G. There are several DNA repair mechanisms that can correct such errors. But in the presence of 5-bromouracil, the rate of misincorporation exceeds the rate of correction. Because T is wrongly incorporated opposite to G, in the next round of replication, when the DNA strand containing T acts as the template, A is incorporated in the new strand that pairs with T. This results in the conversion of GC base pair to AT base pair.

Another compound, 2-aminopurine is a base analog of adenine which brings about mutation in a similar way.

Nitrous Acid

Nitrous acid (HNO_2) is a highly potent mutagen. It can act both on replicating DNA and non-replicating DNA. The amino group of adenine, guanine and cytosine undergo oxidative deamination. NH_2 group is therefore converted to CO group which alters the hydrogen-bonding specificity of the bases (Figure 9.14).

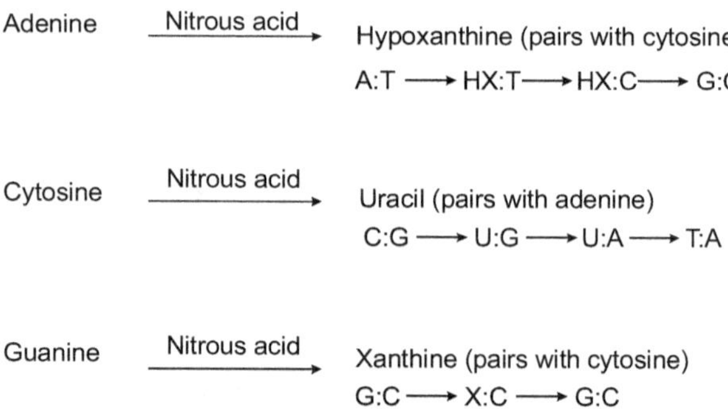

Figure 9.14   Mutagenic action of nitrous acid

Adenine is deaminated to hypoxanthine, which base-pairs with cytosine rather than thymine. Cytosine is converted to uracil, which base-pairs with adenine instead of guanine. Xanthine is formed from guanine deamination. However xanthine also forms base pairs with cytosine like guanine. Deamination of adenine leads to AT $\rightarrow$ GC transitions and deamination of cytosine results in GC $\rightarrow$ AT transition. Therefore, nitrous acid induces transition in both the directions. Thus nitrous-acid-induced mutations can also be induced to revert using nitrous acid again as the mutagen (back mutation or reverse mutation).

Hydroxylamine

Hydroxylamine (NH_2OH) is a hydroxylating agent. It specifically hydroxylates the amino group of cytosine. Hydroxylaminocytosine is formed which can base-pair with adenine to produce GC $\rightarrow$ AT transitions. Because of its specificity, hydroxylamine is useful in identifying the type of transition mutations. Let us consider that nitrous acid has caused a transition mutation. This mutation can be reverted by using nitrous acid itself. But hydroxylamine can bring about reversion only if GC base pair is present at the mutated site. Therefore, hydroxylamine can be used to determine whether the original mutation is AT → GC or GC → AT transition. Both nitrous acid and hydroxylamine act as effective mutagens in prokaryotic systems. The chemical environment in eukaryotic alkylating agent systems is not favourable for their action on DNA.

Alkylating Agents

Alkylating agents such as nitrogen and sulphur mustards, methyl and ethyl methanesulphonates (MMS and EMS) and nitrosoguanidine (NTG) act as potent mutagens. Methyl and ethyl groups are added to the purine and pyrimidine bases and their base-pairing specificity is altered resulting in transition mutation. EMS causes ethylation at the 7-N position and at the 6-O position. Guanine on exposure to EMS forms 7-ethyl guanine which base-pairs with thymine. Such types of mutations trigger the DNA repair mechanism like thymine dimers. Some difunctional alkylating agents with two reactive alkyl groups cross-link DNA strands and induce chromosome breaks and chromosomal aberrations. In general, the mutagenic action of alkylating agents is less specific than base analogs, nitrous acid and hydroxylamine. They also induce different types of mutations—transition, transversion, frameshifts and chromosomal aberrations—with various frequencies which depends on the specific alkylating agent.

Nitrosoguanidine is a potent chemical mutagen (N-methyl N'-nitro-N-nitrosoguanidine) which includes clusters of closely linked mutations in the chromosome during replication. Ethyl methanesulphonate (EMS) is an alkylating agent. It is a potent mutagen used extensively with eukaryotes.

Another mechanism for mutagenesis by alkylating agents is due to error-prone system, such as SOS repair, which becomes inactive due to mutation in one of the genes responsible for the activity of the system.

Depurination is another event resulting from alkylation of bases like guanine. When the purine bases are alkylated in the presence of alkylating agents, the N-glycosidic bond joining the purine nitrogen and deoxyribose is broken. This results in the formation of apurinic site in the DNA, which can be corrected by DNA repair system. But the replication fork sometimes may reach the apurinic site before repair has occurred. In that case, replication stops before the AP site. If the cell has a functional SOS repair system, replication restarts after a pause. There is more chance to add an adenine nucleotide in the new DNA strand opposite to AP site. AP site in the template strand is usually formed by the removal of a purine. Therefore, the new DNA strand should originally have a pyrimidine. After replication, a Pu·Py orientation will be changed to Py·Pu which is an example of a transversion mutation.

Depurination in DNA is also caused by treatment of phages with low pH buffers or at high temperatures.

Acridines and other Intercalating Substances

The acridine dyes like proflavin, acridine orange and a series of compounds called ICR-170, ICR-191, etc. are very powerful mutagens which induce frameshift mutations. Acridines are positively charged which intercalate or get sandwiched between the stacked base pairs in DNA (Figure 9.15). Acridine orange, proflavine and acriflavine are substituted acridines. They are planar, three-ringed molecules and their dimensions are roughly similar to the combination of a purine and a pyrimidine pair. They are therefore able to stack along with a base pair and this insertion occurs between bases in adjacent pairs which is referred to as intercalation. This results in the movement of adjacent base pairs by a distance equal to the thickness of one base pair. They increase the rigidity of the DNA molecule and alter the conformation of the double helical molecule.

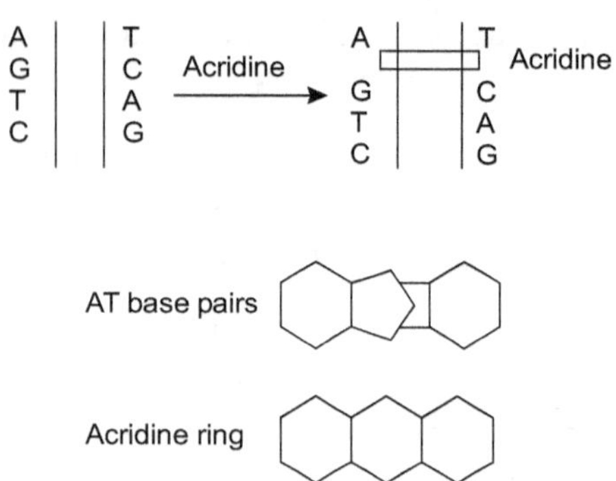

Figure 9.15   Action of intercalating agents

The DNA molecules containing such intercalated acridines during replication undergo addition and deletion mutations which results in the shift of the reading frame.

Mutator Genes

The mutation rate of each gene is dependent to some extent upon the residual genotype. The only effect that some genes seem to exhibit is to increase the mutation rate of another locus. These kinds of genes are called mutator genes. Mutator mutations are one type of mutations that influence the mutability of other genes. Specific mutators are limited to one locus and non-specific mutators are those whose effect is not specific to one locus. These mutations generally occur in genes that control DNA repair.

In *E. coli*, it is identified that there are genes that in a mutant state cause mutations to appear in other genes. For example, DNA polymerase enzyme at times makes errors and also corrects such errors by its $3' \rightarrow 5'$ exonuclease editing function. A mutation in the gene for DNA polymerase is likely to affect its editing function. This gene is referred to as mutator gene which fails to correct the errors in replication. Another example for mutator gene is *dam* gene. This gene product is the enzyme Dam methylase that methylates the parental DNA strand for strand discrimination when mismatch base pair repair is required. The incorrect base in the unmethylated daughter strand is selectively removed. In a *dam⁻* mutant organism, there is no methylation of the parental strand and this may result in the removal of the base present in the parental strand and replacement with a new base that is complementary to that present in the new strand.

A third example for mutator genes can be explained in corn. A dominant gene called "dotted" (*Dt*) is present on chromosome 9 in corn. This gene causes a recessive gene 'a' for colourless aleurone on chromosome 3, to mutate quite frequently to its allele 'A' for coloured aleurone. Plants that are aa Dt have kernels with dots of colour in the aleurone, because in presence of Dt 'a' allele is mutated to 'A' allele. The size of the dot will be large or small depending on how early or late the mutation occurs during seed development.

"ONE GENE"– "ONE ENZYME" HYPOTHESIS

George Beadle in the year 1933, provided the evidence that genes are responsible for the synthesis of enzymes. He along with Boris Ephrussi conducted experiments in *Drosophila*. They proved that mutant genes were responsible for altered eye colour in *Drosophila* which could be associated with errors in biochemical events involving loss of enzyme function. Supported by these findings Beadle and Edward Tatum investigated nutritional mutations in *Neurospora crassa*, which led to the one-gene–one-enzyme hypothesis.

In the early 1940s Beadle and Tatum induced mutations in *Neurospora* and isolated them easily. By inducing mutations, they produced different strains that had genetic blocks of reactions essential for the growth of *Neurospora*.

Beadle and Tatum irradiated asexual conidia (spore) with X-rays to produce mutants and made them to be grown on "complete" medium containing all the necessary growth factors (vitamins and amino acids). Under such growth conditions, a mutant strain was able to grow on minimal medium only on supplementation in the enriched complete medium. All the cultures were then transferred to minimal medium. Growth of the culture on minimal medium indicated that the organisms were able to synthesize all the necessary growth factors themselves and that the culture did not contain any mutation. If no growth occurred, it indicated that the culture contained a nutritional mutation. The steps involved in the induction, isolation and characterization of a nutritional auxotrophic mutation in *Neurospora* are shown in Figure 9.16.

Beadle and Tatum showed that the supplement that restores the growth is the molecule that the particular mutant strain could not synthesize.

The first mutant strain isolated by them required vitamin B_6 in the medium and the second one required vitamin B_1. Using the same protocol, they isolated and studied hundreds of mutants deficient in their ability to synthesize other vitamins, amino acids or other substances.

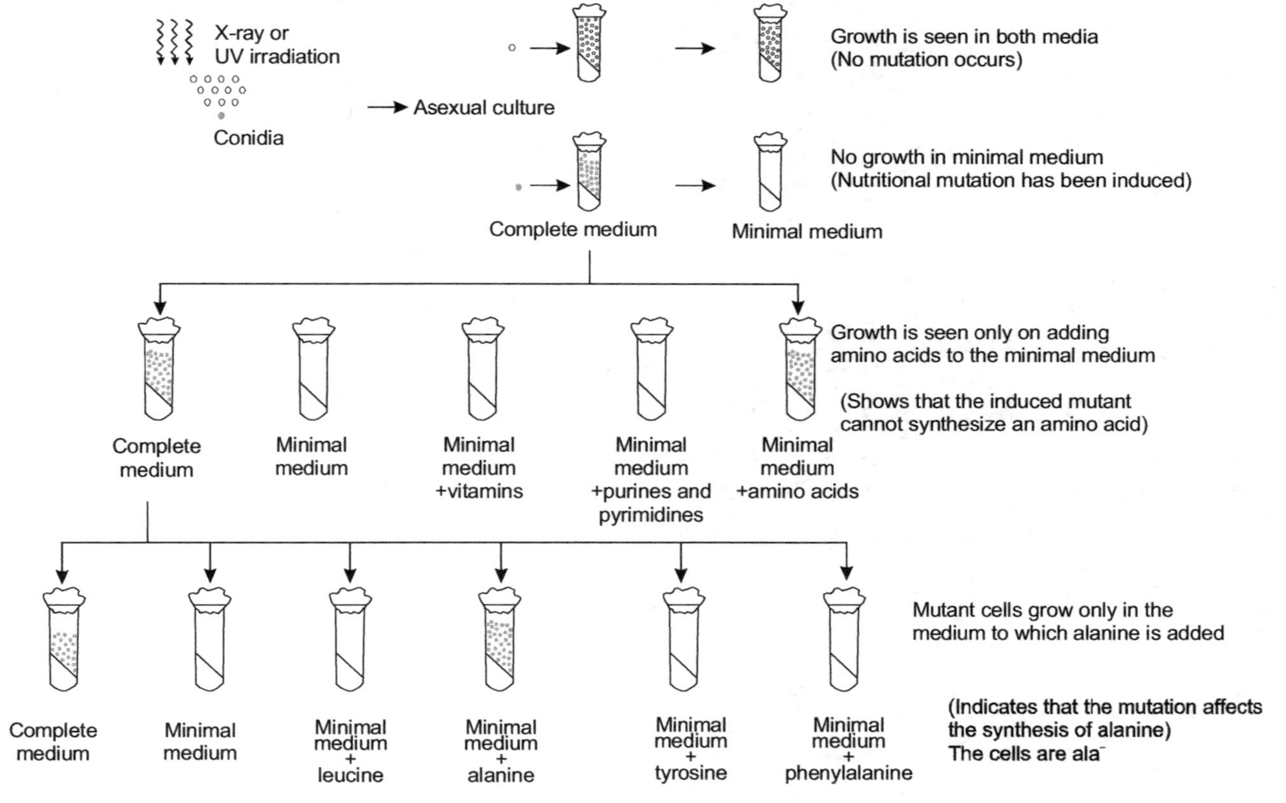

Figure 9.16   Induction, isolation and characterization of a nutritional auxotrophic mutation in *Neurospora*

Beadle and Tatum proved that each nutritional mutation caused the loss of an enzymatic activity that is responsible for an important reaction in the wild type organisms. This fact led to the proposal of a hypothesis called one-gene–one-enzyme. Though this concept was later modified, this formed one of the major principles of genetics.

The one-gene–one-enzyme concept helped in the study of details of metabolism in *Neurospora, Escherichia coli* and other microorganisms.

USE OF MUTANT BACTERIAL STRAINS IN THE ELUCIDATION OF BIOCHEMICAL PATHWAYS

Biochemical pathways include biosynthetic (anabolic) and degradative (catabolic) reactions within a cell. A substrate or precursor is the starting material that is converted to an end product through a number of intermediates. There are defined intermediary steps involved in the pathway and each step is promoted by a definite enzyme. These enzymes are the products of specific genes in the organism. Even if one enzyme is missing or non-functional because of a mutation in the specific gene, the pathway will not be completed. The intermediate compound on which this particular (defective) enzyme is supposed to act will be accumulated in the system. The end product will be obtained, only if the medium is supplemented with the intermediate which has not been formed because of a block in the pathway.

Let us consider a biochemical pathway in a bacterium where A is the substrate that is converted to the end product E, through three intermediate compounds B, C and D. Let E1, E2, E3 and E4 be the four enzymes involved in the pathway.

$$A \xrightarrow{E1} B \xrightarrow{E2} C \xrightarrow{E3} D \xrightarrow{E4} E$$

Let us assume that the gene encoding the enzyme E3 is mutated, so that E3 enzyme becomes non-functional. In this condition, C cannot be converted to D, and E, the end product, cannot be obtained. Compound C will be accumulated. This bacterium is a mutant strain which cannot produce E if the pathway is synthetic or which cannot degrade A if the pathway is a degradative one. The strain can be reverted to normal wild type condition, i.e., E can be formed only if D or E are supplied to the medium.

A total of four different, single mutant strains can be explained with respect to this pathway, in each case one enzyme being non-functional E1 or E2 or E3 or E4 with a mutation in their respective genes. By analysis of the nutrient requirements of these mutants, the order of the pathway can be established.

For example, let us consider that there are four single mutant strains of *Neurospora* which are unable to grow on a minimal medium unless supplemented by one or more of the substances A, B, C, D, E or F. The following table indicates the growth (+) and nil growth (0) of the organism separately in these substances. In addition, it is found that strains 2 and 4 grow, if E and F or C and F are added to the minimal medium. Elucidation of the biochemical pathway from the following informations:

Mutant strain	A	B	C	D	E	F
1	+	0	0	0	0	0
2	+	0	0	+	0	0
3	+	0	+	0	0	0
4	+	0	0	0	0	0

Solution

Mutant strain 1 will grow only if A is added. This shows that the defective enzyme produced by the mutant gene in this strain must act prior to the formation of A and after the formation of B, C, D, E and F. In other words, this mutant brings about a metabolic block in the last step of the sequence and it also shows that A is the end product.

$$(B,C,D,E,F)\xrightarrow{\ 1\ } A$$

Order is unknown

Mutant strain 2 grows only if supplemented either by A or D, or by adding E and F or C and F. It does not grow if B is added. So, the metabolic block in strain 2 is after B but before A. The dual addition of E and F or C and F allow strain 2 to grow. This indicates that there is a split pathway with E and C in one line and F in the other. The position of D is after the block, may be at the initial point of split or after the split.

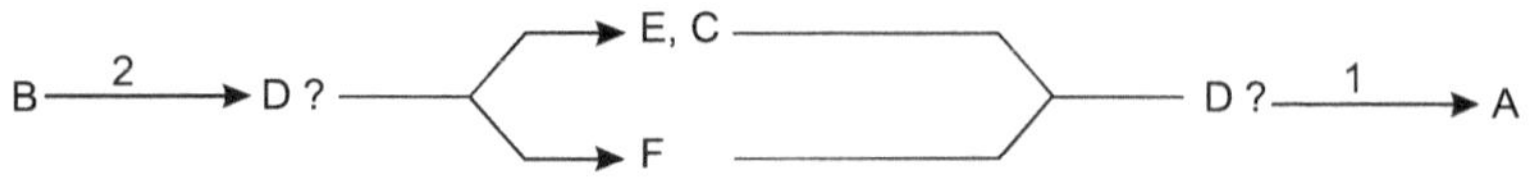

B —2→ D ? → E, C ——————⟩ D ? —1→ A / F ——————

(Order unknown)

Mutant strain 3 grows by adding A or C and not by adding D. Therefore D cannot immediately precede A. The metabolic block in strain 3 must precede the formation of C and not E.

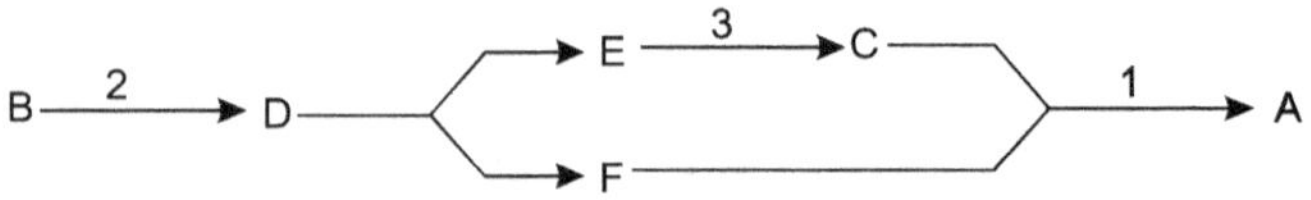

B —2→ D — → E —3→ C —⟩ 1 → A / F ——————

Mutant strain 4 will grow if given dual supplementation of E and F or C and F, but not if D is given. Therefore mutation in strain 4 cannot split D into E and F. Therefore the pathway should be

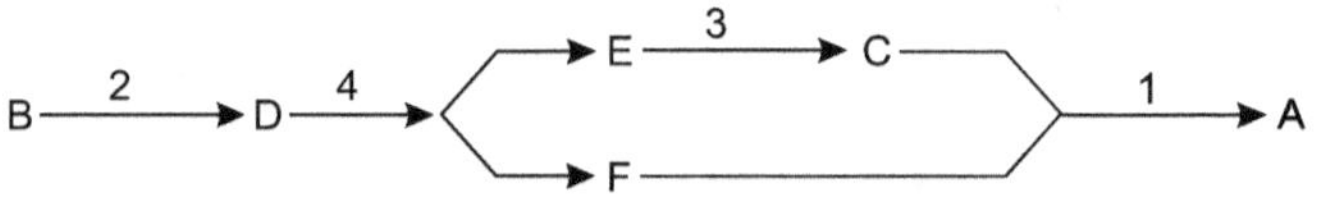

B —2→ D —4→ E —3→ C —⟩ 1 → A / F ——————

The metabolic pathway for the synthesis of the amino acid arginine in *Neurospora* was first attempted by Adrian Srb and Norman Horowitz using seven mutant strains, each requiring arginine for growth (arg⁻).

Given the following responses for single gene mutants for synthesis of arginine, the metabolic pathway leading to arginine synthesis can be elucidated.

Mutant strains	Supplements added to minimal culture medium				
	Citrulline	Glutamic semialdehyde	Arginine	Ornithine	Glutamic acid
Strain 1	+	0	+	0	0
Strain 2	+	+	+	+	0
Strain 3	+	0	+	+	0
Strain 4	0	0	+	0	0

Since none of the four mutant strains respond to the addition of glutamic acid, it can be suggested that glutamic acid is a compound that is present in the metabolic pathway in a position prior to the steps whose actions demand the activity of the four enzymes leading to the formation of arginine. Therefore glutamic acid is the starting material and arginine, the end product.

$$\text{Glutamic acid} \xrightarrow{1,2,3,4(?)} \text{Arginine}$$

Mutant strain 1 responds to the addition of citrulline and arginine. Since arginine is the end product, the position of citrulline is prior to arginine. The defective enzyme produced by strain 1 mutant gene brings a block leading to the formation of citrulline.

$$\text{Glutamic aid} \longrightarrow \text{Citrulline} \longrightarrow \text{Arginine}$$

$$\text{Glutamic acid} \xrightarrow{2,3,4(?)} ? \xrightarrow{1} \text{Citrulline} \longrightarrow \text{Arginine}$$

Mutant strain 2 responds to the addition of citrulline, glutamic semialdehyde, arginine and ornithine. This indicates that the defective enzyme produced by strain 2 mutant gene brings a block leading to the formation of citrulline, glutamic semialdehyde and ornithine. So the pathway should be

$$\text{Glutamic acid} \xrightarrow{2} \text{Glutamic semialdehyde / Ornithine} \xrightarrow{1}$$

$$\text{Citrulline} \longrightarrow \text{Arginine}$$

Mutant strain 3 responds to the addition of citrulline, arginine and ornithine. This shows that the defective enzyme produced by strain 3 mutant gene brings a block leading to the formation of ornithine.

$$\text{Glutamic acid} \xrightarrow{2} \text{Glutamic semialdehyde} \xrightarrow{3} \text{Ornithine} \xrightarrow{1}$$

$$\text{Citrulline} \longrightarrow \text{Arginine}$$

Mutant strain 4 responds only to the addition of arginine indicating that the mutant gene is responsible for a block in the conversion of citrulline to arginine. Therefore the pathway is

$$\text{Glutamic acid} \xrightarrow{2} \text{Glutamic semialdehyde} \xrightarrow{3} \text{Ornithine} \xrightarrow{1}$$

$$\text{Citrulline} \xrightarrow{4} \text{Arginine}$$

CONCEPT OF "ONE GENE"–"ONE POLYPEPTIDE" CHAIN

The concept of one-gene–one-enzyme developed in the early 1940s was not accepted immediately because mutations always did not seem to bring about blocks in pathways and caused the production of inactive proteins. They also seemed to give altered phenotypes like altered eye size, wing shape, etc. in *Drosophila* and different varieties of seed texture, height and fruit size in plants. Simultaneous discoveries of other facts in genetics showed that not all proteins are enzymes, while nearly all enzymes were proved to be proteins at that time. So the concept was modified as one-gene–one-protein.

Later discoveries proved that a gene need not necessarily give a molecule of functional protein. The gene product can also be a polypeptide which may be a subunit of a protein contributing to the quaternary structure of the protein. Therefore the concept was refined as one-gene–one-polypeptide chain.

This can be established by analysing the hemoglobin structure in individuals with sickle-cell anemia. The first direct evidence that genes specify proteins other than enzymes came from the experiment done on mutant hemoglobin molecules in sickle-cell anemia. In normal individuals, the erythrocytes are biconcave and disc-shaped whereas in sickle-cell anemia they take a sickle shape because they become elongated and curved under low oxygen tension due to polymerization of hemoglobin. They become anemic since such abnormal erythrocytes are destroyed more rapidly than normal red cells. Compensatory physiological mechanisms include increased red cell production by bone marrow and accelerated heart function. This leads to abnormal bone size and shape and dilation of the heart. The genotype of normal individual with normal hemoglobin is given as $Hb^A Hb^A$ and individual with sickle cell anemic condition has the genotype $Hb^S Hb^S$. Heterozygous individuals $Hb^A Hb^S$ exhibit sickle-cell trait with less sickling effect. In 1950s Vernon Ingram demonstrated that there is a change in the primary structure of the globin portion of hemoglobin molecule accounting for the formation of HbS, the abnormal sickle-cell hemoglobin. HbS differs from the normal hemoglobin HbA by having a single amino acid change. Valine is substituted for glutamic acid at the sixth position of the β-globin chain, accounting for the peptide difference.

This established the fact that a single gene provides information for a single polypeptide chain.

MUTATIONAL HOTSPOTS

Certain DNA sequences are called mutational hotspots because they are more likely to undergo mutation than others. Hotspots include runs of a single nucleotide and tandem repeats of short sequences (STRPs). Hotspots are found at many sites throughout the genome and within genes. Presence of even very small number of hotspots may account for a disproportionately large fraction of mutations.

Sites of cytosine methylation are highly mutable and this brings about GC → AT transitions. In many organisms like bacteria, maize and mammals about 1% of the cytosine bases are methylated at C-5, giving 5-methyl cytosine by methylating enzymes. During DNA replication, 5-Me-cytosine pairs with guanine and normal replication occurs. Gene activity is found to be reduced in places where cytosine is methylated. For example, the inactivation of X chromosome in females and inactive copies of certain transposable elements in maize are due to cytosine methylation.

Seymour Benzer carried out fine structure mapping of the rII gene in bacteriophage T4. He succeeded in mapping 2400 independent mutations in the rII locus of phage T4. This locus consists of two cistrons rII A and rII B. Mutations were not found to be produced at equal frequencies at all sites within the gene. The 2400 rII mutations were located at only 304 sites. One of these sites accounted for 474 mutations. This site which recorded a very high frequency of mutation is a hotspot.

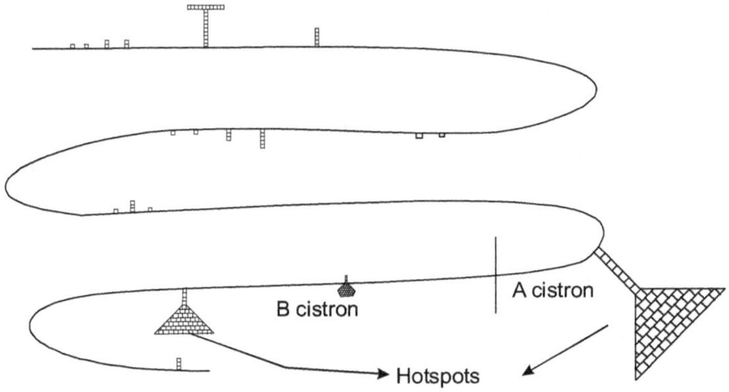

Figure 9.17   rII locus of phage T4 showing hotspots

In Figure 9.17 each small square indicates a separate independent occurrence of a mutation in the rII locus of phage T4.

ASSOCIATION OF MUTATIONS WITH CANCER

In mammals, there is a strong correlation between the accumulation of mutations and cancer. Bruce Ames, a bacterial geneticist developed a simple test to measure the potential of a given chemical compound to promote certain easily detected mutations in a specialized bacterial strain.

In Ames test, mutant strains of *Salmonella typhimurium* are taken in which an enzyme involved in histidine biosynthetic pathway is inactivated. One strain is used to detect base-pair substitutions and three other detect various frameshift mutations. However each mutant strain is unable to synthesize histidine and therefore requires histidine for growth (his⁻ strains). These mutant strains are plated on a histidine-free medium. Only very few cells grow, which would have undergone spontaneous back mutations that permit the histidine biosynthetic pathway to operate. Three identical nutrient plates are taken and inoculated with equal number of such revertant cells. Each plate is made to receive a disc of filter paper containing lower concentrations of a particular mutagen under investigation. The mutation increases the rate of back mutation. Therefore the number of colonies on the plate is increased. The area in the plate immediately around the filter paper disc will be clear, indicating the lethal effect of the mutagen at high concentrations. As the mutagen diffuses away from the filter paper radially, it gets diluted to sublethal concentrations which efficiently brings about back mutation in the *Salmonella* mutant strains (Figure 9.18). Many compounds act as potent mutagens only after entering a cell, where they undergo a variety of chemical transformation. Therefore, sometimes compounds are tested for mutagenicity after first incubating with a liver extract. Sometimes, test compounds may be injected into a mouse where they are modified by liver enzymes and then recovered for assay.

Among the many compounds which are known to be carcinogenic, extensive animal trials have proved that, more than 90% are found to be mutagenic when assayed in the Ames test.

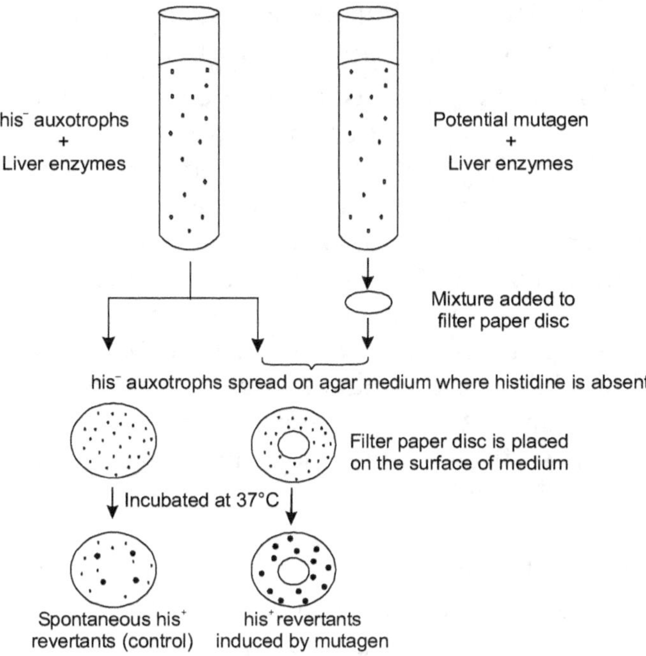

Figure 9.18   The Ames test which screens potential compounds for mutagenicity

BOX 9.3   BLOOM SYNDROME

It is a rare autosomal recessive condition characterized by short stature, facial rash induced by sun exposure, a small narrow head and a predisposition to cancer of all types. Cells from persons with Bloom syndrome exhibit excessive mutations in all genes. Numerous gaps and breaks occur in chromosomes that lead to extensive genetic exchange in cell division. Rates of DNA synthesis is retarded.

The gene for DNA helicase present on chromosome 15 is defective. This unwinding enzyme is essential for DNA synthesis and repair mechanisms. The cells of a person with Bloom syndrome carry two mutated copies of the gene for helicase enzyme.

MUTATIONS IN LIVING ORGANISMS

Mutations that are deleterious to an organism are always maintained at a low frequency in the population because of natural selection.

Mutant organisms cannot compete with wild type individuals, even if they are in a conducive optimal environmental condition.

Mendel's laws of heredity assume equality in survival and reproductive capacity of different genotypes. In all his classical monohybrid and dihybrid crosses while arriving at the F1 and F2 genotypic and phenotypic ratios, Mendel has only highlighted on the variation in phenotypic expression of dominant and recessive alleles and not any difference in their survival or reproducing capacities. Deviations observed from the expected Mendelian ratios therefore would be proportional to the decrease in the survival and/or reproducing ability of the mutant type relative to the wild type. The ability of a mutant to survive and reproduce in competition with other genotypes will be an important phenotypic characteristic from evolutionary point of view.

Mutations in *Drosophila*

In *Drosophila*, white eyed flies are produced by a sex-linked recessive gene (Figure 9.19). If 'w' is the allele for white eye on the X chromosome, white eyed female will be ww and male will be wY (no allele on Y chromosome). White eyed flies may be only 60% as viable as wild type flies. Let us consider a cross between heterozygous wild type female (w^+w) with white eyed male (wY). According to Mendel's concept of zygotic expectation, we except 25% each of white eyed female and male and 25% each of wild type female and male. If only 60% of white eyed flies survive, in contrast to 25 : 25 : 25 : 25 of the offsprings, it will be 25 : 25 :15 :15 or there will be 50 wild type and 30 white eyed females and males.

	Wild type female		White eyed male	
	w^+ w		w Y	
Zygotic expectation	w^+ w	w^+ Y	ww	w Y
	Wild type female	Wild type male	White eyed female	White eyed male

Figure **9.19**   Inheritance of white eye colour in *Drosophila*

Mutations in Plants

Genetic variation is extensive in plants, which could be detected by many techniques, one of which involves analysis of a plant's biochemical composition. For example, the isolation of proteins from maize endosperm, hydrolysis of the proteins and determination of the amino acid composition have revealed that the opaque-2 mutant strain contains high content of lysine than the others which are not mutants. Because maize protein is usually low in lysine, this mutation improves the nutritional value of the plant. Plant geneticists started analysing the amino acid composition of various strains of other grain crops, including rice, wheat, barley and millet. Research work in this line helped in combating malnutrition diseases resulting from inadequate protein or the lack of essential amino acids in the diet.

Tissue culture of plant cell lines in defined medium offers a good method of detection of mutation. The plant cells are treated like microorganisms and their resistance to herbicides or disease toxins is determined by adding the compound of interest (mutagen) to the culture medium. The techniques associated with conditional lethal mutants can be used on plant cells in tissue culture and then applied to the genetics of higher plants, which will provide a means of detection that would not be possible in an intact plant.

TRINUCLEOTIDE REPEAT EXPANSIONS

A class of mutation called trinucleotide repeat expansion which denotes an increase in the number of trinucleotide tandem repeats, was discovered in humans in 1991. A genetic disorder called fragile X syndrome was found to be due to expansion of a region in DNA where the trinucleotide sequence CGG is repeated several times in tandem. Over 10 genetic disorders in humans are now known to be caused by trinucleotide repeat expansions (Table 9.1).

Fragile X syndrome is the most thoroughly studied disorder and offers a good example of some peculiar effects that arise from this type of mutation. Fragile X syndrome is caused by disruption of a gene on the X chromosome called FMR 1 (for "fragile X mental retardation 1") which encodes the protein FMRP. This protein binds to mRNAs in the brain cells and apparently affects the expression of a number of genes essential for proper brain development. Within the untranslated region at the 5' end of the gene is a three-nucleotide sequence (CGG) that is repeated anywhere from 6 to 50 times in most people, with 29 copies being the most common number.

Table 9.1   Some human genetic disorders with trinucleotide repeat expansions

Genetic disorder	Trinucleotide repeat	Symptoms
Fragile X syndrome	CGG	Mental retardation
Fragile X E MR	GCC	Mental retardation
Huntington disease	CAG	Neural degeneration, dementia
Kennedy disease	CAG	Muscle weakness and wasting
Myotonic dystrophy	CTG	Muscle weakness and wasting, cataracts, hypogonadism
Spinocerebeller ataxia	CAG	Ataxic gait, muscular atrophy, diabetes
Dentatorubral–Pallidolugsian atrophy	CAG	Neural degeneration, epilepsy, dementia

People with a normal number of repeats show no symptoms of the disorder. The number of repeats can increase or decrease slightly from generation to generation, but rarely exceeds 50. This normal fluctuation has no effect on the size of the polypeptide because the repeated region appears in the mRNA upstream from the reading frame.

Some people, however, may acquire an abnormally large number of repeats. Repeats ranging from 5 to 230 constitute what is known as a "fragile X premutation". People with the premutation rarely show any symptoms of the disorder, but the descendants of females who carry the premutation have an increased risk of fragile X syndrome. People who have fragile X syndrome often have as many as 700 CGG repeats and sometimes over 1000. When the repeated region expands significantly beyond the premutation size, a "fragile X full mutation" results. The bases in the nucleotides of the repeated region and the promoter become highly methylated, which inhibits transcription of the FMRI gene. Ribosomes do not efficiently translate the few mRNAs that are produced. The amount of FMRP produced under these conditions is inadequate, brain development is inhibited and symptoms of fragile X syndrome appear.

Not all classes of fragile X syndrome are due to trinucleotide repeat expansion. In one case, a missense substitution mutation that substitutes asparagine for isoleucine is responsible. Several deletions of either the entire gene or large portions of it are also known to cause fragile X syndrome.

Several other genetic disorders in humans are due to trinucleotide repeat expansion. Huntington disease is characterized by progressive neurological degeneration that usually does not begin until after the age of 30. Once the degeneration begins, the affected person progressively loses motor function, behaves in unusual ways, becomes demented and eventually dies from the disease, usually within 10–15 years. Huntington disease is usually due to expansion of a CAG repeat in the coding region of the IT 15 gene, which encodes a protein named huntington. Each CAG is in frame and is translated as a glutamine. Unaffected persons have 10–35 repeats that are translated as a string of glutamines. Affected persons have 36–121 repeats and the lengthened string of glutamines alters the protein function.

MUTATIONS AND PROTEIN FUNCTION

The effect of a mutation on protein function depends on several factors: which amino acids in the polypeptide chain are altered, how many amino acids are altered and what specific changes in amino acid sequence arise from the mutation. A same-sense mutation has no effect on the function of a protein because the amino acid sequence remains unchanged. Missense mutations can have a variety of effects, ranging from no loss of function to complete loss of function or, more rarely, a gain of function. The final effect depends on how and where the mutation alters the protein.

A protein's function depends predominantly on the physical shape of the protein, as determined by its primary, secondary, tertiary and quaternary structures, the chemical properties of the protein and the active site of the protein if it is an enzyme. Missense mutations that do not significantly alter any of these will probably have little, if any, effect on the protein's function. For example, valine and isoleucine are both non-polar amino acids, differing from one another by a single methyl group. Their chemical properties are nearly identical. A missense mutation that substitutes valine for isoleucine should have little, if any, effect on protein function. However, a mutation that alters the protein's active site, or alters the tertiary or quaternary structure, may change the protein's ability to carry out its function or, in many cases, eliminate the protein's function altogether. Most frameshift mutations completely eliminate protein function because they alter nearly every amino acid from the point of the mutation onward and usually change the location of the termination codon. Even if a frameshift mutation does not affect the active site of an enzyme, it usually alters the structure of the enzyme drastically, causing the enzyme to lose its function.

Usually, mutations that have an effect on a protein's activity are "loss-of-function mutations", which either reduce the protein's ability to function or eliminate its function altogether. Most of the hundreds of known genetic disorders in humans, such as phenylketonuria, sickle-cell anemia, thalassemias, cystic fibrosis and hemophilias are caused by loss-of-function mutations.

More rarely, a mutation may cause a gene to "gain" function compared to its previous state, usually by producing a greater-than-normal quantity of protein. These are called "gain-of-function mutations". Although gain-of-function mutations are more rare than loss-of-function mutations, they are quite important. Many of the mutations that contribute to development of cancer are gain-of-function mutations in genes that regulate cell growth and division.

Mutations outside of the transcribed region of a gene do not alter the amino acid sequence of a protein. However, depending on where a mutation is located, it may affect the amount of the protein that is produced, thus affecting its level of activity in the cell. For example, a mutation within the promoter region may prevent RNA polymerase from binding and thus prevent transcription, eliminating protein synthesis completely. A promoter mutation may also reduce or, more rarely, enhance the ability of RNA polymerase to transcribe the gene, thus affecting how much protein is produced. In fact, most gain-of-function mutations reside in the promoter or upstream regulatory regions of a gene.

TYPES OF GENOTOXIC AGENTS

Genotoxic agents are of different types:

1. Carcinogen—promotes neoplastic transformation of eukaryotic cells; many carcinogens are also mutagens.
2. Clastogen—induces chromosome fragmentation, i.e., double-strand breaks.
3. Mutagen—promotes mutagenesis either by directly interacting with DNA or by generating metabolic products which do so.
4. Oncogen—induces tumour formation.
5. Supermutagen—an extremely potent mutagen, e.g. Ethyl methanesulphonate.
6. Telogen—causes premature termination of replication, e.g. 2′,3′-dideoxynucleotides.
7. Teratogen—induces development abnormalities, not necessarily a genotoxic agent; includes molecules which mediate their effects without altering DNA structure.

DNA repair mechanisms are essential for survival and integrity and in animals the repair of DNA is very important to overcome the effect of mutations. A number of human diseases result from the loss of DNA repair function or of a cellular response to DNA damage (Table 9.2).

Table 9.2 Human disorders caused due to loss of DNA repair function or cellular response to DNA damage

Disease	Clinical symptoms	Reasons
Ataxia telangiectasia	Facial blood vessel dilation, neuromuscular degeneration, immunodeficiency	Mutation in chr IIq in humans.
Bloom's syndrome	Photosensitivity, immunodeficiency, growth retardation, predisposition to cancer	High frequency of sister chromatid exchange, hypersensitivity to mutagens, delay in joining of Okazaki fragments, DNA ligase deficiency.

(Contd.)

Table 9.2 (Continued)

Disease	Clinical symptoms	Reasons
Cockayne's syndrome	Photosensitivity, neurological disorder, dwarfism, optic defects, disproportionately sized facial features and limbs	Three complementation groups and three genes identified. One is XPB.
Fanconi's anemia	Congenital skin and skeletal abnormalities, loss of blood cells	Multiple chromosome mutation reflecting chromatid breaks. Four complementation groups and a gene FACC identified.
Hereditary nonpolyposis colorectal cancer (HNPCC) Trichothiodystrophy (TTD)	Colorectal tumours, brittle hair, scaly skin, growth defects, mental retardation, photosensitivity	Mutations in genes involved in mismatch repair, TTD involves XP-D genes.
Xeroderma pigmentosum (XP)	Severe photosensitivity, skin cancers, ocular defects, neurological disorders	Seven complementation groups XP-A to XP-G and XP-1 identified. Defect in nucleotide excision repair.

SITE-DIRECTED MUTAGENESIS

Researchers have developed an experimental technique called site-directed mutagenesis to investigate specific genes. They introduced a designed mutation at a prescribed site within a gene of interest. The principle of site-directed mutagenesis is as follows:

Step 1   Analysis of nucleotide sequence of the desired gene

Step 2 Separation of one strand

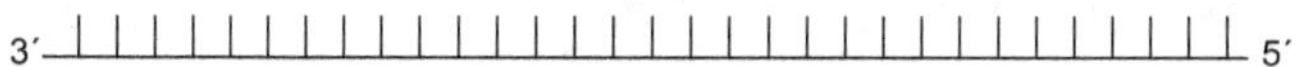

Step 3 Hybridization of this strand with a synthetic oligonucleotide with an altered triplet sequence at a specific site.

Step 4 Synthetic duplex formation

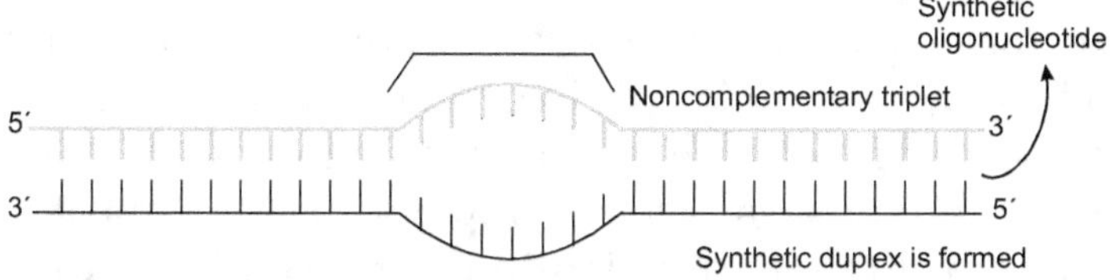

Step 5 Semiconservative replication is promoted so that two copies of gene are obtained—one with original nucleotide sequence and another with altered nucleotide sequence.

Step 6 Transcription and translation promoted to give two gene products—one wild type protein and another mutant protein.

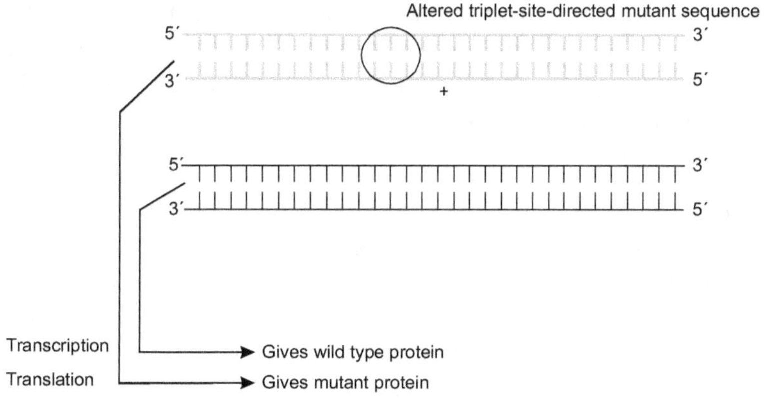

A gene is changed by altering the nucleotide sequence either by deletion or insertion of one or more nucleotides at a particular site. For example, if three nucleotides are inserted, there will be one codon in addition in the mRNA and the final protein will be a mutant protein with one extra amino acid in the corresponding site in the protein. This type of alteration will be helpful in assessing the effect of mutation on gene expression and protein function.

The first step involves sequencing of the nucleotides of the particular gene which can be done either through analysis of the sequence of amino acids in the protein product or through analysis of nucleotide sequence directly in the gene.

Next step includes synthesis of a nucleotide sequence complementary to one of the strands of the gene. To this synthetic nucleotide sequence are added triplet nucleotides at a specific site. This short piece of synthetic DNA is hybridized with the original parent strand to form a partial duplex because of the complementarity along most of its length. On adding DNA polymerase and DNA ligase to the hybrid, the short sequence is extended and the entire gene is got. Both the strands are completely complementary except at the point where an additional trinucleotide is present in the synthetic oligonucleotide. When this DNA is allowed to replicate, two types of DNA are got, one resembling exactly the gene of interest and the other with an added triplet at the predetermined site. When these genes are allowed to be expressed by genetic manipulation, wild type protein (original gene product) and mutant protein are obtained.

A gene with site-directed mutation can be analysed *in vitro*. Insertion of that gene into a living organism helps in the assessment of its function. If the normal gene is removed and replaced by the gene with site-directed mutation, the technique is referred to as **gene knockout**. The organism that is altered genetically is called a **knockout organism**.

In mice, gene knockout technique serves as a model to study human genetic disorders. In such conditions, mice genes comparable to human gene that is responsible for the disorders are isolated, subjected to directed mutagenesis and replaced with non-mutant mouse gene. Human disorders like cystic fibrosis and Duchenne muscular dystrophy have been investigated by this technique.

GENETIC MODIFICATIONS THROUGH MUTATIONS— COMMERCIAL APPLICATIONS OF MUTATION

Among bacteria strain improvement can be done with commercially useful microorganisms for increased product formation. Using mutagenic agents, variations can be obtained in microbial strains. When powerful mutagens like X-rays or nitrogen mustard are used, undesirable effects may be caused. Milder agents like ultraviolet irradiation are preferred. In wild type organisms, the amount of a compound formed is regulated by feedback inhibition of the enzymes involved brought about by a repressor molecule. If mutation is caused in the gene for the repressor, there will be substantial increase in the formation of the required product. Also by abolishing other metabolic activities of the cell that may utilize this compound, depletion in the level of the compound can be prevented.

A primary metabolite, that is, the main product of a reaction sequence can be produced in large amount by increasing the rate of production of the enzymes involved by alteration in the promoter site so as to increase the transcription of the gene. If we consider the production of tryptophan, for example, the whole operon of five genes is repressed by tryptophan, using both a repressor-mediated regulation and attenuation. If the gene for the repressor is inactivated then the level of the enzymes of tryptophan pathway will increase and tryptophan synthesis will increase. Also a mutation in the attenuator region will prevent it responding to the presence of tryptophan leading to increased synthesis of tryptophan.

Overproduction of secondary metabolites like antibiotics is also possible with the help of mutant strains. Most of the commercially important strains of *Penicillium chrysogenum* have descended from a single strain NRRL-1951. This strain produced about 100 units of penicillin per ml. By screening large numbers of spontaneous variants of this strain, a strain B-25, was selected that produced twice the amount of penicillin. X-ray mutagenesis resulted in the mutant X1612 with a yield of 500 units per ml. Further mutation induced by UV irradiation, produced a strain Q-176 with a further increase in yield.

At times, novel metabolites may be produced using mutants that prevent normal production of main products of a pathway and divert the utilization of substrates for novel analogues of the main product. For example, *Streptomyces aureofaciens* usually produces chlortetracycline, but mutants that are blocked at different stages of synthesis accumulate other tetracycline derivatives.

Thus improved and desired variant strains have to be screened and selected. One variant mutant of wheat variety may show resistance to a crop disease and another mutant variety may have good yield. By breeding programmes, these two variants can be crossed, so as to have a new variety crop with both these desirable qualities. Such genetic exchange techniques can be used in combination with gene cloning techniques.

COMMON LESIONS IN DNA

The sequence of nucleotides in DNA can be changed as a result of error in DNA replication and mutations caused by environmental agents. If DNA sequence changes are left uncorrected, both growing and non-growing somatic cells might accumulate so many mutations, that they could no longer lead to the formation of the correct gene product and the cell function gets disrupted. Also the DNA in germ cells with many mutations may pass on the wrong sequence of nucleotides to the offsprings thereby disrupting the correct passage of genetic information from one generation to another.

Common DNA damages are:

1. *Missing bases*   Purines are removed by acid and heat. When DNA is exposed to certain chemical mutagens, the purine and pyrimidine bases undergo substitution reactions and get modified. These altered bases are removed by DNA glycosylase enzymes, and apurinic and apyrimidinic sites (AP sites) are created.

2. *Altered bases*   Ionizing radiations and chemical mutagens like alkylating agents modify the purine and pyrimidine bases which lose their base-pairing specificity.

3. *Incorrect base*   When DNA polymerase enzyme fails to do its proof reading $3' \rightarrow 5'$ exonuclease activity, wrong bases which are incorporated during replication are not corrected. This again results in the formation of mismatch base pairs.

4. *Bulky lesions*   Intercalating agents (e.g. acridines) cause addition or loss of a nucleotide during recombination or replication.

5. *Linked pyrimidines*   UV irradiation results in the formation of cyclobutyl dimers (usually thymine dimers).

6. *Single or double strand breaks*   Ionizing radiations or chemical agents like bleomycin cause break in the phosphodiester bonds. If single strand break is caused, nicks are formed in the DNA whereas double

strand breaks create fragments of DNA either with blunt ends or sticky ends.

7. *Cross linked strands*   Covalent linkage of two strands is caused by bifunctional alkylating agents like mitomycin C.

8. *3'-deoxyribose fragments*   Free radicals cause disruption of deoxyribose structure which leads to strand breaks.

Spicy Questions and Answers

1. **Which of the following single amino acid residue replacements may be due to single-base mutation?**

   (a) **phe to leu**   (b) **lys to ala**

   (c) **ala to thr**   (d) **phe to lys**

   (e) **ile to leu**   (f) **his to glu**

   (g) **pro to ser**

   (a) UUU to UUA   (c) GCU to ACU

   (e) AUU to CUU   (g) CCU to UCU (because of the degeneracy of the genetic code, for the same answers, other possible single base replacements can also be given)

2. **Sickle-cell hemoglobin has a valine residue at position 6 of the β-globin chain, instead of the glu residue found in normal hemoglobin A. What would have been the possible change in the DNA nucleotide sequence for this replacement?**

   Glutamic acid codons are GAA and GAG. Valine codons are GUU, GUC, GUA and GUG.

   GAG codon of glutamic acid would have been changed to GUG codon of valine. In the DNA the coding sequence CTC would have changed to CAC. Another possibility is GAA codon of glutamic acid would have been changed to GUA codon of valine. In the DNA, the coding sequence TTC would have changed to TAC (consider the antiparallel nature of DNA template sequence and mRNA codon sequence).

3. **A normal protein has the following terminal amino acid sequence: ser-thr-lys-leu-COOH. A mutant is isolated with the sequence ser-thr-lys-leu-leu-phe-arg-COOH. What has probably happened to produce the mutant protein?**

The original stop codon is mutated to form leu codon (e.g. UAA would have changed to UUA) and the adjacent triplets read coding for phe and arg and stop codons.

4. **How will you identify whether mutations in bacteria have caused resistance to a particular drug? How can we find out whether a particular drug has caused mutation or has helped in identifying mutation that is already present in the organism?**

Bacteria that have been treated with a mutagen or expected to carry mutations may be introduced into a medium to which the drug is added. If colonies appear, it denotes pre-existing mutations for resistance. This can be confirmed by replica plating technique. Wild-type drug-sensitive cells can be added to two different media—one with the drug, and another without the drug under study and mutation frequency in both can be compared.

5. **At the site of UAA codon in the mRNA, what amino acids can be present, because of suppression by a suppressor tRNA created by a mutant base in the anticodon, assuming only Watson and Crick base-pairing?**

If tRNA is a non-mutant one, it can pair with the codon at two sites. A mutant base in the anticodon will help in the base-pairing of the third site also. Therefore, the amino acids that can be inserted into the suppressed site will be those whose codons differ from UAA in a single base. These codons and the respective amino acids will be AAA (lys), CAA (glu), GAA (glu), UUA (leu), UCA (ser), UAC (tyr) and UAU (tyr).

6. **Two possible anticodons could pair with the codon 5′ UGG 3′. But only one is recognized by this codon. Identify the two anticodons and say why only one is used.**

The two anticodons are 3′ACC 5′ and 3′ACU 5′. But because of wobbling, 3′ACU 5′ can also pair with the codon 5′UGA 3′ which is a stop codeine. In that case it may insert tryptophan in the protein and continue with reading the codon frame, instead of terminating protein synthesis. So, only the anticodon 3′ACC 5′ is used.

1. Define the following:
   i. Mutation
   ii. Aberration
   iii. Mutagenesis

2. Differentiate between:
   i. Adaptive, spontaneous and induced mutation
   ii. Forward and backward mutation
   iii. Point and multiple mutation
   iv. Silent, missense and nonsense mutation
   v. Dominant and recessive mutation
   vi. Homoalleles and heteroalleles
   vii. Transition and transversion mutation

3. What do you mean by chain termination mutation?

4. What are the different types of mutagens?

5. Discuss in detail the molecular mechanism of mutation.

6. How do chemical mutagens cause mutation? Explain with examples.

7. Write notes on alkylating agents.

8. What is intercalation? Name two intercalating agents.

9. What is frameshift mutation?

10. Explain one-gene–one-enzyme hypothesis.

11. How were bacterial mutants useful in the elucidation of biochemical pathways? Discuss with an example.

12. Describe mutational hot spots.

13. Comment on trinucleotide repeats.

14. Explain the types of genotoxic agents.

15. Discuss the commercial applications of mutation.

16. What are the common lesions noticed in DNA?

17. What is replica plating technique?

18. How is mutation associated with cancer?

19. What is suppressor-sensitive mutation?

10

TRANSLATION

The genetic information in the DNA in the form of nucleotides is transcribed into mRNA molecules which carry the codons specifying the twenty standard amino acids. Each triplet codon is translated into the amino acid residues of the polypeptide chain by translation. The ribosomes are called the "work benches" of protein biosynthesis. In prokaryotes, transcription and translation are coupled processes, because they do not have nucleus and the chromosome is transcribed to RNA, and mRNA is translated into protein chain in the cytoplasm. In eukaryotes, transcription and translation are not coupled because there is compartmentalization of these events, transcription taking place in the nucleus and translation in the cytoplasm.

REQUIREMENTS FOR PROTEIN BIOSYNTHESIS

Ribosomes

A ribosome is a multicomponent particle. It possesses several enzymatic activities required for protein synthesis. The ribosomes bring together a single string of mRNA molecule that is transcribed from the structural gene and the charged tRNA molecules in proper position and orientation so that the base sequence of the mRNA is translated into the amino acid residues of the protein.

Chemical composition of prokaryotic ribosomes   A prokaryotic ribosome is a nucleoprotein particle and has two subunits. The intact ribosome and their subunits are named by their sedimentation coefficient values. In prokaryotes, the intact unit is called 70S ribosome. They have two subunits one bigger and another smaller with sedimentation coefficient values 50S and 30S (Refer Figure 7.17). In the laboratory, the stability of 70S ribosome is determined by the Mg^{2+} concentration in the medium. In 0.0005 M $MgCl_2$, the ribosome is completely dissociated into its subunits. In 0.005 M $MgCl_2$, which is nearly the concentration within the bacteria, it is in associated form.

$$30S\ subunit + 50S\ subunit \underset{[Mg^{2+}]\downarrow}{\overset{[Mg^{2+}]\uparrow}{\rightleftharpoons}} 70S\ ribosome$$

At physiological concentration of Mg^{2+}, the 70S ribosome is the preferred form which is favourable for protein biosynthesis. Some of the characteristics of prokaryotic ribosomes are listed in Table 10.1.

Table 10.1   Some of the characteristics of prokaryotic ribosomes

	70S ribosome	50S subunit	30S subunit
Mol. wt.	2.5×10^6	1.6×10^6	0.9×10^6
rRNA molecules present		One 5S rRNA One 23S rRNA	One 16S rRNA
Proteins present		31 proteins (L1, L2,…)	21 proteins (S1, S2,…S21)

Each 50S subunit has one copy of each protein molecule, except for L7 and L12, of which there are four copies in total in one 50S subunit. L26 is not a true component of 50S subunit. The number of copies of L7 relative to L12 depends on the bacterial growth rate. L7 is N-acetylated terminal form of L12. L8 is a complex with one L10 to which the tetrameric L7 and L12 are bound. L26 is actually S20 protein of the 30S subunit. During the dissociation of 70S ribosome, S20 leaves 30S subunit and gets associated with a 50S subunit. Most of the bigger and smaller subunit proteins are basic with 34% of arginine and lysine. This accounts for their strong affinity towards the acidic rRNA molecules.

The size of the three prokaryotic rRNAs—5S, 16S and 23S—are 120, 1541 and 2904 nucleotides. About 70% of the bases in each RNA molecule is internally paired to form stem and loop structures.

Chemical composition of eukaryotic ribosomes   Eukaryotic ribosome has an S value of 80. The two subunits are 40S and 60S (Refer Figure 7.17).

Large number of eukaryotic ribosomal proteins have phosphorylated serine residues. The eukaryotic cell organelles, chloroplasts and mitochondria also have their own ribosomes, which resemble the bacterial ribosomes and this seems to have some evolutionary significance. The characteristics of eukaryotic ribosomes are shown in Table 10.2.

Table 10.2 Some of the characteristics of eukaryotic ribosomes

	80S ribosome	60S subunit	40S subunit
Mol. wt.	4.2×10^6	2.8×10^6	1.4×10^6
rRNA molecules present		One 5S rRNA One 5.8S rRNA One 28S rRNA	One 18S rRNA
Proteins present		49 proteins	33 proteins

Amino Acids and tRNA Molecules

The 20 standard amino acids required for protein biosynthesis are present in the cytoplasmic pool along with the 31 different tRNA molecules present in many copies. During translation, the amino acids should be in activated form. This happens through the charging of tRNA molecules with the amino acids. When an amino acid is attached to a tRNA molecule, the tRNA is said to be aminoacylated. A tRNA molecule with the anticodon for the codon of an amino acid is designated as tRNA^{aminoacid}. For example, the codons for phenylalanine are 5′ UUU 3′ and 5′ UUC 3′. The anticodon which recognizes these two codons is 3′ AAG 5′. The tRNA molecule with this anticodon in position 34, 35 and 36 is designated as tRNA^{phe}.

From the mixture of 20 standard amino acids in the cytoplasm, this tRNA^{phe} will be charged only with phenylalanine and not with any other amino acid and this charged tRNA is called phenylalanyl-tRNA^{phe}. In general, a charged tRNA is designated as aminoacyl-tRNA$^{amino\ acid}$. The term "uncharged tRNA" refers to a tRNA molecule lacking an amino acid.

Selenocysteine is an unusual amino acid present in some proteins. It is a derivative of cysteine in which sulphur is replaced by selenium. This amino acid is identified both in prokaryotes and eukaryotes. The amino acid is incorporated in response to the stop codon UGA. In *E. coli*, four gene products of *selA*, *selB*, *selC* and *selD* genes are required for the incorporation of selenocysteine. The product of *selC* gene is tRNA^{ser} with the anticodon UCA that can recognize UGA codon. Therefore, it is a suppressor tRNA. The enzyme seryl-tRNA synthetase incorporates the amino acid selenocysteine in this tRNA. The products of *selA* and *selD* convert seryl-tRNA^{ser} to selenocysteinyl-tRNA^{ser} through an intermediate phosphoseryl-tRNA^{ser}. The incorporation of selenocysteine into the protein in response to UGA codon requires a protein called SEL B which is the product of *selB* gene. SEL B is homologous to EF-Tu, the elongation factor. Therefore, it replaces this elongation factor and recognizes selenocysteinyl-tRNA^{ser} and the UGA stop codon. This charged tRNA in combination with SEL B competes with the termination factors and brings about the translation of the termination codon.

Aminoacyl-tRNA Synthetases

For each amino acid, there is a specific aminoacyl-tRNA synthetase. For some amino acids specified by more than one codon, more than one synthetase exists. The enzymes are highly specific and a particular enzyme will recognize only a particular tRNA and a particular amino acid with the respective anticodon in the tRNA molecule.

Protein Factors Involved in Translation

In both prokaryotes and eukaryotes, a number of protein factors are involved in protein biosynthesis as indicated in Table 10.3.

Table 10.3 Translational protein factors in prokaryotes and eukaryotes

Prokaryotes	Eukaryotes	Function
Initiation factors		
IF1		Involved in forming
IF2	eIF2	initiation complex
IF3	eIF3, eIF4C ,	
	CBPI (eIF4E)	Involved in cap binding
	eIF4A, eIF4B, eIF4F	Involved in search for first AUG
	eIF5	Helps dissociation of eIF2, eIF3, eIF4C
	eIF6	Helps dissociation of 60S subunit from inactive ribosomes
Elongation factors		
EF-Tu	eEF1α	Delivery of aminoacyl-tRNA to ribosomes
EF-Ts	eEF1β	Helps in recycling factor
	eEF1γ	
EF-G	eEF2	Translocation factor
Release factors		
RF1	eRF	Release of complete
RF2		polypeptide chain
RF3		

STEPS INVOLVED IN PROTEIN BIOSYNTHESIS

Special Properties of the Initiator tRNA Molecules in Prokaryotes and Eukaryotes

The initiator tRNA molecule in prokaryotes is tRNA^{fmet} which differs from other tRNA molecules. It is charged with methionine by methionyl-tRNA synthetase. The same enzyme also charges tRNA^{met} with methionine. The methionine of charged tRNA^{fmet} is immediately recognized by another enzyme, tRNA methionyl transformylase, which transfers a formyl group from N-10, formyl tetrahydrofolate (fTHF) to the NH$_3^+$ group of the methionine to form N-formyl methionine.

$$\text{tRNA}^{fmet} + \text{methionine} \xrightarrow{\underset{\text{synthetase}}{\text{Methionyl-tRNA}}} \text{H}_3\text{N}-\overset{\text{H}}{\underset{\underset{\text{S}}{\underset{|}{\text{CH}_3}}}{\overset{|}{\underset{|}{\text{C}}}}}-\overset{\text{O}}{\overset{\|}{\text{C}}}-\text{O}-\text{tRNA}^{fmet}$$

Methionyl-tRNA^{fmet}

Methionyl-tRNA transformylase (THF → fTHF)

Transformylase does not catalyse the formylation of methionyl-tRNA^{met}. This indicates that there is some structural difference between tRNA^{met} and tRNA^{fmet}.

Another special feature of prokaryotic initiator tRNA is that, a formylated charged tRNA^{fmet} molecule enters the P site (one of the sites in the ribosome which can receive only the initiator aminoacyl-tRNA or peptidyl tRNA referred to as peptidyl site) without prior occupation of the A site (another site in the ribosome referred to as aminoacyl site that is bound by an aminoacyl-tRNA whose anticodon recognizes the mRNA codon present in that site), whereas all other aminoacyl-tRNA molecules bind initially to the A site and then move to the P site only after a peptide bond is formed. This indicates that there should be some structural feature in tRNA^{fmet} that is absent in all other tRNA molecules. It is seen that the 5′-terminal base in tRNA^{fmet} is not hydrogen-bonded to the opposite base in the acceptor stem. Also a hydrogen bond between the uracil adjacent to the anticodon and a ribose in the anticodon loop present in all tRNAs is absent in tRNA^{fmet}. There are also some small differences in the acceptor stem and the DHU loop. The codon–anticodon interaction of tRNA^{fmet} differs from that of other tRNA molecules in that there may be some wobble in the first codon base. The major codon recognized by tRNA^{fmet} is AUG, but tRNA^{fmet} also interacts significantly with the valine codon GUG and weakly with the leucine codons UUG and CUG. This wobble is suggested to result from the lack of the hydrogen bond in the anticodon loop. Therefore, the three codons, GUG, UUG and CUG are occasionally used as start codons and each is recognized by tRNA^{fmet}.

Eukaryotic initiator tRNA is referred to as tRNA^{imet}. In eukaryotes both tRNA^{imet} and tRNA^{met} are charged with methionine, but methionine in initiator tRNA, tRNA^{imet} is not formylated. Therefore, the first amino acid in eukaryotes is methionine and not formyl methionine as in prokaryotes. Also in the *TψC* loop, the sequence AUC or *AψC* is present instead of *TψC*. The eukaryotic tRNA^{imet} like tRNA^{fmet} of the prokaryotes lacks the hydrogen-bonding in the 5′ terminus.

Selection of the Correct AUG Codon for Initiation

The sequence AUG serves both as an initiation codon and the codon for methionine. Since methionine occurs within protein chains, there must be some signal in the base sequence of the mRNA that identifies a particular AUG triplet as start codon.

In prokaryotes, the first AUG is usually 20–30 bases from the 5′ end and this preceding sequence is the leader region. An important feature of the leader region is the presence of a sequence whose consensus is 5′ AGGAGGU 3′, 4–7 bases ahead of the initiator triplet. This sequence only defines the start codon by providing a binding site for the ribosome called Shine-Dalgarno sequence. The 3′-terminal region of the 16S rRNA of several bacteria has a sequence 3′ AUUCCUCCA 5′ which can pair with the mRNA's Shine-Dalgarno sequence. This pairing brings the AUG codon to a start site on the ribosome (Figure 10.1).

Most of the prokaryotic mRNA molecules are polycistronic, which contain sequences specifying the synthesis of several proteins. A polycistronic mRNA molecule must possess a series of start and stop codons. If it encodes four proteins, the sequence will be as follows:

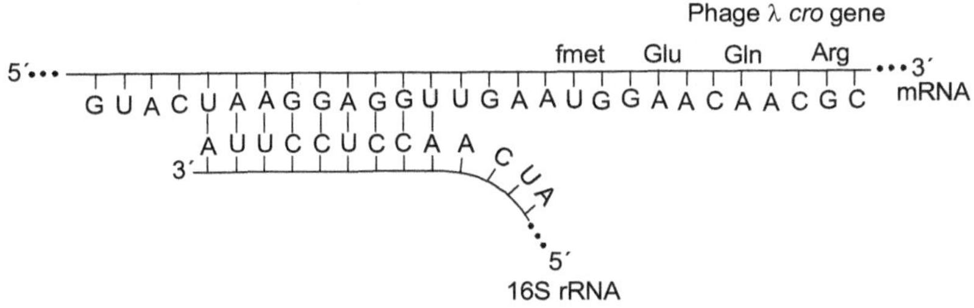

Figure 10.1   Shine-Dalgarno sequence

Start-protein1-stop-start-protein 2-stop-start-protein 3-stop-start-protein 4-stop

In such a polycistronic mRNA, the leader sequence preceding the first start signal may be several hundred bases long. There is usually a sequence called spacer 5–20 bases long between one stop codon and the next start codon. Sometimes, the spacers may have stop codons, so that if the previous stop codon that should actually stop the previous protein synthesis is overlooked due to some reason, the stop codon in the spacer can terminate synthesis. Having stop codons in all the reading frames in the spacer region allows the translational system to have a fresh start at the following AUG that is out of the spacer and is the first codon for the next protein.

The mechanism for initiating synthesis of the first protein molecule in a polycistronic mRNA is not different from that in a monocistronic mRNA. But in case of polycistronic mRNA, for the synthesis of the second protein, protein synthesis must start again after the stop codon of the first protein is reached. The start codon of the second protein is usually near the preceding stop codon, so that reinitiation occurs soon before the ribosome and the mRNA dissociate, or sometimes a second initiation signal may lock the protein-synthesizing system on the second cistron. The number of copies of each protein synthesized from a polycistronic mRNA is not the same. This difference is due to variations in the strength of binding of the ribosomes to the initiation sites for each cistron and also due to the distance between a stop codon and the next start codon.

In eukaryotes, the pattern of recognition of the first AUG codon for initiation and the ribosome-binding sites differ from that in the prokaryotes. In eukaryotes, the initiating tRNA molecule carries methionine and not formyl methionine. The initiator tRNA is designated as $tRNA^{imet}$ that is different from the other $tRNA^{met}$ which can recognize all other AUG codons that are present in the mRNA. The $tRNA^{imet}$ recognizes only the initiator AUG and in rare cases GUG or any other rare start codons.

In eukaryotes, most mRNA molecules are monocistronic. Eukaryotic ribosomes cannot translate prokaryotic polycistronic mRNA. In prokaryotes, related genes are often in adjacent positions. But in eukaryotes the related genes are in different parts of the same chromosome or on different chromosomes, each gene transcribed separately as monocistronic mRNA molecules.

BOX 10.2 PRIONS: A TRANSMISSIBLE DISEASE CAUSED BY INFECTIOUS PROTEINS

Transmissible diseases are usually caused by the genetic materials, DNA or RNA. However, a transmissible disease is caused by an agent without the genetic material. Four human neurological diseases and six similar animal diseases are caused by proteins without DNA or RNA. The human diseases are Kuru, Creutzfeldt–Jacob disease, Gerstmann–Straussler–Scheinker syndrome and a fatal familial insomnia. All these diseases develop slowly, all are fatal and are caused either by the ingestion of a protein from an infected individual or from a mutation of the normal gene. None of these diseases has a cure.

The disease is caused by a protein, similar to one normally produced in the brain of healthy individuals. Prion refers to proteinaceous infectious particle. Prion protein (prp) was isolated. The protein is encoded by a gene on the short arm of chromosome 20. In addition to the infectious form, a familial (inherited) form of the disease results from a mutation of the gene. The normal protein is termed prp^C. The mutated protein is termed as prp^{SC}. prp^C is a glycoprotein present in the membrane surface of the brain cells and some tissues.

A question arises as to how proteins which are not the genetic material transmit the disease to the offspring. One mechanism suggested is that an infective prp^{SC} binds to a normal prp^c producing two infectious prp^{sc} proteins. Two produce four, four produce eight and so on. This resembles the behaviour of the mythical ice-nine, i.e., a single seed caused all of the water on earth, by a chain reaction cascade, to form into a novel type of ice.

In eukaryotic mRNA, the initiator AUG is near the 5′ terminus, 50–100 bases from the 5′ end and is a part of a consensus sequence PuXXAUGG in which X can be any base.

The rate of initiation, i.e., the number of initiation events per unit time varies from one gene to another. This variation depends on the sequence of bases surrounding AUG. For example, PuXXAUGG is the most efficient one and if the sequence is XXXAUGPy it becomes inactive for initiation. In most eukaryotes, PuXXAUGG was found to be ACCAUGG and was referred to as Kozak sequence, named after Marilyn Kozak who identified this sequence.

The actual mechanism for recognizing the 5′ end of eukaryotic mRNA is based on the recognition of the 5′ cap of the mRNA by the cap-binding protein with recruitment of other initiation factors and the small subunit of the ribosome. Then the small subunit moves down mRNA. The ribosome scans the mRNA until it recognizes the initiation codon and this model is referred to as the **scanning hypothesis.**

In several cases in eukaryotes, a process called **shunting** occurs, in which the first AUG does not serve as the initiation codon. As scanning begins, it bypasses a region of mRNA upstream of the initiation codon—the leader otherwise called **5′-untranslated region (5′-UTR)**—in favour of the correct start AUG codon.

In some cases, very small genes called **open reading frames (ORFs)** are present in this region of mRNA and play some role in shunting. Some ORFs are translated and then the main gene is translated by the same ribosome by a process called **reinitiation**. Such a type of translation is seen in the genes of some plants and animal viruses.

Attachment of an Amino acid to a tRNA Molecule

Charging of tRNA with the amino acid takes place in two steps:

Step 1:

$$\text{Amino acid} + \text{ATP} \rightleftharpoons \text{Aminoacyl-AMP} + \text{PPi}$$

$$\text{H}_2\text{N}-\underset{\underset{R}{|}}{\overset{\overset{H}{|}}{C}}-\overset{\overset{O}{\|}}{C}-\text{OH} + \text{ATP} \rightleftharpoons \text{H}_2\text{N}-\underset{\underset{R}{|}}{\overset{\overset{H}{|}}{C}}-\overset{\overset{O}{\|}}{C}-\text{O}-\underset{\underset{O^-}{|}}{\overset{\overset{O}{\|}}{P}}-\text{ribose-adenine} + \text{PPi}$$

Step 2:

$$\text{Aminoacyl AMP} + \text{tRNA} \rightleftharpoons \text{Aminoacyl-tRNA} + \text{AMP}$$

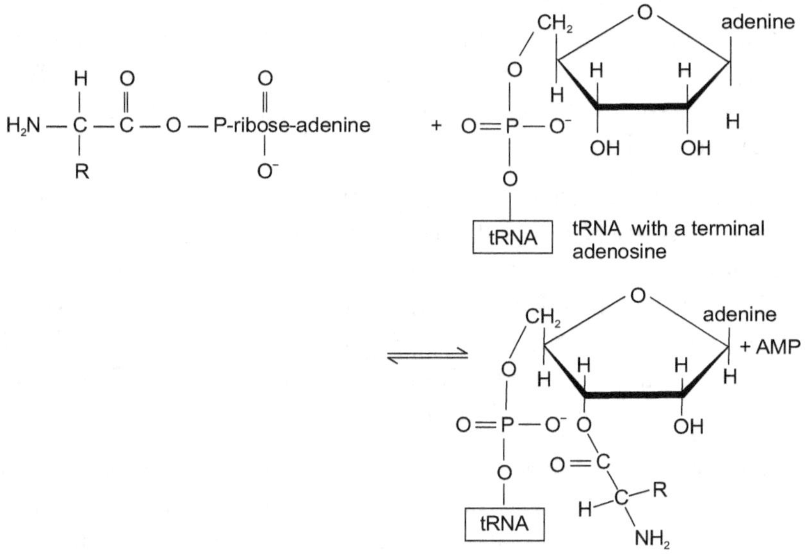

$$H_2N-\underset{R}{\overset{H}{C}}-\overset{O}{\overset{\|}{C}}-O-\text{P-ribose-adenine} \quad + \quad O=\underset{O}{\overset{}{P}}-O^- \quad \text{(tRNA with a terminal adenosine)}$$

tRNA with a terminal adenosine

$$\rightleftharpoons \quad + \text{ AMP}$$

The overall reaction is

Amino acid + ATP + tRNA + H_2O $\rightleftharpoons$ Aminoacyl-tRNA + AMP + 2Pi

Both step 1 and step 2 are mediated by the enzyme, aminoacyl-tRNA synthetase.

An individual aminoacyl-tRNA synthetase is specific not only for a single amino acid but also for a particular tRNA, in order to maintain the fidelity of protein synthesis. The interaction between aminoacyl-tRNA synthetases and tRNAs is referred to as the "second genetic code".

Initiation

Initiation of protein biosynthesis is a three-stage process. Intact ribosomes do not directly bind mRNA to start polypeptide synthesis. Initiation is a complex process in which the two ribosomal subunits and the initiator tRNA molecule assemble on a properly aligned mRNA to form a complex. This assembly process requires the participation of initiation factors that are not permanently associated with the ribosome. They are soluble proteins and are different from the permanent ribosomal proteins. Three initiation factors IF1, IF2 and IF3 are present in *E. coli* (Figure 10.2). The three stages of initiation are:

1. Once a cycle of polypeptide synthesis is over, the 30S and 50S ribosomal subunits remain associated as inactive 70S ribosomes. IF3 binds to the 30S subunit to promote the dissociation of the subunits. IF1 increases the dissociation rate by assisting the binding of IF3.

2. mRNA and IF2 form a ternary complex with GTP, and formyl-X methionyl-tRNA^{fmet} subsequently binds to the 30S subunit in either order. IF3 also functions in assisting the 30S subunit in binding the mRNA although it is not involved in the selection of the initiation codon. One 30S subunit, one mRNA molecule, a charged tRNA^{fmet} and the three initiation factors with GTP comprise the 30S pre-initiation complex.

3. IF3 is then released and the 50S subunit joins the 30S initiation complex in a manner that stimulates IF2 to hydrolyse its bound GTP to GDP + Pi. It is an irreversible reaction and conformationally rearranges the 30S subunit and releases IF1 and IF2 for participation in further initiation reactions.

Therefore, initiation results in the formation of formyl-methionyl-tRNA^{fmet}–mRNA–ribosome complex in which formyl-methionyl-tRNA^{fmet} occupies the P site of the ribosome while the A site is ready to accept an incoming aminoacyl-tRNA (Figure 10.2). The tRNA^{fmet} is the only tRNA that directly enters the P site. All other tRNAs enter the P site only through the A site during chain elongation.

Eukaryotic initiation resembles the overall prokaryotic process. But eukaryotes have many more initiation factors as indicated in Table 10.3. In some eukaryotic systems, there are more than 10 factors, many with multiple subunits. It is difficult to distinguish them from the ribosomal proteins. They are differentiated from prokaryotic initiation factors by adding 'e' prefix.

The most striking difference between eukaryotic and prokaryotic ribosomal initiation occurs in the second stage of the process, the binding of mRNA, eIF2, GTP and methionyl-tRNA^{imet} to the 40S ribosomal subunit. Another eukaryotic initiation factor eIF4E is a cap-binding protein which helps the 40S subunit in binding at or near the eukaryotic mRNA's 5'cap. This subunit then migrates downstream until it encounters the first AUG.

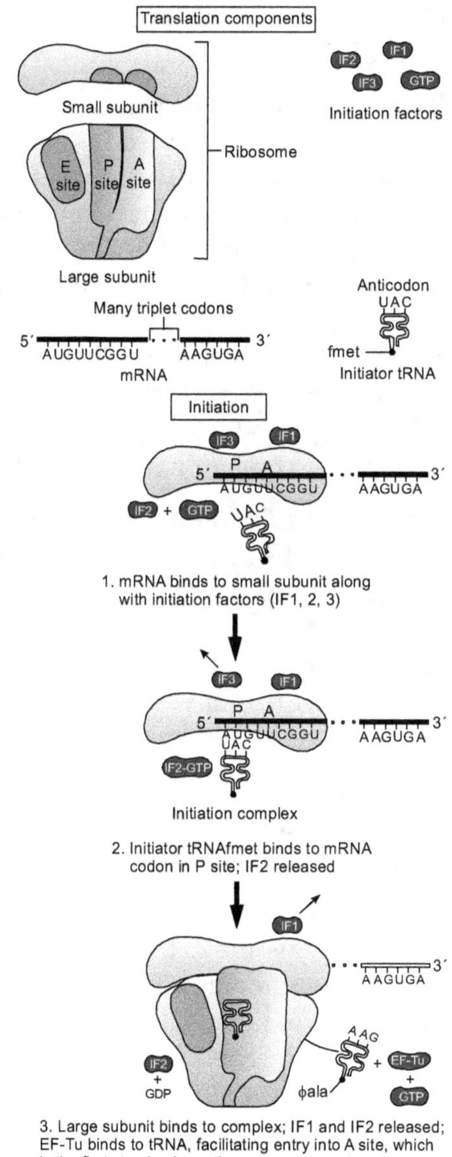

Figure 10.2   Initiation of protein biosynthesis

Elongation

Each ribosome has two tRNA binding sites, which overlap the 30S and 50S subunits. These sites are called the aminoacyl or A site and the peptidyl or P site. Each site consists of a collection of segments of S and L proteins and 23S rRNA.

Three non-ribosomal proteins in prokaryotes—the elongation factors, EF-Tu, EF-Ts and EF-G—are required for chain elongation. These proteins together comprise about 5–10 per cent of the soluble protein of the cell. EF-Tu is a carrier protein that is required for the binding of charged tRNA to the A site of an active 70S ribosome. It acts in a multistep process. First, EF-Tu and GTP form a binary complex. Binding of GTP causes a conformational change in EF-Tu, which allows a charged tRNA molecule to bind to the binary complex to form a ternary complex. In the ternary complex, EF-Tu is situated near the CCA terminus of the acceptor stem of the tRNA molecule. But the anticodon loop is completely free for pairing with the codon.

The ternary complex is able to bind to the A site. As a result of codon–anticodon binding, the aminoacyl-tRNA undergoes a small conformational change that exposes the $T\psi C$ G sequence in the $T\psi C$ loop. This allows pairing of the tRNA to a sequence AAGC in the 5S RNA, thereby stabilizing binding to the A site. Once the ternary complex is in the A site, a ribosomal protein having GTPase activity causes hydrolysis of the GTP bound to EF-Tu to yield bound GDP and Pi.

The GDP · EF-Tu complex now lacks the ability to bind to the charged tRNA and it gets dissociated from the tRNA. This leaves the aminoacyl-tRNA in the A site. Removal of EF-Tu is also necessary for peptide bond formation.

GDP binds more strongly to EF-Tu than does GTP. Therefore, some means is required either to regenerate GTP in the binary complex or to dissociate the complex. Another elongation factor EF-Ts displaces GDP from EF-Tu–GDP complex and forms EF-Tu · EF-Ts complex. The released GDP joins the internal pool of metabolic intermediates and is reconverted to GTP. Once GTP is available again, it displaces EF-Ts from EF-Tu · EF-Ts complex and forms EF-Tu · GTP complex. EF-Ts that is released is again available to complex with EF-Tu. These reactions can be summarized as follows:

$$\text{EF-Tu} \cdot \text{GDP} + \text{EF-Ts} \rightleftharpoons \text{EF-Tu} \cdot \text{EF-Ts} + \text{GDP}$$

$$\text{GDP} + \text{Pi} \longrightarrow \text{GTP}$$

$$\text{EF-Tu} \cdot \text{EF-Ts} + \text{GTP} \rightleftharpoons \text{EF-Tu} \cdot \text{GTP} + \text{EF-Ts}$$

At this stage, both the P site and A site, are occupied by aminoacyl-X tRNA and a peptide bond can form. The equivalent elongation factors in eukaryotes are eEF1α, eEF1βγ and eEF2 (Table 10.3).

Chemistry of peptide bond formation and transpeptidation   Earlier it was thought that the blockage of the NH_2 group of fmet by the formyl group was responsible for peptide bond formation between the COOH group of fmet and the NH_2 group of the adjacent amino acid (Figure 10.3). But in eukaryotes the first amino acid methionine is not formylated and even then protein biosynthesis takes place in the correct direction. Therefore it is considered that the relative orientation of the two amino acids in the A and P sites determine the event of peptide bond formation between them.

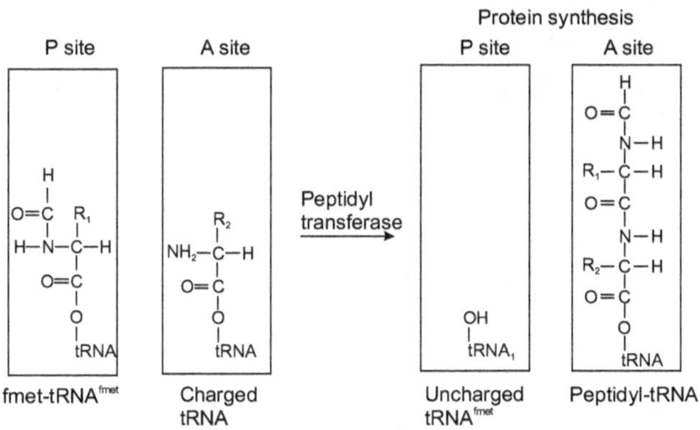

Figure 10.3   Peptide bond formation

Peptide bond is formed by an unusual enzyme complex called peptidyl transferase. The active site of peptidyl transferase consists of portions of several proteins of the 50S subunit. The nascent polypeptide chain is lengthened at its C terminus by one residue and transferred to the A site tRNA. This process is called transpeptidation.

Translocation   Formation of the peptide bond is accomplished by cleavage of the bond connecting fmet and tRNA^fmet by the enzyme tRNA deacylase (tRNA hydrolase in eukaryotes) which is a ribosomal component. Deacylated tRNA binds very poorly to the P site. Therefore this tRNA leaves the ribosome immediately after peptide bond formation. Also, the binding of the peptidyl-tRNA molecule that is now in the A site is weakened. The binding of peptidyl-tRNA to the

P site is always strong. Therefore there is a tendency for the peptidyl-tRNA to move from the A site to the P site. This movement is called translocation. A third elongation factor EF-G controls this process. As with other elongation factors, EF-G also first binds with GTP and then GTP is hydrolysed to GDP which is later released. The elongation factors in eukaryotic system are shown in Table 10.3. Before the repetition of elongation, the tRNA that is attached to the P site, which is now uncharged, moves transiently through a third site on the bacterial ribosome called the E site or exit site and then leaves the ribosome (Figure 10.4).

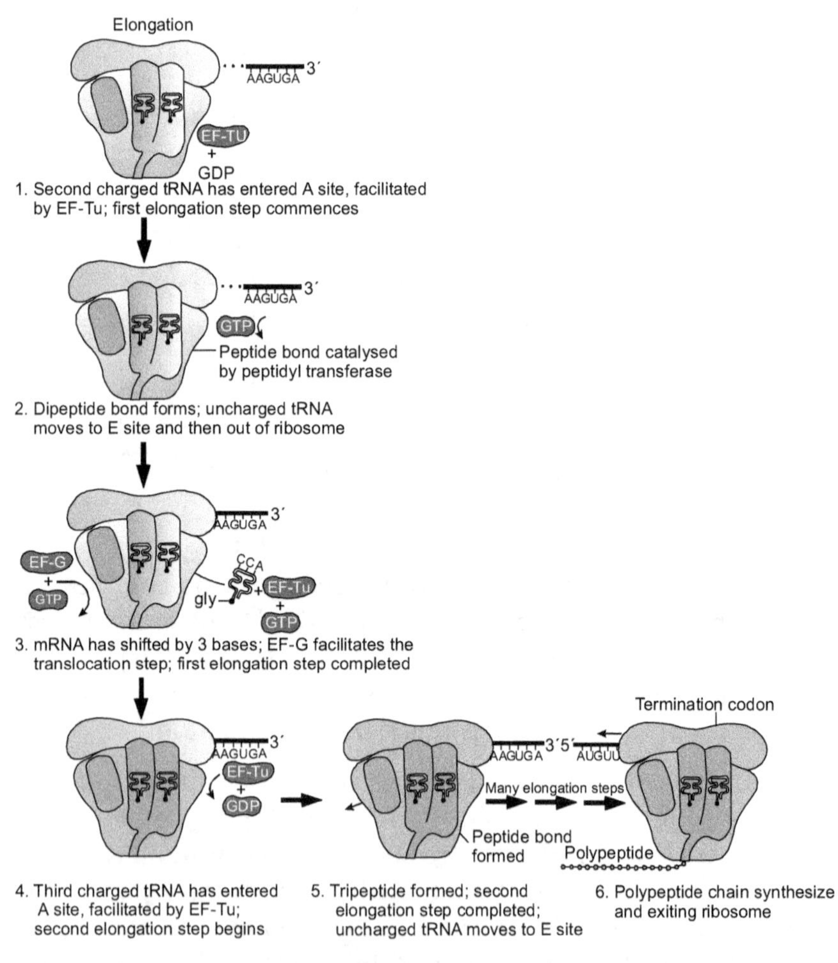

Figure 10.4 Elongation and termination of protein biosynthesis

Termination

Termination of translation is signalled by one or more of three triplet codes in the A site, UAG, UAA or UGA. They are the stop codons, otherwise called termination codons or nonsense codons. They do not specify any amino acid and they cannot recognize any anticodon of the tRNA in the A site. The completed polypeptide is therefore still attached to the terminal tRNA at the P site and A site is now empty. The termination codon signals the action of GTP-dependent release factors which cleave the polypeptide chain from the terminal tRNA, releasing it from the translation complex. In prokaryotes, there are three release factors RF1, RF2 and RF3. RF1 is specific for UAA and UAG stop codons and RF2 for UGA and UAA codons. RF3 stimulates the action of RF1 and RF2. Once the polypeptide is cleaved, the uncharged tRNA is shifted to the E site and then released from the ribosome. Then the ribosomal subunits get dissociated. Eukaryotes have only one RF that does the function of both RF1 and RF2, which is referred to as eRF.

When the release factors bind at the termination codon, they induce peptidyl transferase to transfer the growing polypeptide to a water molecule rather than to another amino acid.

BOX 10.3   ANTISENSE OLIGONUCLEOTIDES

The chemotherapies for cancer and AIDS are often associated with toxic side effects. Conventional therapeutic drugs target both normal and diseased cells. A promising candidate as an alternative has emerged—the antisense oligonucleotide. Usually RNA is obtained from only one DNA strand by transcription and this RNA is referred to as sense RNA. Sometimes, it is possible for the other DNA strand to be copied into RNA which is referred to as antisense RNA. Because of complementarity, the sense RNA and antisense RNA can form a duplex structure. This may prevent the binding of the ribosome to initiate protein synthesis, or the double-stranded RNA is easily attached by intracellular ribonucleases. This results in a block in gene expression. *In vitro* production of antisense RNA and targeting them into the cells seemed to be promising to combat cancer. Antisense oligonucleotide approach was used to treat cytomegalovirus-induced retinitis in AIDS patients.

DIRECTION OF POLYPEPTIDE CHAIN GROWTH AND TRANSLATION OF mRNA

The protein molecules have an amino terminus and a carboxyl terminus. Synthesis begins at the amino terminus. If a protein has a sequence

$$H_2N\text{-met-ser-ala-glu-arg-COOH,}$$

methionine is the starting amino acid and arginine the last amino acid.

Translation of mRNA occurs in $5' \rightarrow 3'$ direction. Let us consider a synthetic mRNA with the nucleotide sequence 5′ AAAAAA——AAAC 3′ which is used in an *in vitro* protein synthesizing system that does not require an AUG start codon. The codons AAA and AAC correspond to lysine and asparagine respectively. The protein obtained from this mRNA will have the amino acid sequence

$$H_2N\text{-lys-lys-lys…asn-COOH}$$

indicating that the N-terminal lysine is the first amino acid and the C-terminal asparagine is the last amino acid. This is a proof to show that mRNA is translated from 5′ to 3′ end.

INHIBITORS OF PROTEIN BIOSYNTHESIS

Many inhibitors of protein synthesis are known. They have been very useful in studying the events in translation because a particular inhibitor is likely to block a step in the complex process of protein synthesis. Many antibacterial agents called antibiotics have been isolated from microorganisms and were used both clinically and as reagents for unravelling the details of protein synthesis, and RNA and DNA synthesis. Some of the antibiotics have only limited clinical application because they inhibit the growth of both animal and bacterial cells and are considered as toxic agents.

Inhibitors of Protein Synthesis in Prokaryotes

The inhibitors of protein biosynthesis in prokaryotes are listed in Table 10.4 and chemical structures of some of the inhibitors are given in Figure 10.5.

Puromycin　Puromycin resembles the 3′ end of an aminoacyl-tRNA. It binds to the A site of the bacterial ribosome. Peptidyl transferase creates a bond between the nascent peptide attached to the tRNA in the P site and an α-amino group of puromycin present in the A site. Elongation can no longer occur. The peptide chain is released prematurely and protein synthesis terminates.

Table 10.4 Inhibitors of protein synthesis in prokaryotes

Inhibitor	Action
Streptomycin, Neomycin, Kanamycin	Binds to S12 of 30S subunit and inhibits the binding of tRNA[fmet] to P site. If already initiation has taken place, it leads to misreading of mRNA
*Chloramphenicol (Chloromycetin)	Inhibits peptidyl transferase activity of 70S ribosome
Tetracycline	Inhibits binding of charged tRNA to 30S particle
Erythromycin	Binds to 50S subunit and prevents formation of 70S ribosome. But it is not effective on 70S ribosome
*Puromycin	Causes premature chain termination as it is an analogue of charged tRNA
*Fusidic acid	Binds to EF-G · GTP. Even then hydrolysis of this complex and translocation occur. But EF-GDP is not released from the ribosome. So the ribosome cannot bind another aminoacyl-tRNA
Kasugamycin	Inhibits binding of tRNA[fmet]
Lincomycin	Inhibits the activity of peptidyl transferase
Kirromycin	Inhibits the release of EF-Tu
Thiostrepton	By inhibiting EF-G, prevents translocation

* They are active in eukaryotes also.

Streptomycin   Streptomycin interferes with the binding of tRNA[fmet] to the P site and thereby inhibits the initiation of synthesis of a polypeptide chain. When its concentration is very less, it causes misincorporation of amino acids because it alters anticodon–codon recognition and induces misreading of the code.

Chloramphenicol   Chloramphenicol is active against both bacterial and mammalian cells. It inhibits peptidyl transferase in both bacterial and mitochondrial ribosomes. Cytoplasmic ribosomes are unaffected. This antibiotic is of clinical use administered for serious infection.

There are other antibiotics also acting on prokaryotic systems (Table 10.4).

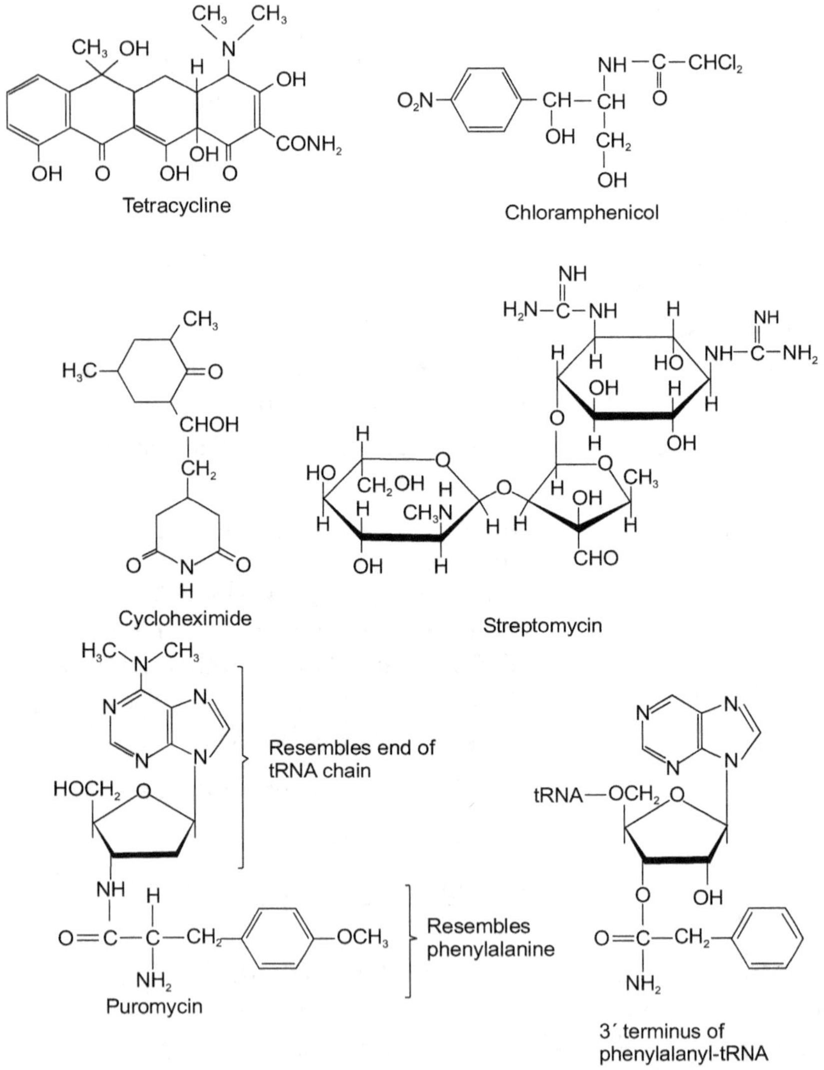

Figure 10.5   Selected inhibitors of protein biosynthesis

Inhibitors of Translation Effective in Eukaryotes

Two important inhibitors that are effective only in eukaryotes are cycloheximide and diphtheria toxin. Cycloheximide inhibits the peptidyl transferase activity of the eukaryotic ribosomes.

Diphtheria toxin is an enzyme coded for by a bacteriophage that is lysogenic in the bacterium *Corynebacterium diphtheriae*. It catalyses a reaction in which NAD^+ adds an ADP ribose group to a specially modified histidine in the translocation factor eEF2, the eukaryotic equivalent of EF-G. Since the toxin is a catalyst, minute amounts can irreversibly block a cell's protein synthetic machinery. Pure diphtheria toxin is found to be one of the most deadly substances known.

Many other antibiotics are also known to inhibit protein synthesis in eukaryotes (Table 10.5).

Table 10.5 Inhibitors of protein synthesis in eukaryotes

Inhibitor	Action
Abrin, ricin	Inhibits binding of aminoacyl-tRNA.
Diphtheria toxin	Enzymatic catalysis of a reaction between NAD^+ and eEF2 to yield an inactive factor. Inhibits translocation.
Chloramphenicol (also called chloromycetin)	Inhibits peptidyl tranferase of mitochondrial ribosomes. Inactive against cytoplasmic ribosomes.
Puromycin	Causes premature chain termination by acting as an analogue of charged tRNA.
Fusidic acid	Inhibits translocation by altering an elongation factor.
Cycloheximide (also called actidione)	Inhibits peptidyl transferase.
Pactamycin	Inhibits positioning of $tRNA^{imet}$ on the 40S ribosome.
Showdomycin	Inhibits formation of the $eIF2\text{-}tRNA^{imet}\text{-}GTP$ complex.
Sparsomycin	Inhibits translocation.

BOX 10.4   PROTEASOMAL COMPLEX

Proteasomes are large protein complexes in eukaryotic and prokaryotic cells. Their main function is to degrade unneeded or damaged proteins by proteolysis. In structure, proteasome is a large barrel like complex with a core of four stacked rings around a central pore. Each ring has seven proteins with protease activity. The proteasomal degradation pathway is essential for many cellular processes, including the cell cycle, regulation of gene expression and responses to oxidative stress. Three scientists Avram Hershko, Aaron Ciechanover and Irwin Rose shared the 2004 Nobel Prize in chemistry for their work on the proteasomal complex system related to ubiquitin-mediated protein degradation". The most common form of the complex is called 26S proteasome which contains one 20S core particle structure and two 19S regulatory caps. Proteasome inhibitors have been identified which have effective antitumour activity inducing apoptosis. One such compound developed was bortezomib used in the treatment of multiple myeloma, pancreatic cancer and B-cell-related cancers like non-Hodgkin's lymphoma. Another inhibitor, ritonavir, was developed to target HIV infection.

COMPLEX TRANSLATION UNITS

During translation, the mRNA molecule does not pass through just one ribosome. After about 25 amino acids have been joined together in a polypeptide chain, the initiation AUG codon site of the mRNA molecule is completely free of the ribosome and a second initiation complex forms with another set of ribosomal subunits. When the second ribosome moves a distance similar to that traversed by the first one, a third ribosome is able to attach. This process continues until the mRNA is covered with ribosomes—one 70S ribosome per 80 nucleotides. This large translation unit is called a polyribosome or polysome.

It is thus seen that the ribosomes move along the mRNA in the $5' \rightarrow 3'$ direction. There is a gradient in the length of the nascent polypeptide chain increasing in the $5' \rightarrow 3'$ direction, because a ribosome near the 3' end has been synthesizing protein for a greater time than the one near the 5' end. The total number of ribosomes per polysome is proportional to the length of the polypeptide chain and inversely related to the spacing between adjacent ribosomes (Figure 10.6). In prokaryotes that involve polycistronic mRNA, polysomes with 40–50 ribosomes are involved in translation. This accounts for the synthesis of the required number of protein copies from a single mRNA molecule.

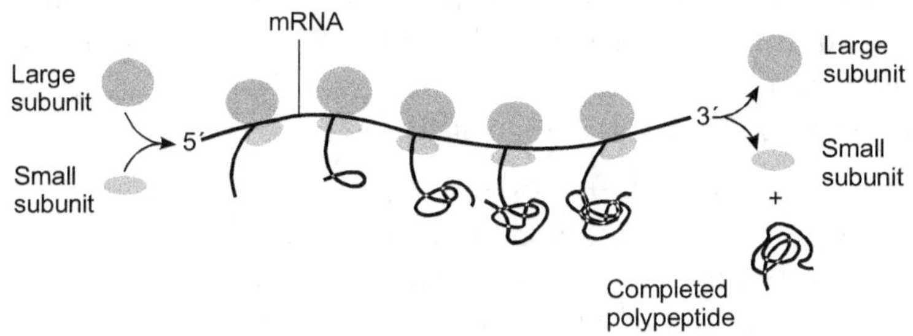

Figure 10.6 Protein biosynthesis in polyribosomes

In *E. coli*, the polypeptide elongation rate is found to be 17 amino acids/second and the mRNA elongation rate is 55 nucleotides/second. The mRNA half-life is 1.3–1.8 minutes.

ENERGETICS OF PROTEIN BIOSYNTHESIS

Tremendous energy is required to synthesize a protein true to the information stored in the mRNA. Formation of each aminoacyl-tRNA uses two high-energy phosphate groups. One GTP is hydrolysed in the initiation step that is bound to IF2. In every elongation step, one GTP bound to EF-Tu and another GTP in the translocation step are hydrolysed. A total of $2n - 1$ high-energy bonds are utilized in the elongation step. In the termination step, one GTP is hydrolysed. Therefore the total number of high energy groups hydrolysed are $2n + 1 + (2n - 1) + 1$, i.e., more than 4 high-energy bonds per amino acid added in the peptide chain. This represents 4×30.5 kJ/mol $= 122$ kJ/mol of phosphodiester bond energy required to generate a peptide bond.

POST-TRANSLATIONAL MODIFICATIONS

Though the genetic code specifies only the 20 standard amino acids, analysis of the mature proteins reveals the presence of many structural variants of the standard amino acids. Apart from the modifications of the amino acid side chains, in many proteins, parts of the originally synthesized polypeptides are removed during the process of maturation. These events constitute post-translational modifications of the proteins.

1. The N-terminus and C-terminus amino acids are removed or may be modified. For example, in bacteria the initial N-terminal formylmethionine residue is removed enzymatically. In prokaryotes, in some proteins, deformylase removes the formyl group from N-terminal formylmethionine and leaves methionine as the N-terminal amino acid. Usually, the amino group attached to the initial methionine residue is removed and the amino group of the N-terminal residue is chemically modified.

2. At the N-terminal end of some proteins, a sequence of up to 30 amino acids called signal sequence direct the protein to the location where it functions. The signal sequence determines the final destination of a protein in the cell and the process is called protein targeting. Once their function is over, these signal sequences are removed enzymatically from the polypeptide. For example, secretory proteins and plasma membrane proteins are to be initially transported into the lumen of the endoplasmic reticulum which is promoted by the signal sequence. These sequences are removed after the polypeptides are transported even prior to achieving their functional status as proteins.

3. Sometimes individual amino acids may be modified. For example, the phosphate group may be added to the hydroxyl groups of certain amino acids like tyrosine as in glycogen phosphorylase to activate the enzyme. The phosphate groups create negative charge on the amino acid that may ionically bond with the adjacent molecules. Phosphorylation is important in regulating many cellular activities that are promoted by kinase enzymes. The milk protein casein has many phosphoserine groups that bind Ca^{2+}.

   Extra carboxyl groups are added in some proteins. For example, the blood-clotting protein prothrombin contains a number of γ-carboxy glutamate residues in its amino terminal region. These carboxyl groups bind Ca^{2+}, required to initiate the clotting mechanism.

   In some proteins methyl groups may be added to the amino acid residues. Monomethyl and dimethyl residues occur in some muscle proteins and in cytochrome *c*. The calmodulin of most organisms contain one trimethyl residue at a specific position. In some proteins, the carboxyl groups of some Glu residues undergo methylation, removing their negative charge.

Other modifications include acetylation, glycosylation, hydroxylation, nucleotidylation, and ADP-ribosylation (as in eEF-2 to modulate the protein activity).

Carbohydrate side chains are sometimes attached covalently to the polypeptide chain to produce glycoproteins, like the antigenic determinants that specify the antigens in the ABO blood type system in humans. In some glycoproteins, the carbohydrate side chain is attached enzymatically to Asn residues to form N-linked oligosaccharides, in others to Ser or Thr residues to form O-linked oligosaccharides. Many proteins that function extracellularly, as well as the lubricating proteoglycans that coat mucous membranes, contain oligosaccharide side chains.

A number of eukaryotic proteins are modified by the addition of groups derived from isoprene. A thioether bond is formed between the isoprenyl group and a Cys residue of the protein. Proteins modified in this way include the Ras proteins, which are the products of the ras oncogenes and proto-oncogenes and G proteins and lamins, which are the proteins in the nuclear matrix. Sometimes these isoprenyl groups help to anchor the protein in a membrane.

Several side-chain modifications covalently bond cofactors to enzymes to increase their catalytic efficiency. For example N^{ε}-lipoyllysine is linked to dihydrolipoyl transacetylase and 8α-histidylflavin is linked to succinate dehydrogenase.

4. Many proteins are involved in a variety of biological processes, but are initially synthesized as inactive precursors. These are activated under proper conditions by limited proteolysis. The polypeptide precursors are correctly trimmed. For example, the primary translation product of insulin is preproinsulin which contains a 24-residue signal peptide preceding the 81-residue proinsulin molecule. The signal peptide is removed by signal peptidase, as the nascent chain is transported into the lumen of the endoplasmic reticulum. Proinsulin folds, and two disulphide bonds cross-link the ends of the molecule. Before secretion, a trypsin-like enzyme cleaves after a pair of basic residues. Then a carboxypeptidase B-like enzyme removes these basic residues to generate the mature form of insulin. Trypsinogen and chymotrypsinogen are converted

to the active forms trypsin and chymotrypsin, by tryptic cleavages of specific peptide bonds. Inactive proteins that are activated by removal of polypeptides are called proproteins, whereas the excised polypeptides are termed propeptides.

In the case of collagen biosynthesis, the collagen propeptides are necessary for proper procollagen folding. The N terminal and C terminal propeptides of procollagen are respectively removed by amino and carboxyl procollagen peptidases.

5. The polypeptide chains are often complexed with metals. Hemoglobin containing four iron atoms along with four polypeptide chains is a good example.

Some proteins are synthesized as segments of polyproteins, polypeptides that contain the sequences of two or more proteins. Examples include most polypeptide hormones, proteins synthesized by many viruses including those causing polio and AIDS, and ubiquitin, a highly conserved eukaryotic protein involved in protein degradation. Specific proteases post-translationally cleave polyproteins to their component proteins, by recognizing specific cleavage sites.

These post-translational modifications of the proteins are important in achieving the functional status specific to any given protein. They finally assume a correct three-dimensional structure related to their specific function which is referred to as protein folding. For many proteins, folding is dependent on members of a family of ubiquitous proteins called **chaperones**. They are also called as **molecular chaperones** or **chaperonins**.

Spicy Questions and Answers

1. **Why is it possible for a prokaryotic mRNA to contain more than one reading frame, while eukaryotic mRNAs can contain only a single reading frame?**

   Prokaryotic mRNAs can carry several separate reading frames because the small ribosomal subunit binds to a Shine-Dalgarno sequence in a single mRNA, one for each reading frame. In eukaryotes, all small ribosomal subunits must attach at the 5' end of the eukaryotic mRNA and there are not many binding sites downstream for ribosome binding.

Therefore there is only one binding site, only one initiation codon and only one reading frame in eukaryotes.

2. **If a portion of the DNA that encoded the 5.8S rRNA in one large rRNA precursor gene were deleted in a mammalian cell, will the cell die due to lack of protein synthesis?**

No, because there are many copies of the gene in the chromosome.

3. **Inosine is a modified nucleoside in the tRNA anticodon. However, a tRNA with the anticodon 3′ ACI 5′ is not found. Why is this so?**

The anticodon 3′ ACI 5′ would pair with the termination codon UGA. This would form the codon to encode cysteine instead of termination of translation, violating the genetic code.

4. **What is the anticodon in the single tRNA that carries tyrosine?**

The tyrosine codons are 5′ UAU 3′ and 5′ UAC 3′. The anticodon is 3′ AUG 5′.

5. **Synthetic mRNAs are prepared using U and G in the ratio 5:1. These synthetic mRNAs incorporate cysteine but not alanine in a cell-free translation system. Charged cysteinyl-tRNA^{cys} are treated with Raney nickel, which is a catalyst that removes sulphur from cysteine, converting cysteine to alanine. The product got is alanyl-tRNA^{cys}. When this is added to the cell-free translation system, alanine is incorporated in place of cysteine. What does this result suggest about the function of tRNA?**

It shows that the anticodon and not amino acid in the tRNA is involved in recognizing the mRNA codon.

6. **Predict the amino acid sequences of peptides formed in response to the following mRNA sequences. Assume that the reading frame starts with the first codon in each sequence.**

   (a) GGUCAGUCGCUCCUGAUU

   (b) UUGGAUGCGCCAUAAUUUGCU

   (c) CAUGAUGCCUGUUGCUAC

   (d) AUGGACGAA

   (a) gly-gln-ser-leu-leu-ile

   (b) leu-asp-ala-pro

   (c) his-asp-ala-cys-cys-tyr

   (d) met-asp-glu (in eukaryotes)
       fmet-asp-glu (in prokaryotes)

7. **From a given sequence of amino acid residues in a protein, can we predict the base sequence of the mRNA that coded it? Give reasons for your answer.**

   No, we cannot predict the exact base sequence of mRNA even if we know the protein's amino acid sequence, because one amino acid may have more than one codon. Only in case of amino acids (methionine and tryptophan) which have one codon each, the sequence of bases can be predicted. In most of the amino acids, the third base of the codon will be different and in these positions of mRNA, the bases cannot be predicted correctly.

8. **Determine the energy cost (in terms of high-energy phosphate groups expended) required for the biosynthesis of the β-globin chain of hemoglobin with 146 amino acid residues?**

   In synthesizing a protein with 146 amino acid residues, the number of high-energy phosphate groups required is 583, based on 4 per amino acid residue added minus one.

9. **The following is the base sequence of a strand of DNA TACGTCTCCAGCGGAGATCTTTTCCGGTCGCAACTGAGGTTGATC. The strand is transcribed from left to right and codes for a small peptide.**

   (a) **Identify the 5′ and 3′ end of the DNA strand.**

   (b) **Give the sequence of the complementary DNA strand.**

   (c) **What is the sequence of the transcript?**

   (d) **Give the sequence of the amino acids of the peptide.**

   (e) **What is the direction of translation?**

   (f) **Identify the N-terminal and C-terminal ends of the peptide.**

   (a) Since transcription proceeds from left to right, the sequence of bases in the DNA given in the problem has 3′-OH group on the left end and 5′-phosphate in the right end.

   (b) 5′   ATGCAGAGGTCGCCTCTAGAAAAGGCCAGCGTTGACT
       CC AACTAG 3′

   (c) 5′ AUGCAGAGGUCGCCUCUAGAAAAGGCCAGCGUUG
       ACU CCAACUAG 3′

   (d) met-gln-arg-ser-pro-leu-glu-lys-ala-ser-val-asp-ser-asn-stop

   (e) Translation proceeds from 5′ to 3′ end of mRNA. In this case, translation also goes from left to right.

   (f) Methionine is the N-terminal amino acid and asparagine is the C-terminal amino acid.

10. The amino terminus of a wild-type enzyme has the amino acid sequence

    met-leu-his-tyr-met-gly-asp-tyr-pro

    Two mutant types of the enzyme were identified which made the enzyme inactive. The following are the details of the amino acid sequence of these two mutant types:

    Mutant 1: met-ile-thr-leu-tyr-gly-asp-tyr-pro

    Mutant 2: only a tripeptide was got.

    (a) **What single base change would have accounted for producing these two mutants?**

    (b) **What is the sequence of the tripeptide produced by mutant 2?**

    (a) When comparing the amino acid sequence of mutant 1 and the wild type, the change occurs in the wild type frame leu-his-tyr-met, which has become gly-asp-tyr-pro by mutation. This shows that a frameshift mutation has occurred in this frame with insertion (+)/deletion (–) of base each at the ends of this frame.

    Base sequence of wild type: AUG-UUA-CAU-UAU-AUG—

    Base sequence of mutant 1: AUG(+)AUU-ACA-UUA-UAUG(–)—

    In mutant 2, the fourth codon of tyrosine UAU is mutated to UAG or UAA, so that the sense codon is altered to a stop codon.

    (b) The resulting tripeptide in mutant 2 is met-leu-his.

11. **How many aminoacyl-tRNA synthetases are there? What do they use for recognition signals?**

    There are twenty aminoacyl-tRNA synthetases each for a particular amino acid. Any region of the tRNA can act as the recognition signal, but the anticodons play an important role.

12. **Why is the E site necessary in translocation?**

    E site is required for the binding of the deacylated tRNA that moves out of P site. E site enables stop codon recognition in the A site and stops the elongation step. The E site has three functions.

    1. The E-site-induced low-affinity A site plays an important role for the selection of the correct aminoacyl-tRNA.

    2. The low affinity of the E site induced by the occupation of the A site is the active mechanism that rejects the deacylated tRNA from the E site.

    3. A cognate tRNA at the E site seems to be essential for maintaining the reading frame.

Review Questions

1. What are the basic requirements for protein biosynthesis?

2. Discuss the chemistry of prokaryotic and eukaryotic ribosomes.

3. Explain the charging of tRNA molecules.

4. What is meant by activation of amino acids in translation?

5. List and explain the functions of the protein factors involved in the various steps of translation.

6. Explain the interaction between mRNA, tRNA and rRNA during protein biosynthesis.

7. Discuss the energetics of protein synthesis.

8. Bring out the special features of initiator tRNA molecules in prokaryotes and eukaryotes.

9. How is the first AUG recognized as the start codon during mRNA scanning in prokaryotes and eukaryotes?

10. What is meant by Shine-Dalgarno sequence? Give its significance.

11. How is an initiation complex formed during translation?

12. Discuss the role of elongation factors in peptide bond formation.

13. What is meant by "second genetic code". Give its importance in maintaining the fidelity of protein synthesis.

14. What is the role of GTP in translation?

15. Explain i) translocation ii) open reading frame iii) polyribosomes.

16. List the inhibitors of protein biosynthesis in prokaryotes and eukaryotes and explain their mechanism of action.

17. Compare and contrast prokaryotic and eukaryotic translation.

18. Write an essay on the steps involved in protein biosynthesis.

19. Explain post-translational modification of nascent polypeptide chain with suitable examples.

11

REGULATION OF GENE EXPRESSION IN PROKARYOTES

For a bacterium to survive, it is not necessary that all the genes should be transcribed at all times. To conserve energy and resources, bacteria regulate the activity of their genes so that only those gene products required for the functioning of the cell are produced. For example, if a particular amino acid is present in the surrounding medium, it would be a waste of energy and resources for the bacterium to synthesize that particular amino acid. Regulation of gene expression allows bacteria to respond to changes in their environment, mainly to the presence or absence of nutrients.

Bacteria regulate gene expression in order to control the amount of gene product, i.e., the protein, present in the cell. The steady-state concentration of a gene product is determined by the balance between the rate of synthesis and the rate of degradation of the expressed protein.

Gene expression in prokaryotes is regulated at three important levels—transcriptional level, mRNA turnover period and translational level.

In *Escherichia coli*, the number of genes is approximately 3,000–4,000 and the number of proteins found at a given time is 600–700. This means that all the genes are not always expressed. Only the genes whose products are required at a particular time are active at that time. Also in a bacterial cell, there are about 10,00,000 protein molecules and the number of copies of various protein molecules varies from 10 to 5,00,000. This suggests that those genes which are active at a particular time also do not produce the proteins at the same rate. The genes synthesize their products in quantities required by the cell and at a time when they are needed by the cell/organism. This again shows that some regulatory mechanisms operate for the expression of genes in prokaryotes.

There are at least six potential points at which the amount of protein in a cell can be regulated

1. synthesis of the primary transcript
2. post-transcriptional processing of mRNA (in eukaryotes)
3. mRNA degradation
4. protein biosynthesis
5. post-translational modification of protein
6. protein degradation

Some gene products are required all the time and their genes are expressed at a more or less constant level in all the cells of a species or organism. The genes for the enzymes that catalyse the steps in central metabolic pathways like citric acid cycle fall in this category. These genes are referred to as housekeeping genes. Unregulated expression of such genes is called **constitutive gene expression**. The amounts of other gene products rise and fall in response to molecular signals. Gene products that increase in concentration under prescribed molecular circumstances are referred to as **inducible** and the process of increasing the expression of such genes is called **induction**. The expression of many genes encoding DNA repair enzymes, for example, is induced in response to DNA damage. Here DNA damage is the molecular signal for induction.

TERMINOLOGY IN REGULATION OF GENE EXPRESSION

Structural genes   Structural genes are nucleotide sequences of DNA that serve as templates for the synthesis of RNAs—tRNA, rRNA and mRNA. The average length of structural genes specifying proteins in prokaryotes is about 1000 base pairs. In eukaryotes it may be about 10,000 base pairs on an average due to the presence of introns.

Cistron   A cistron is a gene defined on the basis of a *cis-trans* complementation test. It is defined as a segment of DNA that codes for a complete polypeptide chain. Two mutations in the same cistron (*cis* configuration) cannot complement each other, whereas two mutations in different cistrons may complement each other.

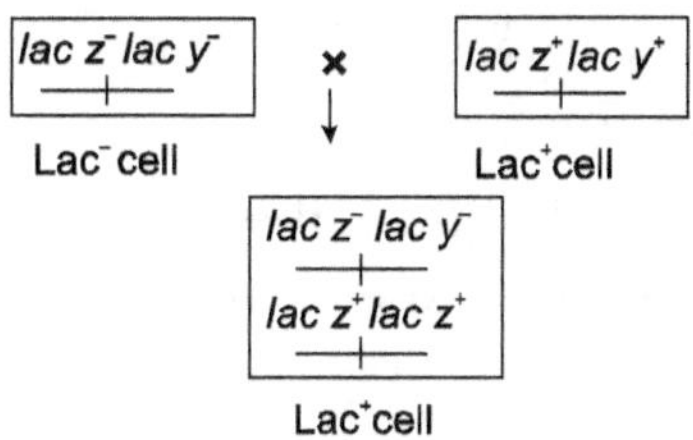

(a) *Cis*-configuration: Complementation occurs

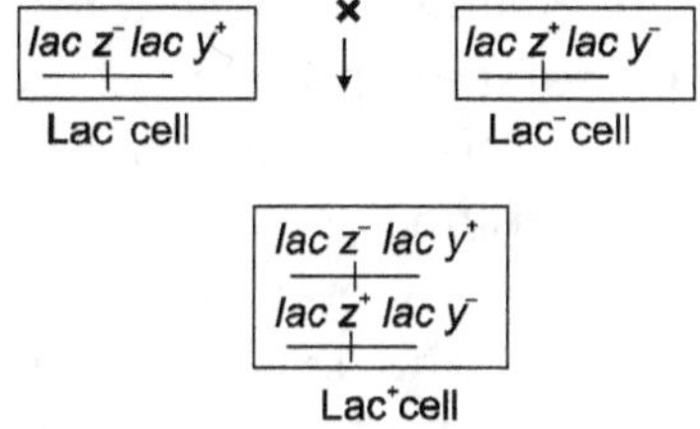

(b) *Trans*-configuration: Complementation occurs

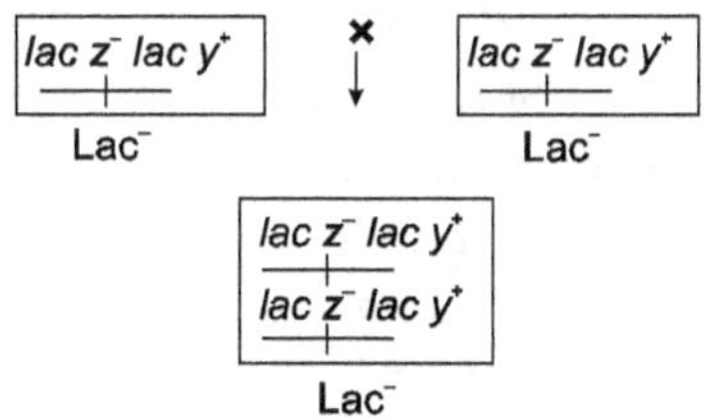

(c) *Trans*-configuration: No Complementation

Figure 11.1 Mutations in *cis*- and *trans*-configuration

In Figure 11.1a, two *E. coli* cells Lac⁻ and Lac⁺ (in which mutations are there in *lac z* and *lac y* genes) when crossed, because of complementation, the resulting partial diploid can metabolize lactose since the two mutations in *cis*-configuration are on the same chromosome. *lac z* gene is responsible for the production of the enzyme β-galactosidase and *lac y* for the enzyme β-galactoside permease. The first enzyme is required for the metabolism of lactose, a galactoside (to break down lactose into glucose and galactose) and the second enzyme for the transport of lactose inside the cell.

Figure 11.1b illustrates *trans*-configuration, i.e., the cell at the left is defective only in its ability to make β-galactosidase and the cell at the right is defective in its ability to make β-galactoside permease. So, neither cells can utilize lactose. When crossed, the resulting hybrid cell contains one unmutated copy of each gene and therefore can metabolize lactose, because of complementation.

If there is no complementation in the hybrid as in Figure 11.1c, we can conclude that the two mutations in the two original cells are in the same gene or cistron. But it will not be possible to say which gene is defective, *lac z* or *lac y*.

Controlling sites   There are short nucleotide sequences of DNA, usually 15 to 30 base pairs in length, adjacent to the structural genes (preceding them) that control the expression of these structural genes. These sites are present in *cis*-configuration and not in *trans*-configuration to them. Controlling sites are categorized as promoter sites (promoters), operator sites (operators), initiator sites (initiators), attenuator sites (attenuators) and transcriptional terminator sites (terminators). In most cases, controlling sites are bound by regulatory proteins or RNA polymerases.

Regulatory proteins   Regulatory proteins are encoded by regulatory genes. They affect the expression of structural genes by binding to the controlling sites near the structural genes. They either activate transcription or repress it. A regulatory protein that stimulates gene transcription is called an **activator**. Activators are of two types—transcriptional factors and transcriptional activators. Proteins that terminate transcription are referred to as **terminators**. **Repressors** are proteins that inhibit initiation of transcription when they bind to those controlling sites called operators.

In general, the activators stimulate the binding of RNA polymerase to the promoter site on the DNA at the beginning of the structural genes and repressors inhibit RNA polymerase binding.

Induction and repression   Gene products that decrease in concentration in response to a molecular signal are referred to as repressible and the decrease in gene expression is called **repression**. For example, the presence of large amounts of an amino acid leads to repression of the genes for the enzymes catalysing the biosynthesis of that amino acid in bacteria. The system is said to be turned off when repressed.

Gene products that increase in concentration in response to a molecular signal are referred to as inducible and the increase in gene expression is called **induction**. For example, presence of lactose as the sole carbon source in its environment for a bacterium causes it to synthesize the enzymes required for the utilization of lactose that are not required if glucose or any other carbon source is available for the organism as energy source. The system is said to be turned on when induced.

Both induction and repression of gene expression can be regulated at the transcriptional level.

Types of responses to regulatory signals   There are three types of responses to a regulatory signal (Figure 11.2).

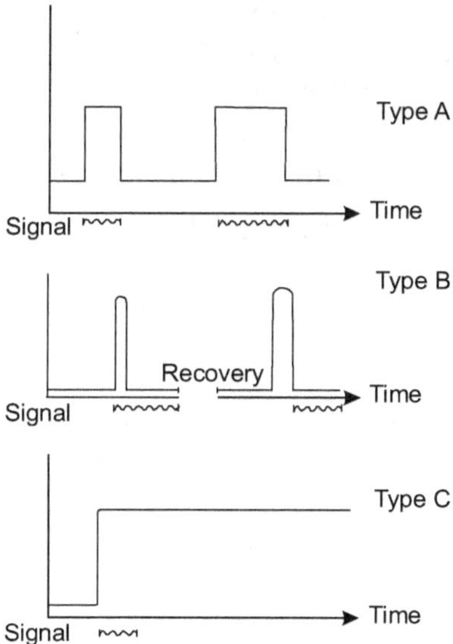

Figure 11.2   Three types of responses to a regulatory signal

A type A response is characterized by an increased rate of gene expression that is dependent upon the continued presence of the inducing signal. On removing the signal, the rate of gene expression diminishes to its basal level. Once the signal reappears, the rate is further increased. Such responses are seen in prokaryotes when the intracellular concentration of a nutrient is suddenly changed and in higher organisms in response to hormones or growth factors.

A type B response shows a transient increase in the rate of gene expression even when there is continued presence of the signal. After the regulatory signal has terminated and the cell is allowed to recover, a second transient response to another signal is observed. This type of response is seen during the development of an organism, when a gene product is produced only transiently even though the signal persists.

In type C response, there is increased rate of gene expression that persists indefinitely even after the termination of the signal. This type of response occurs during the development of differentiated function in a tissue or organ.

Coordinate regulation   In the bacterial system, when several enzymes act in sequence in a single metabolic pathway, usually all or none of these enzymes are produced. This is known as coordinate regulation which results from the control of the synthesis of polycistronic mRNA molecules encoding all of the gene products that function in the same metabolic pathway. Here, the genes for all the enzymes involved in the metabolic pathway are present clustered close to one another on the chromosome and have a common promoter region.

When transcription is in the "off" state, there always remains a basal level of gene expression. This may amount to only one or two transcription events per cell generation leading to very little protein synthesis.

Feedback or end-product inhibition   Usually the activity of enzymes involved in a pathway is regulated by the concentration of the end product of that pathway. For example, in biosynthetic pathways, if the concentration of the final compound is high, the compound will inhibit the first step of the reaction sequence which is referred to as **feedback inhibition** or **end-product inhibition**.

There are several common patterns of regulation of transcription. These depend on the type of metabolic activity of the system being regulated. For example, in a catabolic (degradative) pathway, the concentration of the initial substrate determines whether the enzymes involved in the pathway are synthesized or not. In an anabolic (biosynthetic) pathway, the final product is the regulatory substance.

Negative regulation and positive regulation   Regulation of gene expression is of two types, negative regulation and positive regulation. In negative regulation, an **inhibitor molecule** is present in the cell which turns-off transcription. An anti-inhibitor, called an **inducer** is required to turn the system on. In positive regulation, an **effector molecule**, which may be a protein, a small molecule or a molecular complex activates a promoter. Negative and positive regulation are not mutually exclusive and some systems are both positively and negatively regulated and need two regulators. The types of regulation—positive or negative—as decided by the binding of the regulator to DNA is given in Table 11.1.

Table 11.1   Types of regulation

Binding of regulator to DNA	Positive regulation	Negative regulation
Binding	Turn-on	Turn-off
No binding	Turn-off	Turn-on

A degradative system may be regulated either positively or negatively. In a biosynthetic pathway, the final product of the pathway often negatively regulates its own synthesis. That is, absence of the final product encourages synthesis and presence of the product inhibits its synthesis.

BOX 11.1   RNA THERMOMETERS

RNA thermometers are thermosensors that regulate gene expression by temperature induced changes in RNA conformation. Naturally occurring RNA thermometers exhibit complex secondary structures which are believed to undergo a series of gradual structural changes in response to temperature shifts. At low temperature the mRNA adopts a conformation that makes the Shine-Dalgarno sequence within the 5′-UTR and prevents ribosome binding and translation. At elevated temperatures, the RNA secondary structure melts locally and gives the ribosomes access to the ribosome binding site. RNA thermometers do not require binding of a metabolic ligand to induce the conformational change and they respond directly to the temperature. Biological processes controlled by RNA thermometers include bacterial pathogenesis, heat-shock responses and the life cycle of bacteriophages. Researchers have developed minimum sized synthetic RNA thermometer which can be used to induce or repress gene expression by a simple temperature shift. Efficient RNA thermometers could be built from a single small RNA stem-loop structure masking the ribosome binding site, providing useful RNA—the only tools for the regulation of gene expression in prokaryotes.

OPERON MODEL IN PROKARYOTES

OPERON HYPOTHESIS

Francois Jacob and Jacques Monad described the operon model in 1961, based on the regulation of lactose metabolism by the intestinal bacterium *E. coli*. They proposed the model to explain the induction or repression of enzyme synthesis. Operon is a unit of coordinated control of protein synthesis. In prokaryotes, structural genes tend to be organized in clusters controlled from a single regulatory site (Figure 11.3). The cluster of genes, collectively known as an **operon**, usually produces products with related functions, e.g. the enzymes involved in a single metabolic pathway. The regulatory site, linked to the gene cluster it controls, is located in the upstream region near the 5' side of the gene and is called *cis*-**acting element**. Certain molecules bind to such regulatory sites and control the transcription of the gene cluster. Such molecules are called *trans*-**acting elements**, as they are not physically linked to the genes they regulate. The arrangement of the regulatory sites and the structural genes they regulate is an operon.

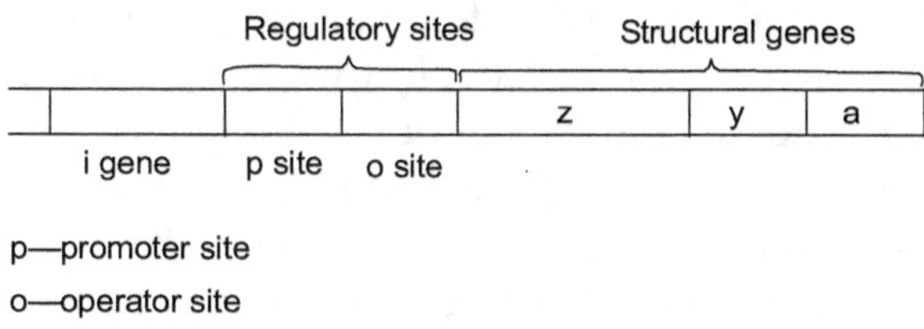

Figure 11.3   The lactose operon

TYPES OF CONTROL OF OPERONS

Operons may be inducible or repressible depending on whether they are switched-on or off by a particular metabolite. Negatively controlled inducible operons are not normally transcribed because a specific repressor protein is bound to the operator. Induction occurs when an inducer, a small molecule, binds to the repressor, altering its conformation so that it now dissociates from the operator and allows transcription to proceed.

Table 11.2   Types of control of operons

Types of operon	State of regulatory molecules	Binding to operator	Transcription	Examples
Negative inducible	Repressor free	+	–	lac
	Repressor binds to inducer	–	+	gal
Positive inducible	Apo-activator free	–	–	ara
	Apo-activator binds to co-activator	+	+	mal
Negative repressible	Apo-repressor free	–	+	trp
	Apo-repressor binds to co-repressor	+	–	arg

Positively controlled operons become active when a co-activator, a small molecule, binds to an apo-activator (protein) altering its conformation so that the complex can bind to a site near the operon and allow transcription to be initiated. Negatively controlled repressible operons are normally transcribed, but when a co-repressor, a small molecule, binds to the apo-repressor (protein) this complex binds to the operator resulting in inhibition of transcription. These types of positive or negative/inducible or repressible regulation decided by the binding of the regulatory molecule to the operator and turning on or turning off transcription is summarized in Table 11.2.

LACTOSE OPERON IN *E. COLI*

Lactose metabolism is regulated by the lactose operon in *E. coli*. β-galactosidase is an enzyme that hydrolyses lactose, a β-galactoside to galactose and glucose (Figure 11.4). The lac operon is classified as an inducible operon, because the pathway's (lactose utilization) substrate induces transcription of the operon.

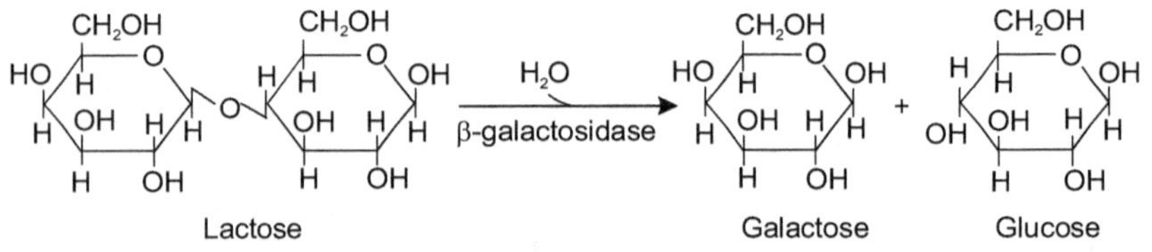

Figure 11.4 Hydrolysis of lactose to galactose and glucose by the enzyme, β-galactosidase

The structural gene for β-galactosidase is designated as *lac z* which is clustered with two other genes, *lac* y gene whose gene product is β-galactoside permease that is responsible for the transport of lactose into the cell and *lac* a gene whose gene product is thiogalactoside transacetylase that transfers an acetyl group from acetyl-CoA to β-galactosides like lactose. Acetylation gives an advantage for utilizing certain non-metabolizable analogues of β-galactosides, because it results in detoxification and excretion.

The structural genes for these three enzymes along with the lac promoter and lac operator that is a regulatory region, are collectively referred to as lac operon. This genetic arrangement of the structural genes and their regulatory genes allows for the coordinate expression of the three enzymes involved in lactose metabolism. All the three linked genes are transcribed together into one large mRNA molecule called polycistronic mRNA, which is a characteristic feature of prokaryotes.

These polycistronic mRNA molecules contain multiple independent translation start (AUG) and stop (UAA) codons for each cistron. Thus, each protein (enzyme) is translated separately and they are not processed from a single large precursor protein.

When *E. coli* cells are presented with lactose or some specific lactose analogs, the expression of the activities of β-galactosidase, galactoside permease and thiogalactoside transacetylase is increased 10–100-fold. As shown earlier, this is a type A response. Upon removal of the signal, i.e., the inducer, the rate of synthesis of these three enzymes get decreased. Since there is no significant degradation of these enzymes in bacteria, the level of β-galactosidase as well as that of the other two enzymes will remain the same unless they are diluted out by cell division.

When *E. coli* is exposed to both lactose and glucose as carbon sources, the organism will first metabolize glucose and temporarily stop growing until the genes of the lac operon are induced and produce the lac enzymes to metabolize lactose. Although lactose is present from the beginning of the bacterial growth phase, lac enzymes are not produced, i.e., the lac structural genes are not expressed until glucose present in the medium is exhausted. This phenomenon was first thought to be attributable to repression of the lac operon by some catabolite of glucose. Hence it was termed **catabolite repression**. Later studies showed that catabolite repression is in fact mediated by a **catabolite gene activator protein (CAP)** along with cyclic AMP.

Negative Regulation of lac Operon

The mechanism of repression and de-repression of the lac operon explains the negative regulation of lac operon. The expression of the normal *lac* i gene of lac operon is constitutive. It is expressed at a constant rate. The gene product is lac repressor protein which has four identical subunits each with a molecular weight of 38,000. The repressor protein has high affinity for the operator locus. The operator site is a region of double-stranded DNA 27 base pairs long with twofold rotational symmetry in a region that is 21 base pairs long. At any one time, only two subunits of the repressor appear to bind to the operator. The binding occurs mostly in the major groove. The operator locus is between the promoter site, the binding site for DNA-dependent RNA polymerase (that is responsible for the transcription of the structural genes) and the *lac* z gene (the structural gene for β-galactosidase). When the repressor protein binds to the o-site, it prevents the transcription of the lac structural genes *lac z*, *lac y* and *lac a*, because it blocks the RNA polymerase from transcribing these genes.

The lac operon is not always transcribed when there is a scarcity for glucose. Failure of RNA polymerase binding to the p-site occurs, when repressor protein is bound to the o site, only in conditions when glucose is present in the medium. This does not mean that the lac operon is transcribed whenever glucose is scarce. There is no point in doing so unless lactose is present.

The lac repressor is an allosteric protein. In the absence of lactose in the environment, it has strong affinity for the operator, but if lactose is present, a small amount of lactose enters inside the cell via the basal level of permease and it is hydrolysed when a small amount is converted to the lactose isomer, allolactose, which acts as an inducer (Figure 11.5).

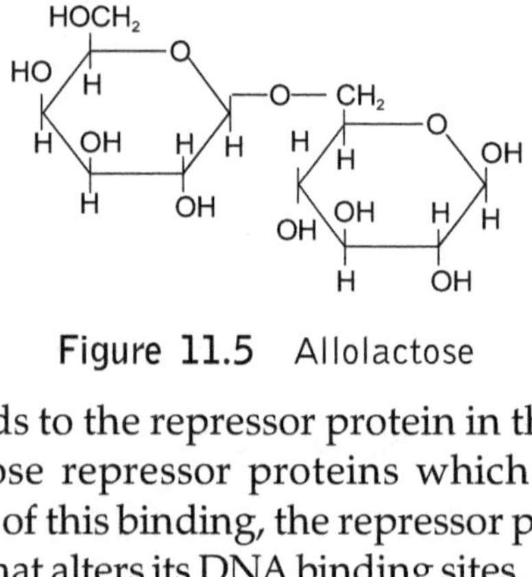

Figure 11.5   Allolactose

Allolactose binds to the repressor protein in the inducer binding site, including those repressor proteins which are bound to the operators. Because of this binding, the repressor protein undergoes a structural change that alters its DNA binding sites, causing the protein to separate from DNA. When the repressor protein leaves the operator, it will not prevent the binding of RNA polymerase on the promoter site. This will help initiation of transcription of the structural genes. The lac enzymes are produced that will then permit lactose to enter the cell and convert the lactose molecules into glucose and galactose. Therefore, it is found that in negative regulation of lac operon, gene expression is turned off when the repressor protein is bound to the o site which is referred to as repressed state and gene expression is turned on when allolactose, the inducer, binds to the repressor protein and relieves it from the o site, thereby permitting RNA polymerase to bind to the promoter site. This is known as **de-repressed state**.

When lactose is no longer available, all the remaining lactose molecules and the allolactose molecules, that are in the inducer binding sites are broken down into glucose and galactose. With no allolactose in the inducer binding site, the repressor proteins resume their original conformation, again bind to the operators and prevent further transcription. Both the lac mRNA transcribed from lac operon and the lac enzymes translated from lac mRNA are highly unstable. Within minutes after transcription of the lac operon ceases, the lactose-metabolizing enzymes are rapidly degraded. The enzymes are no longer needed because lactose is no longer available.

In spite of the elegant nature of this system, we can notice a paradox. When lactose is present in the environment, it should first enter into the cell and also it should be converted to allolactose which is required for de-repression. Entry of lactose into the cell requires β-galactoside permease and conversion of lactose to allolactose requires β-galactosidase. Both the enzymes are encoded by the lac operon which cannot be transcribed until lactose enters inside the cell and is converted to allolactose. A question arises here as to how lactose can enter inside the cell and be converted to allolactose when the genes for these functions are turned off. The answer to this paradox is that in the absence of lactose, prevention of transcription is not completely effective. Lac operon is occasionally transcribed and a small amount of the lac enzymes is produced. These low levels of permease and β-galactosidase produced are sufficient to take inside the cell a few molecules of lactose and convert them to allolactose to turn-on the lac operon.

Mutations in lac Operon

Mutation in lac i gene   Each subunit of the repressor protein has two active sites: DNA-binding site and inducer binding site. A mutation in the *lac* i gene may affect either one of these sites or both. A mutation that eliminates the function of the DNA-binding sites prevents the repressor protein subunits from binding to the operators irrespective of whether lactose is present or absent. Even though allolactose can enter the inducer-binding site, this has no effect on operon transcription because the repressor protein cannot bind to the operator under any conditions. The mutant repressor protein is thus unable to prevent transcription and RNA polymerase transcribes the *lac* operon at the same rate in the presence and absence of lactose. A mutation that may eliminate the function of the DNA-binding sites in the lac repressor protein creates *lac* i⁻ mutant alleles, the superscript minus sign indicating that functional *lac* i gene product is absent.

A mutation that eliminates the function of the inducer-binding sites on the repressor protein has an opposite effect compared to the mutation that eliminates the function of the DNA-binding sites. Since the DNA-binding site is intact, the repressor protein binds to the operators and prevents the transcription of lac structural genes. But allolactose cannot bind to the repressor protein because the inducer-binding site is mutated. So the repressor protein remains permanently bound to the operator and permanently turns off the expression of *lac* genes. Such a mutant is designated as *lac* iˢ. 's' refers to "super-repressed" condition.

Mutation in operator sites   Actually the operator consists of three sites designated as o_1, o_2 and o_3. o_1 is the **principle operator** that is present downstream from the promoter. o_2 and o_3 are called **auxiliary operators,** because they interact with a protein that binds first to the principle operator. o_2 is located within the *lac z* gene and o_3 upstream from o_1 (Figure 11.6(a)).

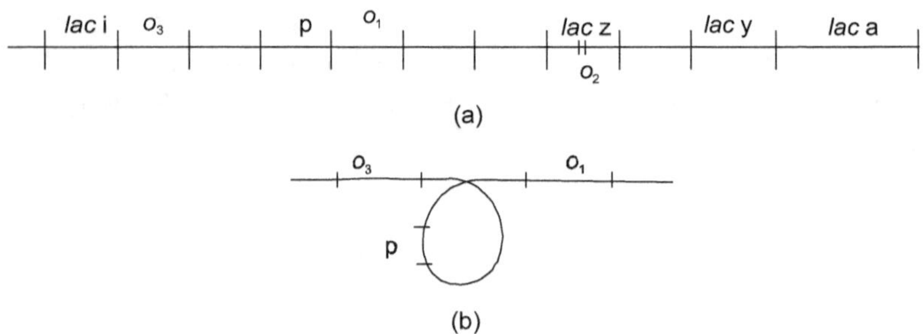

Figure 11.6   (a) lac operator sites (b) Looping out of p site

Out of the four subunits of the repressor protein, two subunits bind to o_1 and o_2 or o_3 and then bind to the other two subunits. The p site loops out and is not available for the binding of RNA polymerase (Figure 11.6(b)).

The operator sites are characterized as having palindromic sequences capable of binding to the DNA-binding sites of the subunits of the repressor molecules.

Any mutation within the operators will affect the binding of the repressor protein. A mutation in o_1 site prevents the repressor from binding and allows RNA polymerase to transcribe the lac operon constantly. This is called as **constitutive mutation**. Constitutive mutations in o_1 are referred to as *lac o*$_1^C$. In some other mutations in *o* site, repressor binds more strongly to *o* site and allolactose is not successful in removing the repressor protein from the operator. Under such conditions, even in the presence of lactose, transcription of *lac* genes is very much reduced.

Mutations in the lac structural genes   If *lac z* gene is mutated, the amino acid sequence of β-galactosidase is altered. The enzyme now loses its activity. However, we may expect the *lac* y and *lac* a genes to be transcribed and translated normally. But it is not so. A lac z⁻ mutation may eliminate production of β-galactoside permease and β-galactoside transacetylase as well as β-galactosidase. This is because in the absence of β-galactosidase, lactose is not converted to the true inducer allolactose and therefore the gene products of the other two genes, i.e., *lac* y and *lac* a are also not produced. This is an example of polar mutation, a mutation that affects the genes downstream from it as well as the gene in which it occurs.

If *lac* y gene is alone mutated, it may not produce functional β-galactoside permease. Therefore, lactose cannot enter into the cell and the cell cannot utilize lactose as an energy source. Though *lac z* gene is normal and is able to produce functional β-galactosidase, the phenotype of the cell is lac⁻ since it cannot utilize lactose.

Mutation in *lac* a gene may eliminate functional β-galactoside transacetylase. This enzyme is not essential for lactase metabolism. Cells with mutations in this gene are capable of utilizing lactose. Therefore *lac* a⁻ mutation does not cause lac⁻ phenotype.

If there is a frameshift mutation in the *lac* z gene, a termination codon is shifted out of frame and polypeptide chain synthesis crosses the *lac* y gene fraction. *lac* y gene is not translated since translation initiation site is blocked. This is also an example of polar mutation.

trans and cis effects of mutations in merozygotes   Francois Jacob and Jacques Monad developed the operon model by doing experiments with merozygotes which are bacterial cells with an extra copy of part of their genome present within the F factor. Certain terminology was used to describe the findings. *i, o, z, y* and *a* stand for *lac* i, o_1, *lac* z, *lac* y and *lac* a respectively. Superscript plus indicates their functional status. Superscript minus indicates their non-functional status. Superscript C designates a constitutive mutation. Since a merozygote has two copies of the entire lac operon, the genes on the F factor are separated from those on the chromosome by a slash. For example $F'i^+o^+z^+y^+a^+/i^+o^+z^-y^+a^+$ shows that F' plasmid is wild type and *lac* z^- gene is present in the chromosome in a merozygote (Figure 11.7).

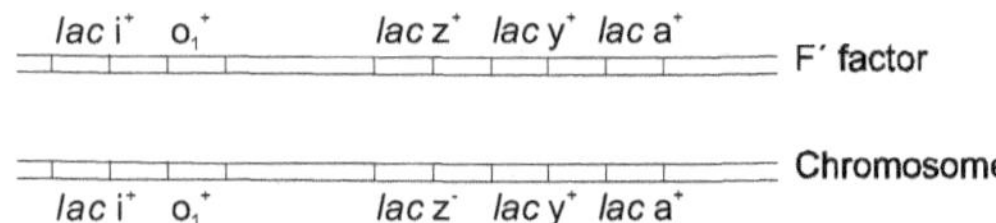

Figure 11.7 Representation of wild type F′ plasmid and lac z⁻ chromosome

Any gene or DNA sequence that affects the expression of a gene located on the same DNA molecule is called a *cis*-acting element.

A DNA sequence that affects genes on a different DNA molecule than the one on which it is located is called a *trans*-acting element (Figure 11.8).

Most *cis*-acting elements are not genes but are DNA sequences to which regulatory proteins bind. Most *trans*-acting elements are protein-encoding genes that regulate the expression of other genes.

A cell with the chromosomal genotype $i^-o^+z^+y^+a^+$ transcribes lac operon constitutively because there is no functional repressor protein to block transcription. If an F' factor with the genotype $F'i^+o^+z^-y^-a^-$ is introduced to make the merozygote $F'i^+o^+z^-y^-z^-/i^-o^+z^+y^+z^+$, the lac operon in the chromosome starts functioning normally, as if the chromosome contained a functional *lac* i gene. Here I⁺ in F' factor is *trans*-acting. In fact, *lac* i gene is both *cis*- and *trans*-acting because the repressor protein it encodes may bind to operators on the same DNA molecule as well as to operators on a different DNA molecule in the same cell.

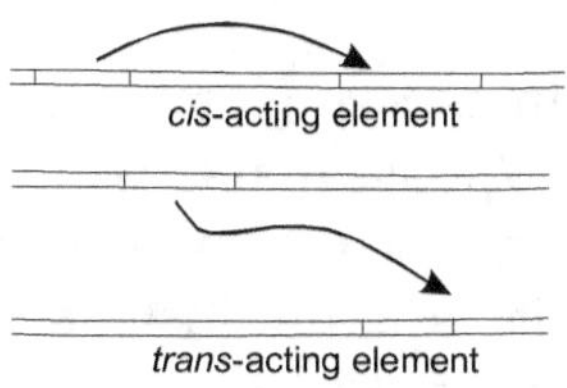

Figure 11.8   *cis* and *trans*-acting elements

Constitutive operator mutations are *cis*-acting. A *lac o*$_1^C$ mutation affects only the expression of genes on its own molecule. For example, the merozygote F'$i^+o^cz^-y^-a^-$/$i^+o^+z^+y^+a^+$ does not produce the enzymes encoded by the chromosomal lac operon constitutively. Instead, the chromosomal lac operon functions as an inducible operon. o^c constitutive mutation in F' factor has no effect on the chromosomal lac operon.

However, the merozygote F'$i^+o^+z^-y^-a^-$/$i^-o^cz^+y^+a^+$ produces functional enzymes constitutively because the o^c mutation is in the same DNA molecule as the functional enzyme groups.

Dominance of o^c mutations   A striking property of O^c mutations is that in certain cases, they are dominant. The expression of the partial diploids with different combination of i and o mutants is indicated in Table 11.3.

Table 11.3   Expression of partial diploids with various combinations of I and O mutants

Genotype	Constitutive (C) or inducible (I)
F'o^cz^+/o^+z^+	C
F'o^+z^+/o^cz^+	C
F'i^-z^+/i^{+z+}	I
F'i^+z^+/i^-z^+	I
F'o^cz^+/i^-z^+	C
F'o^cz^-/o^+z^+	I
F'o^cz^+/o^+z^-	C

Gratuitous Inducers

Lactose is rarely used in experiments to study the expression of *lac* genes, because β-galactosidase that is synthesized will act on lactose and reduce the level of lactose which may complicate the analysis of the functioning of the lac operon. Instead, two sulphur-containing analogues of lactose, isopropyl thiogalactoside (IPTG) and thiomethyl galactoside (TMG) are used (Figure 11.9), which are effective inducers without being substrates of β-galactosidase. Inducers having this property are called gratuitous inducers and this type of induction is called gratuitous induction.

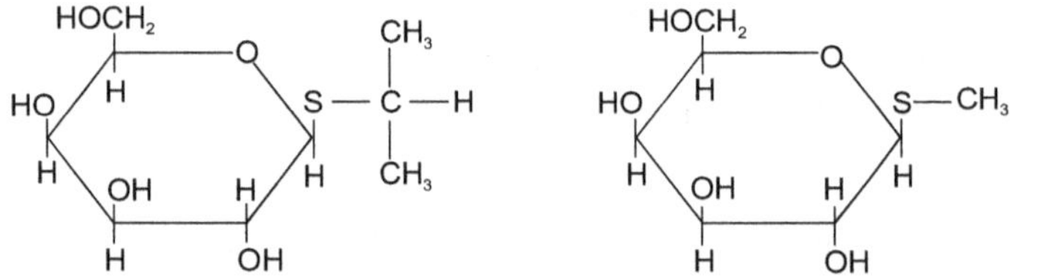

Isopropyl thiogalactoside (IPTG)   Thiomethyl galactoside (TMG)

Figure 11.9 Isopropyl thiogalactoside (IPTG) and thiomethyl galactoside (TMG)

Positive Regulation of lac Operon

If an *E. coli* cell is present in an environment with an ample supply of glucose and lactose, it would not be energetically efficient for the cell to be induced by lactose to produce the lac enzymes, because glucose is already present. Another molecule called catabolite activating protein (CAP) represses the expression of lac operon when glucose is present. This inhibition is referred to as **catabolite repression.**

RNA polymerase binding to the promoter site is enhanced by CAP. In the absence of glucose and under inducible conditions, CAP exerts positive control by binding to the CAP site, facilitating RNA polymerase binding at the promoter and promoting transcription. For maximal transcription, the repressor protein must be bound by lactose or allolactose, so as not to repress operon expression and CAP must be bound to the CAP binding site.

When glucose is present in the environment, CAP binding is inhibited. In order to bind to the promoter, CAP must be bound to another molecule—cyclic adenosine monophosphate (cAMP). The level of cAMP is dependent on the activity of the enzyme adenylate cyclase that converts ATP to cAMP (Figure 11.10). It is therefore clear that glucose inhibits the activity of adenylate cyclase, thereby causing a decline in the level of cAMP which is now not available to form a complex with CAP.

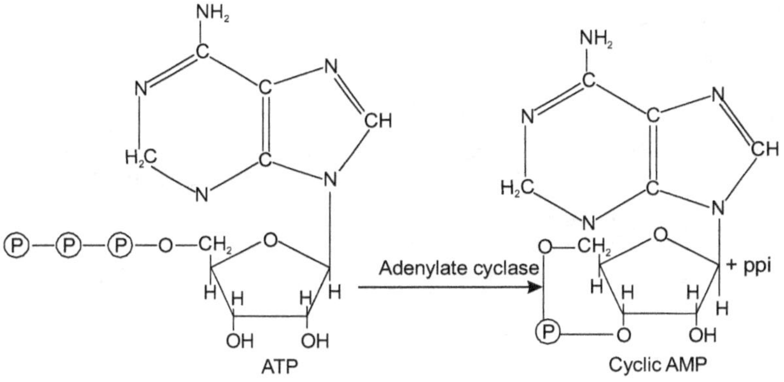

NH_2
CH
H_2C
P—P—P—O—CH_2
H H
H H
OH   OH
ATP
Adenylate cyclase
NH_2
CH
H_2C
O—CH_2
H H
H H
P—O   OH
+ ppi
Cyclic AMP

Figure 11.10 Formation of cAMP from ATP

CAP is a dimer which gets inserted into adjacent regions of a specific nucleotide sequence in the promoter site. Neither cAMP–CAP complex nor RNA polymerase has a strong affinity to bind to *lac* promoter DNA. Also they do not have a strong affinity to bind to one another. When both of them are present together in the presence of *lac* promoter DNA, a tightly bound complex is formed and this binding is called **cooperative binding**.

With both positive and negative regulation operating on it, lac operon is not transcribed in the absence of lactose regardless of glucose concentration (Figure 11.11a), because the repressor protein is bound to the operator. The lac operon is also not transcribed when glucose concentration is high, even if lactose is present, because CAP does not bind to DNA in the presence of glucose (Figure 11.11b). Therefore, for lac operon to be transcribed, lactose must be present and the concentration of glucose must below (Figure 11.11c).

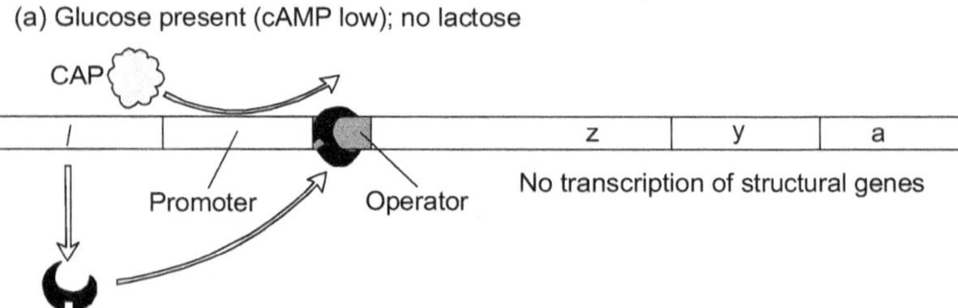

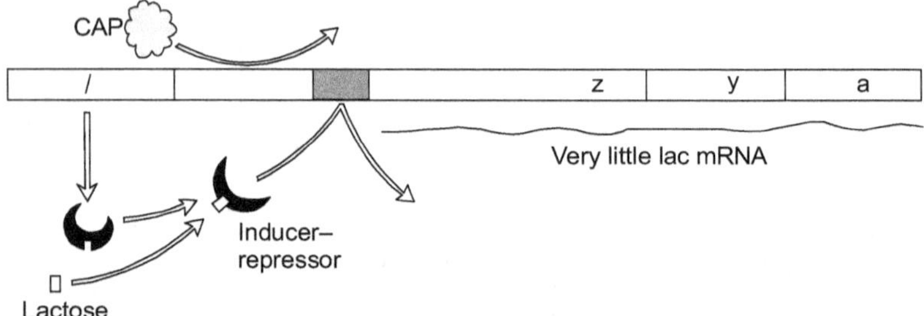

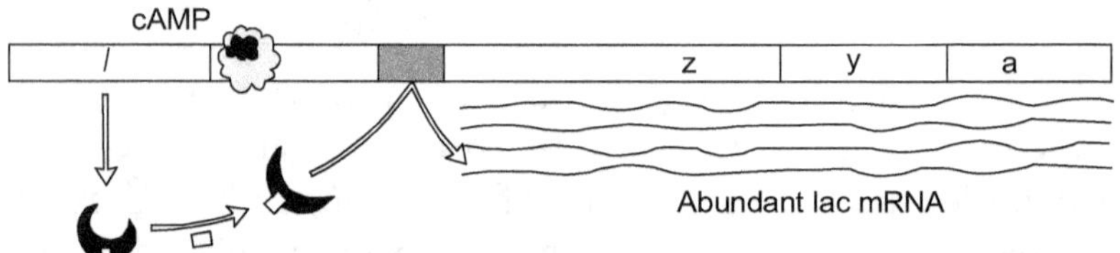

Figure 11.11 Summary of combined positive and negative regulation of the lac operon

ARABINOSE OPERON

Like lactose operon, arabinose operon is also an example for both positive and negative regulation. Ara operon is an inducible operon and is unique because the same regulatory protein is capable of exerting both positive or negative control. So, gene expression will be either induced or repressed.

The regulatory protein of the *ara*C gene acts as either an inducer (in the presence of arabinose) or a repressor (in the absence of arabinose).

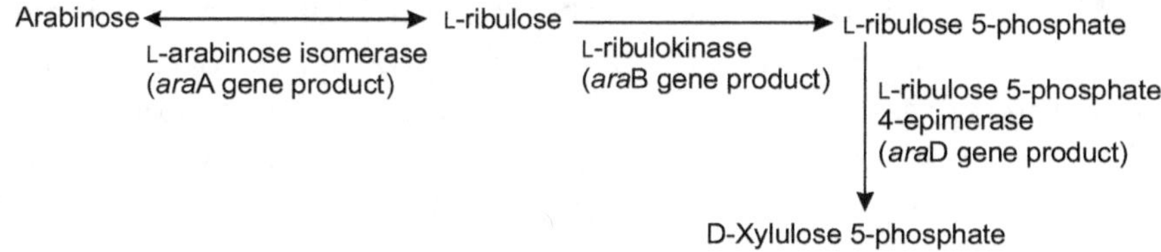

Figure 11.12   Utilization of arabinose

Arabinose operon consists of three structural genes, *ara*B, *ara*A and *ara*D whose gene products are three enzymes L-ribulokinase, L-arabinose isomerase and L-ribulose 5-phosphate 4-epimerase required for the utilization of arabinose. The pathway of arabinose breakdown using these three enzymes is indicated in Figure 11.12. The transcription of these three structural genes is controlled by a regulatory protein AraC encoded by the *ara*C gene, which interacts with two regulatory regions *ara*I and *ara*o$_2$, the inducer and operator sites respectively of the operon. An additional operator region o$_1$ is also present, but is not involved in the regulation of the *ara* structural genes (Figure 11.13(a)).

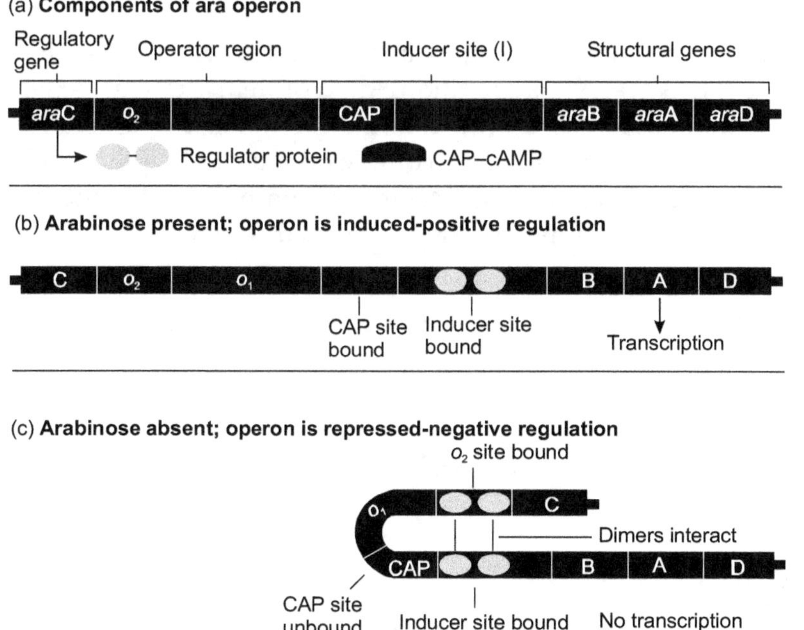

Figure 11.13   Genetic regulation of the ara operon

The ara i and ara o$_2$ sites can be bound individually or coordinately by the *Ara* C protein. In lac operon, the product of the regulatory gene, the repressor protein, functions in a negative manner, turning off transcription of the operon and the catabolite activator protein (CAP) exerts a positive control over the lac operon by stimulating transcription of the operon. In ara operon, the regulatory protein, *Ara* C, exhibits both negative and positive regulatory effects on the transcription of the structural genes of the operon depending on the environmental conditions.

The three structural genes of ara operon are cotranscribed as a single polycistronic mRNA initiated at a promoter (referred to as P$_{BAD}$) near the operator site. Active transport of arabinose into the cells is carried out by the products of genes *ara*E, *ara*F and *ara*G. These genes are present at sites away from ara operon. AraC, the regulatory protein, is produced from a transcript that is initiated by a promoter (referred to as P$_c$) adjacent to *ara*C gene. P$_c$ promoter is 100 nucleotide pairs away from P$_{BAD}$ and the two promoters initiate transcription in opposite directions, one transcribing *ara*C and the other *ara*B, *ara*A and *ara*D genes.

In the presence of arabinose and cAMP, the ara operon is induced (Figure 11.13(b)). In this condition, Ara C protein acts as an activator of transcription. The Ara C protein forms a complex with arabinose. This complex and the cAMP–CAP complex open the loop by binding at their *ara*I sites. This permits RNA polymerase to bind at the P$_{BAD}$ site and initiate transcription of the ara structural genes.

In the absence of arabinose and cAMP, AraC protein acts as a negative regulator (a repressor) preventing the expression of the structural genes (Figure 11.13(c)). As a dimer, the AraC protein binds at both the *ara* i site and *ara* o$_2$ site. These two dimers then bind to each other to form a DNA loop. When the loop structure is formed, it prevents the binding of RNA polymerase at the adjacent promoter P$_{BAD}$ of the operon.

GALACTOSE OPERON

Galactose is another sugar, used by bacterial cells as a carbon source. The sugar is metabolized by three enzymes, galactokinase, galactose transferase and galactose epimerase (Figure 11.14).

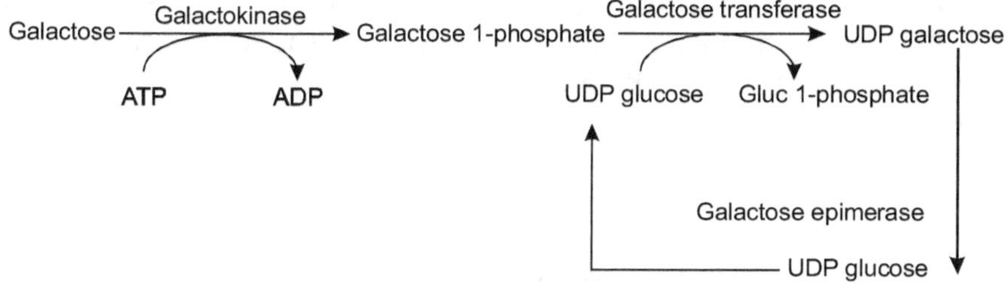

Figure 11.14   Catabolism of galactose

The regulation of the gal operon, like the lac operon, involves a repressor and cAMP. But unlike lac operon which exhibits turn on-off regulation, gal operon exhibits two states of regulation namely turn on at high level and turn on at low level.

The three structural genes of gal operon are gal *K*, gal *T* and gal *E* for the three galactose-metabolizing enzymes. The operon includes an operator (*gal* O), promoter (*gal* P) and a repressor gene (*gal* R). One difference from lac operon is that the repressor gene in gal operon is very far from the cluster of o site, p site and the three structural genes. The second difference is that there are two operators—*gal* O$_e$ present upstream and adjacent to *gal* P (as in lac operon), and the other, *gal* O$_i$, 95 base pairs downstream from *gal* O$_e$ and within the *gal* E gene (Figure 11.15(a)).

In the presence of glucose, galactose operon is expected to be non-inducible. However, it is inducible when glucose is present. Gal operon is found to have two promoters. The mRNA molecules synthesized from these two promoters differ in length by only 5 nucleotides. Both transcripts contain the information of all the three structural genes, *gal* K, *gal* T and *gal* E and are identical. There is difference only with respect to the base sequences of the promoter–operator region. This shows that the two mRNA molecules have different start sites designated as S1 and S2.

Transcription from S1 occurs only when glucose is absent or only if cAMP–CRP is present. Transcription from S2 occurs when glucose is present. cAMP–CRP is not required for initiation at S2. Actually cAMP–CRP inhibits transcription from S2. This proves that glucose does not inhibit induction of gal operon because one of the start sites does not require cAMP–CRP and thus remains active even though glucose is present.

Two cAMP–CRP binding sites have been identified, one from –25 to –50 and the other from –50 to –68. Binding to the second site requires binding of cAMP–CRP and RNA polymerase to the first site (Figure 11.15(b)).

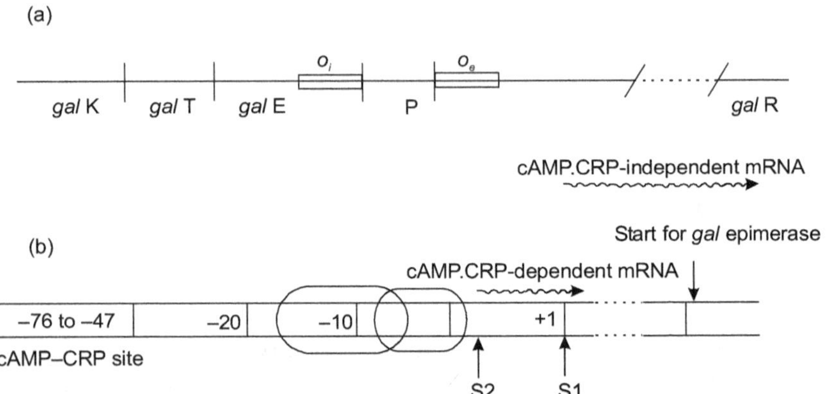

Figure 11.15   The operator–promoter region of the gal operon

The reason for having two transcription initiation sites in gal operon is based on the dual role of galactose in cellular mechanism. Galactose not only serves as a carbon source. It is required as the precursor molecule, UDP galactose, in the synthesis of *E. coli* cell wall. If exogenous galactose is not available, UDP galactose is formed from UDP glucose by the enzyme galactose epimerase, the product of *gal* E gene. This means that epimerase enzyme should be synthesized always. Another *gal* E gene is present away from gal operon. Alternatively, background constitutive synthesis of this enzyme in small amounts is required.

If S1 was the only promoter, since its activity depends on cAMP–CRP, epimerase could not be made constitutively, when glucose is present. If S2 was the only promoter, galactose may fully de-repress the operon even when glucose is present which will be wasteful. Thus, for the sake of both necessity and economy, a cAMP–CRP-independent S2 promoter is needed for the background constitutive synthesis and a cAMP–CRP-dependent promoter S1 is needed to regulate high-level synthesis.

TRYPTOPHAN OPERON

The tryptophan operon is required for the synthesis of the amino acid tryptophan. It is an example of a negatively controlled repressible operon.

A negatively controlled operon is one that is inhibited by a regulatory protein such as a repressor. A repressible operon is one that is inhibited when an effector molecule is present.

Tryptophan operon has five structural genes: *trp* E, *trp* D, *trp* C, *trp* B and *trp* A that code for enzymes involved in the conversion of chorismic acid to tryptophan (Figure 11.16).

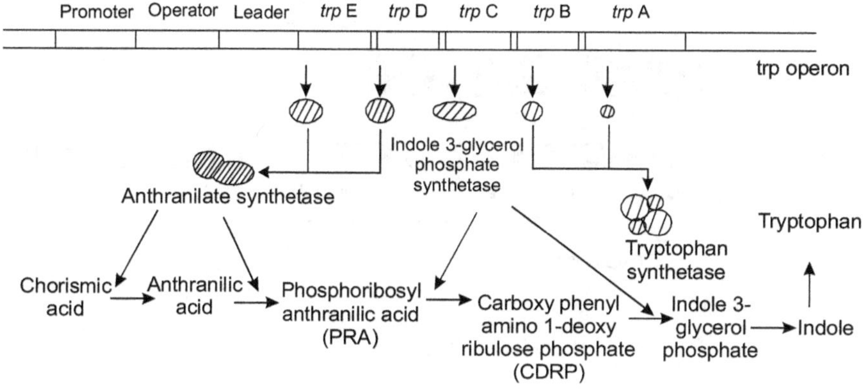

Figure **11.16**   The *trp* operon

The controlling sites lie next to *trp* E and consist of a promoter site *trp* P, an overlapping operator site, *trp* O and a leader region *trp* L that codes for a leader peptide and for an RNA attenuator (Figure 11.17(a)).

The enzymes that synthesize tryptophan are produced only when tryptophan is not readily available in the cell's environment. Therefore, tryptophan is the regulatory factor and is also the product of the pathway. Thus, unlike lactose, which induces transcription of the lac operon, tryptophan represses transcription of trp operon.

This clearly describes the two types of operons—inducible operons and repressible operons. For inducible operons (e.g. lac operon), high concentration of the substrate of the biochemical pathway induces transcription. For repressible operons (e.g. trp operon) high concentration of the product of the pathway represses transcription. However, lac operon is both inducible and repressible, because high concentration of the substrate (lactose) induces transcription and high concentration of the product (glucose) represses transcription.

Like lac operon, trp operon is regulated by a repressor protein, an inactive regulatory protein referred to as aporepressor that binds to the operator. The trp repressor protein is encoded by *trp* R located far from trp operon on the chromosome. When tryptophan concentration is high, trp repressor protein binds to DNA and represses transcription. When tryptophan concentration falls, the repressor protein comes out of the DNA-binding site and permits transcription of the trp operon (Figure 11.17(b) and (c)).

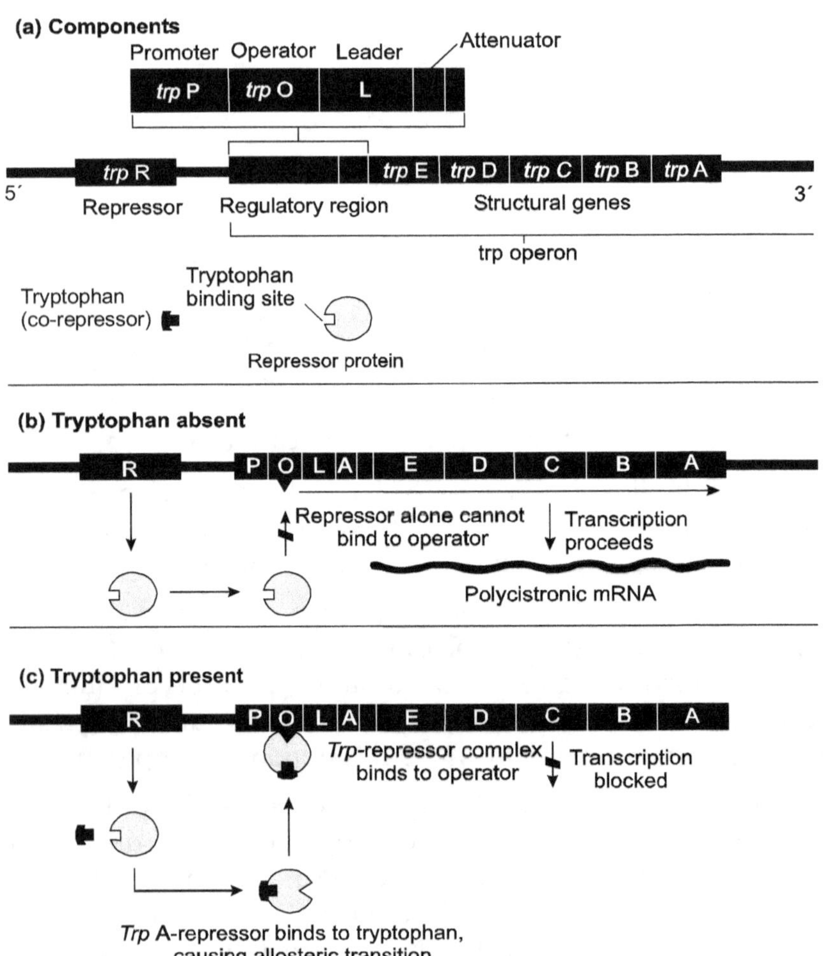

Figure 11.17　Regulation of the tryptophan operon

The repressor protein has a domain that is specific for tryptophan binding. When tryptophan is present in the cell, it binds to the repressor protein, and brings a conformational change in the repressor protein that increases its affinity for trp operator. Therefore, increased level of tryptophan turns off transcription of trp operon. Hence, tryptophan is referred to as a co-repressor, which converts the inactive **aporepressor** to active repressor called **holorepressor**.

In the presence of tryptophan-repressor protein complex, transcription of trp operon is reduced about 70-fold. When the level of tryptophan comes down, tryptophan separates from the repressor protein, and the repressor protein comes out of the operator, thereby promoting transcription.

Trp repressor protein acts not only on trp operon. First of all, it regulates its own synthesis. When the level of trp repressor protein increases, it binds to an operator site adjacent to the promoter of *trp* R gene (that transcribes and translates the repressor protein) and represses transcription. When the level of trp repressor protein comes down, the protein leaves the operator site and allows transcription of *trp* R gene. This form of regulation, where the product of a gene regulates the gene itself is called **autogenous regulation**.

Along with the repressor protein, another mechanism called attenuation also reduces transcription of the trp operon. A 70-fold reduction in transcription is caused by repression and a 10-fold reduction by attenuation. Together a 700-fold reduction is caused in presence of tryptophan.

Charles Yanofsky, Kevin Bertrand and co-workers observed that even when tryptophan is present and trp operon is repressed, initiation of transcription of the leader sequence occurs. But that mRNA synthesis is usually terminated at a point about 140 nucleotides along the transcript. This process is called **attenuation**. This effect severely diminishes the genetic expression of the operon.

In the absence of tryptophan or in the presence of tryptophan in low concentration, the repressor is inactive and does not bind to the operator. Transcription is initiated, but is not terminated and therefore proceeds through the leader sequence and the structural genes. As a result, a polycistronic mRNA is got and this is translated to give the enzymes involved in tryptophan biosynthesis.

In the leader sequence is a segment of DNA that, when transcribed, gives rise to an RNA molecule that immediately folds back on itself and forms a hairpin loop. The loop region is followed by a poly(U) sequence, similar to the poly(U) sequence found usually in the 3' end of termination region of many prokaryotic RNA transcripts. In contrast to the termination of most regular transcripts, attenuation does not require the rho termination factor. Attenuation is a *rho*-independent process.

In the absence (or low concentration) of tryptophan, attenuation is bypassed. The leader sequence that is transcribed must be translated to form the hairpin. The leader transcript has two UGG triplets encoding tryptophan. Even though this leader sequence is not a part of the structural genes, following its transcription, an initial AUG sequence prompts translation by ribosomes. When there is enough level of tryptophan, charged tRNA^{trp} is also present. As a result, translation proceeds and the termination hairpin is formed. If cells are starved of tryptophan, charged tRNA^{trp} is unavailable and the ribosome is blocked during translation of triplets that request for charged tRNA^{trp}. Therefore, the termination hairpin structure is not formed in the leader transcript and attenuation is overcome and transcription and translation of the entire set of structural genes proceeds.

The phenomenon of the attenuation appears to be a mechanism common to other bacterial operons that regulate the enzymes essential to the biosynthesis of amino acids. In addition to tryptophan, operons involved in threonine, histidine, leucine and phenylalanine display attenuators in their leader sequences. As in tryptophan operon, each of these amino acid operons has multiple codons for these amino acids for regulation, which prompt "stalling" of translation if the respective amino acid is missing. For example, the leader sequence in the histidine operon codes for seven histidine residues in a row. The threonine operon leader sequence calls for eight threonine residues. When these amino acids are present, stalling does not occur, a hairpin structure is formed and attenuation occurs.

The mechanism of attenuation in trp operon is shown in Figure 11.18.

(a) Selected regions of leader transcript and potential base pairing

(b) High concentrations of tryptophan: Attenuation occurs

(c) Low concentrations of tryptophan: Attenuation is overcome

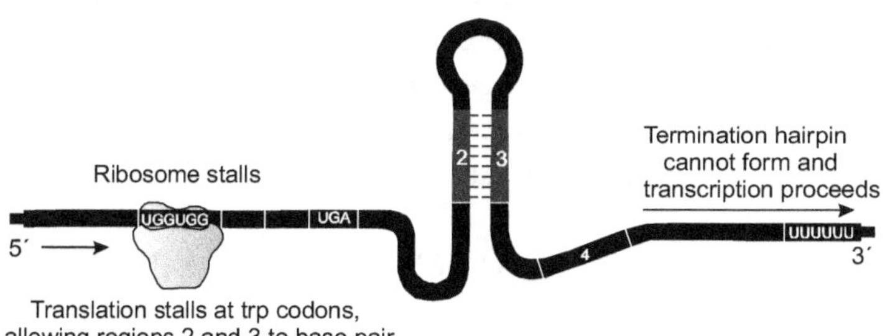

Figure 11.18 Involvement of the leader sequence of the mRNA transcript of the trp operon of *E. coli* during attenuation

REGULATION OF GENE EXPRESSION IN PHAGES

Phage genes are subjected to the same types of regulation as the genes in the bacterial chromosome. The best studied case of phage gene regulation is the competition between lysis and lysogeny in *E. coli* cells that are infected with the temperate phage lambda λ.

When wild type λ phage particles infect *E. coli* cells on a culture plate, turbid plaques appear, instead of clear plaques indicating that some cells are lysed and others are lysogenic. Because both lysis and lysogeny take place in the same plaque, it is clear that there is some regulatory mechanism to determine the choice between lysis and lysogeny.

The phage DNA has a gene *cI* which encodes a repressor protein called pCI. This protein represses transcription of the genes for the lytic pathway, resulting in lysogeny. When *cI* gene is not expressed, pcI repressor protein is not produced and the genes of the lytic pathway are expressed, resulting in lysis. pcI binds to operators and exerts negative regulation of the operons that initiate the lytic pathway.

The name cI stands for "clear I", because mutations in *cI* gene cause clear, rather than turbid, plaques. This indicates that all the cells within the plaque have lysed and none are lysogenic. Lambda phages that are mutant for cI cannot enter lysogeny because the lytic pathway is not repressed.

When λ phage first enters a cell, the *cI* gene is not immediately expressed. If it were, then lysogeny would be immediately established in all the infected cells. In some cells, lysis occurs and new phage particles are produced to infect other cells. In other cells, lysogeny occurs and the cells do not die. When both lysis and lysogeny occur in a bacterial colony, a turbid plaque appears.

EXPRESSION OF LYSIS AND LYSOGENY IN LAMBDA PHAGE

When λ phage infects a cell, the first proteins produced are pN, pCro, pCIII, pCII and pCI.

The *cI* gene is flanked on either side by promoter–operator complexes. On the left side, there is a single operator O_L and a single promoter P_L. On the right side is the *cI* gene's own promoter P_M and three other promoters OR^3, OR^2 and OR^1 followed by another promoter P$_R$. L, R and M stand for left, right and maintenance (Figure 11.19). P$_L$ and P$_R$ are the promoters for genes that initiate the lytic pathway as well as genes that stimulate transcription of *cI* gene to initiate lysogeny. The genes that succeed in being expressed first determine whether lysis or lysogeny is established in the cell.

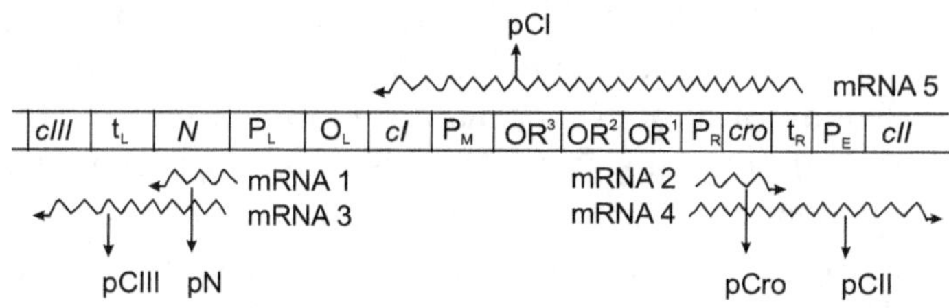

Figure 11.19   Expression of lysis and lysogeny in lambda phage

When lambda first enters the cell, its DNA gets circularized and the bacterial RNA polymerase binds to P$_L$ and P$_R$. Transcription is initiated and *N* gene is transcribed on the left side and *cro* gene on the right side. Both the genes have transcription termination at their ends t$_L$ and t$_R$, thus mRNA$_1$ and mRNA$_2$ are got. *N* gene product, pN is an antiterminator protein allowing transcription to proceed crossing t$_L$ and t$_R$ terminator. Thus mRNA$_3$ and mRNA$_4$ are got whose products are pCII and pCIII (p stands for protein). These two proteins are essential for lysogeny maintenance. The cII protein, pCII is very unstable and can be readily degraded by a protein called HflA encoded by a bacterial gene. The *cIII* gene product, pCIII protects pCII from degradation by HflA. pCII assists RNA polymerase in binding to another promoter, P$_E$ where E stands for establishment of lysogeny. Only if pCII is produced, this promoter becomes active. RNA polymerase transcribes a long mRNA$_5$ passing through *cro* gene, PR, OR^1, OR^2, OR^3, P$_M$ and *cI* genes. But only one gene is transcribed in this process, i.e., *cI*. Therefore, the product of mRNA$_5$ is only pCI protein.

Once pCI is produced, it binds to O$_L$ and the three O$_R$ and blocks RNA polymerase from binding to P$_R$ and P$_L$. Thus, *cro*, *N*, *cII* and *cIII* genes are now not transcribed. Another gene, int, codes for an enzyme called integrase. Once repression of P$_R$ and P$_L$ is established, *int* gene is expressed and integrase is produced which helps in the integration of phage DNA into the host chromosome. Thus lysogeny is established. For the continuous maintenance of lysogeny, pCI must be continuously produced. If pCI production is stopped, their binding to P_R and P_L will not occur and these promoters will become active once again and the cells will enter lytic pathway.

There should be some mechanism that regulates the production of pCI protein. pCI itself provides both positive and negative regulation of the *cI* gene. pCI has greater affinity for OR^1 and OR^2, than for OR^3. If there is less amount of pCI available, it binds preferentially to OR^1 and OR^2 and OR^3 remains vacant. In this situation, pCI prevents RNA polymerase from binding to P_R, which prevents transcription of *cro* and other genes downstream. Also pCI activates RNA polymerase bound to P_M, stimulating transcription of the *cI* gene and continued production of pCI.

When the amount of pCI is high, OR^3 also is occupied by pCI. This inhibits RNA polymerase from binding to P_M, stopping the transcription of *cI* gene. This is an example of autogenous regulation, where the product of a gene regulates its own production. This type of regulation ensures that the correct amount of pCI is constantly available for the maintenance of lysogeny.

Simultaneously we have to consider the effect of *cro* gene product also. pCro protein binds to the same operators as pCI. But pCro has greatest affinity for OR^3 and inhibits the transcription of *cI* gene. If pCI is not produced, lysis is favoured. If the level of pCro is increased, it binds also to OR^1 and OR^2 and also to O_L. This reduces transcription of genes under the regulation of these operators, including *cII*. Therefore lysogeny cannot be initiated or reinitiated. pCro also regulates its own production by autogenous regulation. If pCro succeeds in binding to the operators against pCI, lysis occurs. If pCI succeeds in binding to the operators against pCro, lysogeny results.

When the lysogenic cells are in a turbid plaque, they are surrounded by λ phage particles from their lysed neighbour. The λ particles do not simply infect the lysogenic cells and make them enter into lytic pathway because the lysogenic cells are immune to lysis by further infection. This immunity is due to the presence of pCI in lysogenic cells. λ DNA that enters a lysogenic cell is immediately inactivated because, pCI produced by the prophage already present in the cell binds to OR^1, OR^2 and O$_L$ on the infecting DNA. The infecting DNA has no chance to express any gene but *cI*. So lysis is immediately prevented. pCI is therefore behaving as a *trans*-acting protein.

When the cell is exposed to dangerous environment like ultraviolet radiation, induction occurs, when an integrated λ prophage excises itself from the chromosome and enters the lytic pathway. When the bacterium is exposed to UV rays, rec A protein is activated and acquires a proteolytic activity, causing the cleavage of certain proteins including pCI. When pCI is cleaved, it becomes inactive and cannot bind to the operators. Now pCro is free to bind to OR^3 and initiates lytic pathway.

Regulation of gene expression during the lytic cycle of bacteriophages is quite different from the operon concept of regulation. Viral genes are expressed in genetically preprogrammed 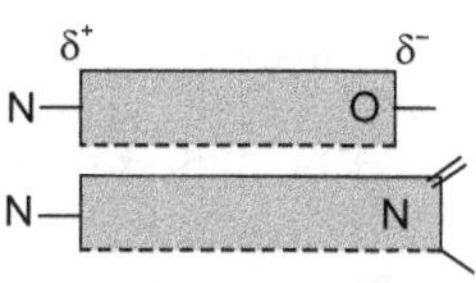 sequences as in higher organisms. One set of phage genes called "early genes" is expressed immediately after infection. The products of these early genes turn off the expression of the early genes and turn on the expression of the next set of genes and so on. Two to four sets of such genes are involved depending on the virus. This regulation of sequential gene expression during phage infection occurs primarily at the transcriptional level.

Spicy Questions and Answers

1. **How will you find out whether the genes for the utilization of arabinose are under inducible control in *E. coli*?**

    Inducible means the genes that produce enzymes required for the sugar arabinose utilization are not active in the absence of the inducer, arabinose. On adding arabinose, the enzymes are produced. Therefore the contents of the cells have to be assayed before and after adding arabinose to the medium. Estimation of arabinose will help us to find that the bacterial cell is incapable of metabolizing arabinose before induction but capable of metabolizing it afterward. If the cells were capable of metabolizing arabinose in both cases, we can say that arabinose utilization is constitutive. If the cells were not capable of utilizing arabinose in both cases, we can conclude that the bacterium is incapable of using the sugar arabinose as an energy source.

2. **What are the phenotypes of the following partial diploids for the lac operon in *E. coli* in the presence and absence of lactose?**

(a) **(F′)i$^+$o$^+$p$^+$z$^-$ / i$^-$o$^+$p$^+$z$^+$ (chromosome)**

(b) **(F′)i$^+$o^{c}p$^+$z$^+$ /i$^+$o$^+$p$^-$z$^+$ (chromosome)**

(a) Functional repressor in the plasmid (i$^+$) will bind to both DNAs and hence transcription does not occur in the absence of lactose and will be induced in presence of lactose.

(b) F′plasmid always will transcribe, because the repressor cannot bind to the operator. But chromosomal DNA can never make RNA because it is p$^-$.

3. **Differentiate between positive and negative control.**

In negative control, the regulatory molecule interferes with transcription while in positive control, the regulatory molecule stimulates transcription.

4. **For the following *lac* genotypes, predict whether the structural genes are constitutive, permanently repressed, or inducible in the presence of lactose:**

i$^+$o$^+$z$^+$

i$^-$o$^+$z$^+$

i$^+$o^{c}z$^+$

i$^-$o$^+$z$^+$/F′I$^+$

i$^+$o^{c}z$^+$/F′o$^+$

i^{s}o$^+$z$^+$

i^{s}o$^+$z$^+$/F′i$^+$

inducible, constitutive, constitutive, inducible, constitutive, repressed, repressed.

5. **Predict the levels of gene activity of lac operon, lac repressor status and CAP protein status under the following cellular conditions:**

	Lactose	Glucose
(a)	–	–
(b)	+	–
(c)	–	+
(d)	+	+

(a) Operon is off because the repressor is bound to the operator and also even if CAP is bound to its binding site, it will not over ride the action of the repressor.

(b) Repressor is now bound to lactose and the operon is on and also with no glucose CAP is bound to its binding site, thus enhancing transcription.

(c) Repressor is bound to the operator and since glucose inhibits adenylate cyclase, CAP protein will not interact with its binding site. Lac operon is therefore turned off.

(d) Though lac repressor is not bound to the operator site, because glucose is also present, CAP will not interact with its binding site. So, *lac* genes are not expressed.

6. How can inducible and repressible enzymes be distinguished?

They can be distinguished by studying the synthesis or lack of synthesis of the enzyme in cells grown on chemically defined media. If the enzyme is synthesized only in the presence of a metabolite, or a set of metabolites, it is inducible. If it is synthesized in the absence of or not in the presence of a particular metabolite or set of metabolites, it is said to be repressible. Induction is characteristic of catabolic or degradative. Repression is characteristic of anabolic or biosynthetic pathway.

7. What is the biological significance of catabolite repression?

It refers to the use of glucose as the sole source of carbon when this sugar is available even when other less efficient carbon sources are present.

Review Questions

1. With suitable examples, distinguish between regulation based on negative mechanism and that based on positive mechanism.

2. Justify the statement that lac operon is an example for both positive and negative regulation.

3. Explain the different levels of gene expression in prokaryotes.

4. Define an operon and explain the different elements in an operon.

5. Write notes on

   i. Catabolite repression

   ii. Gratuitous induction

6. Discuss the regulation of expression of lysis and lysogeny in lambda phage.

7. Demonstrate the mechanism of positive and negative regulation of arabinose operon.

8. How does regulation of gal operon differ from lac operon?

9. A repressible operon is inhibited when an effector molecule is present. Explain this with the example of trp operon.

10. Differentiate between repressed state and derepressed state of lac operon.

11. What are *cis*-acting and *trans*-acting elements? What is their role in the regulation of gene expression?

12. Differentiate between inducible operons and repressible operons with suitable examples.

13. Comment on the various mutations possible in lac operon. Add a note on their consequences.

12

REGULATION OF GENE EXPRESSION IN EUKARYOTES

Multicellular eukaryotes exhibit differential regulation of gene expression. For example, the pancreatic cells produce insulin which is not produced by the retinal cells and the retinal cells produce retinal pigment which is not produced by the pancreatic cells. Thus eukaryotes express a subset of genes in one cell type and a different subset of genes in another cell type. This does not mean that, this regulation is accomplished by eliminating unused genetic material. Actually, there are several mechanisms which activate specific portions of the genome and repress the expression of other genes. The expression of a gene at the wrong time, in the wrong cell type or in abnormal amounts may lead to deleterious phenotypes or even death.

The main reasons for the difference in the regulation of gene expression between prokaryotes and eukaryotes are as follows:

1. The eukaryotic genome is highly complex. The DNA that has the genetic information is present associated with histones to form chromatin. Chromatin may be in condensed or extended form to form heterochromatin or euchromatin.

2. The genetic information is present in a single chromosome in prokaryotes whereas in eukaryotes there are many chromosomes present enclosed within the nucleus.

3. In eukaryotes, there is compartmentalization of transcription and translation because of the presence of the nucleus. Because of this, as in prokaryotes, attenuation control is not possible.

4. The entire post-transcriptional modification occurs in the nucleus in eukaryotes before getting transported to the cytoplasm.

5. Eukaryotic mRNA molecules have longer half-life. In prokaryotes, when sufficient copies of a protein are synthesized, transcription is stopped and mRNA gets degraded.

6. Because eukaryotic mRNA is highly stable, they also have translation control level.

7. Even though eukaryotic cells contain a complete set of genes, different cell types use different sets of overlapping genes to make different proteins.

8. Introns are present in eukaryotic genes. A large fraction of DNA nucleotides are untranslated.

LEVELS OF REGULATION OF EUKARYOTIC GENE EXPRESSION

There are six levels at which eukaryotic gene expression is regulated (Figure 12.1).

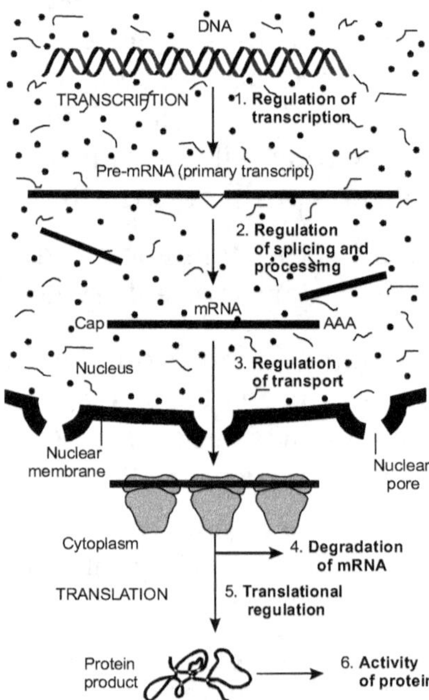

Figure 12.1   Levels of regulation of eukaryotic gene expression

1. Control at transcriptional level
2. Control at post-transcriptional level
3. Regulation of transport of mRNA from the nucleus to the cytoplasm
4. Stability of mRNA molecules
5. Control at translational level
6. Control at the level of post-translational modification

EUKARYOTIC CHROMOSOMES AND REGULATION OF GENE EXPRESSION

In chromosomal regions that have been activated for transcription, a variety of structural changes can be seen. The most obvious is an increased sensitivity of the DNA to nuclease-mediated degradation. The nucleosomes usually comprising of about 200 base pairs are unaffected by the nucleases. The hypersensitive sites found on or away from the 5' end of the transcribed gene or near the 3' end or even within the gene are highly sensitive to DNases. These sites are also the binding sites for known regulatory proteins. The relative absence of nucleosomes in these regions may facilitate binding of the regulatory proteins.

Transcriptionally active chromatin tends to be deficient in histone H1 and the other core histones have a greater tendency to be modified by acetylation or by the attachment of ubiquitin.

Transcription of a eukaryotic gene is strongly repressed when its DNA is condensed within chromatin. The DNA containing the β-globin gene cluster is in active chromatin in the reticulocyte but in inactive chromatin in muscle cells. The presence of nucleosomes and histone complexes with DNA strands provides a barrier against the ready association of transcription factors with specific DNA regions. The dynamics of the formation and disruption of nucleosome structure are therefore an important feature of regulation of eukaryotic gene expression. Transcription-associated structural changes take place in chromatin, which is referred to as **chromatin remodelling**. These include the action of some enzymes involved in acetylation and deacetylation of the histones in the nucleosomes and use of chemical energy of ATP to remodel the nucleosomes.

Each of the core histones, H2A, H2B, H3 and H4 has two distinct structural domains. A central domain is involved in histone–histone interaction and the wrapping of DNA around the nucleosome. A second, lysine-rich amino-terminal domain is positioned near the exterior of the assembled nucleosome particle. The lysine residues are acetylated by histone acetyl transferases (HAT). Cytosolic HATs are referred to as type A enzymes which acetylate newly synthesized histones before the histones are transported into the nucleus. The subsequent assembly of the histones into chromatin is facilitated by additional proteins: CAF1 for H3 and H4 and NAP1 for H2A and H2B.

When chromatin is being activated for transcription, the nucleosomal histones are further acetylated by the nuclear HATs referred to as type B enzymes. Acetylation of multiple lysine residues in the amino-terminal domains of histones H3 and H4 can reduce the affinity of the entire nucleosome for DNA. Acetylation may also prevent or promote interactions with other proteins involved in transcription or its regulation. When transcription of a gene is no longer required, the acetylation of nucleosomes in that vicinity is reduced by histone deacetylases in a gene-silencing process that restores the chromatin to a transcriptionally inactive state.

BOX 12.1   RNA SILENCING

RNA silencing refers to gene silencing effects by which the expression of one or more genes is down-regulated or entirely suppressed by the introduction of an antisense RNA molecule. An example is RNA interference in which endogenously expressed microRNA or exogenously derived small interfering RNA includes the degradation of complementary messenger RNA.

MicroRNA (miRNA) are single stranded RNA of 21–23 nucleotides in length, encoded by genes that are transcribed from DNA but not translated into protein. They are processed from primary transcripts pri-miRNA to short stem-loop structures called pre-miRNA and finally to functional miRNA. Short interfering RNA (siRNA) or silencing RNAs are double-stranded 20–25 nucleotide pair long molecules. A third class of RNA called Piwi-interacting RNA (piRNA) is expressed in mammalian testes and somatic cells. They form complex with Piwi proteins (a regulatory protein responsible for maintaining incomplete differentiation in stem cells and stability of cell division rate). piRNAs have been linked to transcriptional gene silencing of retrotransposons and other genetic elements in spermatogenesis.

Chromatin remodelling also requires protein complexes that actively move or displace nucleosomes and they hydrolyse ATP in this process. An enzyme complex called SW1/SNF (protein complex involved in switching of mating type and sucrose nonfermentation) found in eukaryotic cells contains at least 11 polypeptides which together create hypersensitive sites in the chromatin and stimulate the binding of transcription factors. NURF is another ATP-dependent enzyme complex that remodels chromatin along with SW1/SNF.

REGULATION OF GENE EXPRESSION AT TRANSCRIPTIONAL LEVEL

The Promoter Regions in Eukaryotes

The promoters consist of nucleotide sequences that serve as the recognition regions for the binding of RNA polymerase. They are located immediately adjacent to the genes they regulate and are considered as a part of the gene.

Eukaryotic promoters require the binding of a number of protein factors to initiate transcription. Both positive and negative regulatory elements are found in eukaryotic cells. But positive mechanisms predominate in most of the systems. Every eukaryotic gene requires some sort of activation to be transcribed.

The signals in DNA which control transcription are of several types. Two types of sequence elements are promoter-proximal. One of these defines where transcription is to commence along the DNA and the other determines how frequently this event should occur. Most mammalian genes have a TATA box that is usually located 15–30 bp upstream from the transcription start site. The TATA box binds a 30-kDa TATA binding protein (TBP), which in turn binds several other proteins called TBP-associated factors (TAFs). The complex of TBP and TAF is referred to as TFIID. Binding of TFIID to the TATA box sequence marks the first step in the formation of the transcription complex on the promoter. Few genes lack a TATA box when an initiator sequence (Inr) directs RNA polymerase II to the promoter and in so doing provides basal transcription starting from the correct site. The proteins that bind to Inr to direct pol II binding include TFIID and one or more TAFs. Promoters that have both a TATA box and an Inr may be stronger than those with one of these elements.

Sequences further upstream from the start site determine how frequently the transcription event occurs. These are the GC and CAAT boxes. Each of these boxes binds a protein, Spl (rich in glutamine) in the case of the GC box and CTF (rich in proline) by the CAAT box. The frequency of transcription initiation is a consequence of these protein–DNA interactions, whereas the protein–DNA interaction at the TATA box ensures the fidelity of initiation.

Role of Transcription Factors and Formation of Transcription Complex

Eukaryotic transcription is mediated by three classes of RNA polymerases—I, II and III—which require the presence of specific proteins, the transcription factors. Three classes of transcription factors are involved in class II gene transcription—those transcribed by pol II to make mRNA.

The first class of transcription factors, TF A, B, D, E, F and H are required for basal transcription and are therefore considered as **basal elements**. Some of these factors are composed of many subunits. These general transcription factors (GTFs) are abbreviated as TFIIA, TFIIB, etc. As mentioned earlier, TFIID is the only factor among them capable of binding to the TATA box, which has TBP and TAF. TFIID directs the assembly of the other components by protein–DNA and protein–protein interactions.

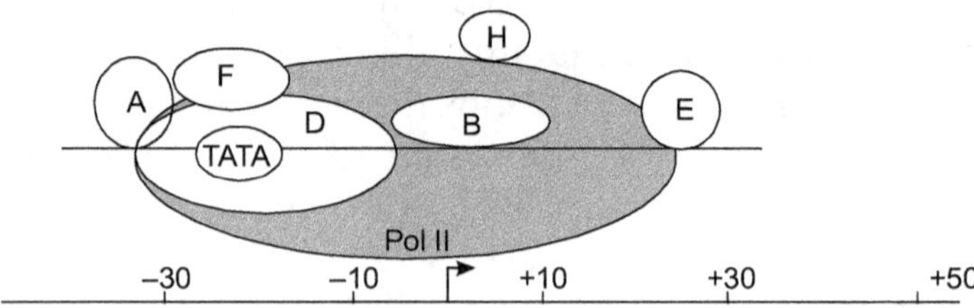

Figure 12.2 The eukaryotic basal transcription complex

TBP binds to the TATA box in the minor groove of DNA. Most transcription factors bind in the major groove. TBP is also an important component of class I and class III initiation complexes even if they do not contain TATA boxes. Among the other factors, first TFIIB binds to TFIID/promoter complex. Once a stable ternary complex consisting of

TFIIB, TFIID and promoter is formed, the pol II / TFIIF complex binds to the promoter followed by TFIIA, TFIIE and TFIIH.

The complex consisting of RNA polymerase II and the different transcription factors at the promoter site is shown in Figure 12.2.

Eukaryotic pol II consists of 14 subunits. It has a series of heptad repeats of the sequence tyr-ser-pro-thr-ser-pro-ser at the carboxyl terminal of the largest pol II subunit. This carboxyl terminal repeat domain (CTD) has 26 repeated units in yeast and 52 units in mammalian cells. CTD is a substrate for several kinases. It is the binding site for a group of proteins called Srb or mediator proteins. Pol II is activated when serine and threonine residues in CTD are phosphorylated and inactivated when CTD is dephosphorylated. If pol II lacks CTD tail, it cannot activate transcription.

TBP supports basal transcription but not the augmented transcription provided by certain activators (a transactivator) like Sp1 bound to GC box. TFIID supports both basal and enhanced transcription by Sp1, Oct1, AP1, CTF, ATF, etc. The TAFs are essential for this activator-enhanced transcription and hence are referred to as coactivators. RNA polymerase III that transcribes tRNA genes and 5S rRNA genes recognizes a promoter that is internal to the gene to be expressed in the downstream region. In eukaryotic tRNA genes two internal separated blocks of sequences (A and B) exist that act as an intragenic promoter. These regions are responsible for the formation of DHU and T ψ C arm in tRNA. The optimal distance between A and B block in tRNA gene for promoter function is found to be 30–40 base pairs. Transcription start point occurs between 10 and 16 base pairs upstream from the A block. For the transcription of 5S rRNA gene, a specific transcription factor is required that binds to an intragenic promoter for that gene.

Enhancers Control Chromatin Structure and the Rate of Transcription

Specific additional DNA sequences apart from the promoter region may be on either side of a gene, at some distance from the gene or even within the gene. They are referred to as the *cis*-regulators as they are present adjacent to the structural genes they regulate as opposed to *trans*-regulators like the binding proteins which regulate a gene on any chromosome. The enhancers are sites which interact with multiple regulatory proteins, transcription factors and increase the efficiency of transcription initiation or activate the promoter. Positive and negative gene regulators have binding sites in the enhancers. Therefore enhancers resemble the operator regions in prokaryotes. But they are more complex in structure and function.

An example of an enhancer located within the gene it regulates is found in the immunoglobulin heavy chain gene where an enhancer is located in an intron between two coding regions. This enhancer is active only in the cells in which the immunoglobulin genes are expressed. This indicates that tissue-specific gene expression can be modulated through enhancers. Downstream enhancers are found in the human β-globin gene and the chicken thymidine kinase gene. In chickens, an enhancer located between the β-globin and ε-globin genes works in one direction to control the ε-globin gene during embryonic development and in the opposite direction to regulate the β-globin gene during adult life.

Enhancers differ from promoters in a number of ways. Promoter sequences are essential for basal-level transcription whereas enhancers are necessary for full level transcription. Also enhancers are responsible for time- and tissue-specific gene expression. The transcription factors first bind to enhancers and alter the configuration of chromatin. By bending or looping the DNA, the transcription factors bring distant enhancers and their promoters into direct contact in order to form complexes with transcription factors and polymerases (Figure 12.3).

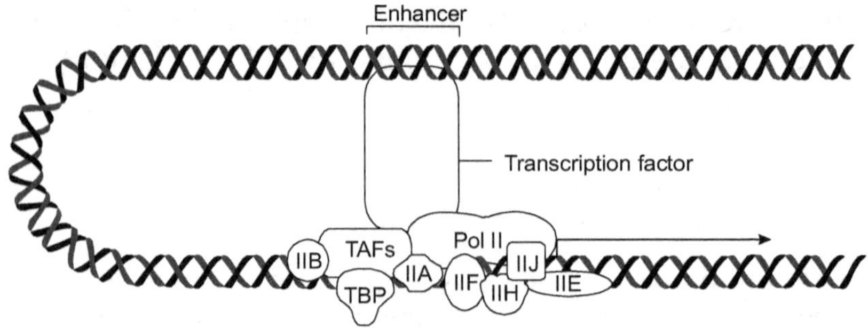

Figure 12.3 DNA looping and binding of transcription factors to the enhancers

In the new configuration, transcription is stimulated to a higher level, increasing the overall rate of RNA synthesis.

BRITTEN–DAVIDSON'S MODEL FOR REGULATION OF GENE EXPRESSION

The model is also referred to as **gene battery model**. The model assumes the presence of four classes of sequences.

1. **Producer genes** comparable to structural genes of prokaryotic operon

2. **Receptor site** comparable to operator sites in operons

3. **Integrator gene** comparable to regulator gene and is responsible for the synthesis of an activator RNA that may or may not give rise to proteins before it activates the receptor site

4. **Sensor site**, which regulates the activity of integrator gene which can be transcribed only when the sensor site is activated. The sensor sites are recognized by agents like hormones and proteins which change the pattern of gene expression. For example, a hormone–protein complex or a transcription factor may bind to a sensor site and cause the transcription of integrator.

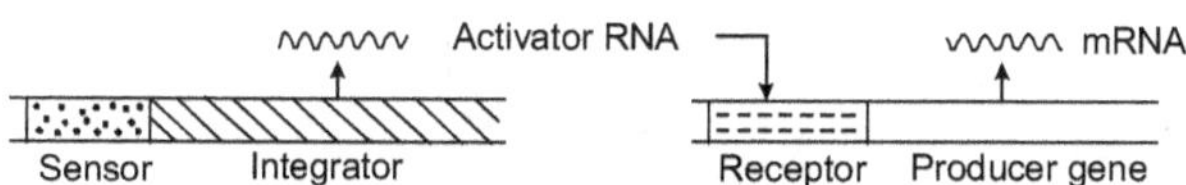

Figure 12.4   The interrelationship of the four classes of sequences in Britten–Davidson's model

In this model, the integrator gene and producer gene are involved in RNA synthesis, whereas the receptor site and sensor site are sequences that help only in the recognition without taking part in RNA synthesis. Also it is proposed that the integrator genes and receptor sites may be repeated a number of times, to control the activity of a large number of genes in the same cell. Repetition of receptor ensures that same activator recognizes all of them and several enzymes of one pathway are synthesized simultaneously. The interrelationship of the four classes of sequences in Britten–Davidson's model is shown in Figure 12.4.

Sometimes, the same gene may be required to be transcribed in different developmental stages. This can be achieved by multiplicity of receptor sites and integrator genes as shown in Figure 12.5 and 12.6. Each producer gene may have several receptor sites, each responding to one activator RNA. A single activator RNA can recognize several producer genes and different activators may activate the same gene at different times. An integrator gene may also occur in cluster with the same sensor site.

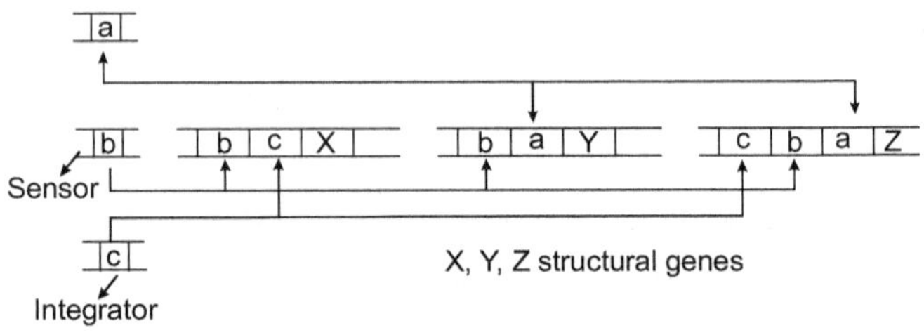

Figure 12.5   Redundancy of receptor sites

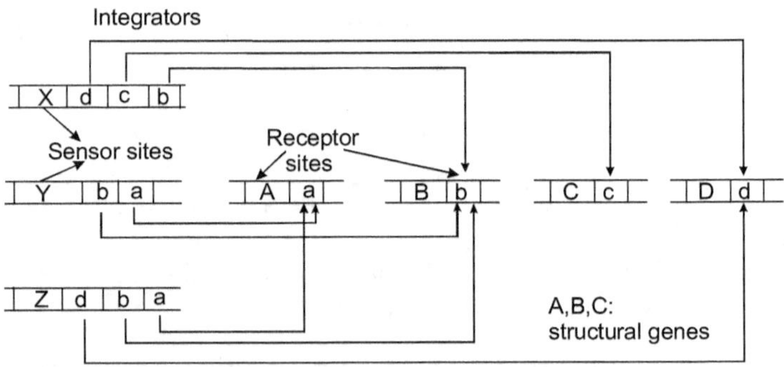

Figure 12.6   Redundancy of integrator genes

In Britten–Davidson model, if a set of structural genes are controlled by one sensor site, it is termed a **battery**. If one sensor site is associated with several integrators, it may cause transcription of all integrators at the same time thus causing transcription of several producer genes through receptor sites. The repetition of integrator and receptor sequences account for the repeated sequences in eukaryotic DNA molecules.

DISCRETE DNA-BINDING DOMAINS FOR THE REGULATORY PROTEINS INVOLVED IN TRANSCRIPTIONAL CONTROL

The regulatory proteins involved in gene expression have recognition surfaces on the DNA with which they interact. Most of the protein–DNA contacts that impart specificity in binding are hydrogen bonds. The DNA binding domains of regulatory proteins tend to be small with 60 to 90 amino acid residues. Four structural motifs that play a major role in DNA binding have been found in many regulatory proteins: the **helix-turn-helix motif/helix-loop-helix motif, the zinc finger, the leucine zipper** and the **HMG box**.

The Helix-turn-Helix/Helix-loop-Helix

This DNA binding motif forms the physical basis for protein–DNA interactions for many prokaryotic and eukaryotic regulatory proteins. It consists of two short α-helical segments 7 to 9 amino acid residues long, separated by a β-turn (about 20 amino acids in total) (Figure 12.7(a)). This structure is not stable by itself, but it represents the reactive portion of the larger DNA-binding domain. One of the two α-helices is referred to as the recognition helix. It contains many of the amino acids that interact with the DNA in a sequence-specific way. When bound to DNA, the recognition helix is positioned in or nearby in the major groove. For example, the MyoD protein which appears to be the primary signal for differentiation of muscle cells, binds DNA most tightly when it forms a heterodimer with the ubiquitously expressed E2A protein. In prokaryotes, the lac repressor, CAP–cAMP receptor protein and tryptophan repressor are some examples that have this DNA-binding motif.

Another structural motif found in some transcription factors is the helix-loop-helix, (HLH) a stretch of two helical regions of amino acids separated by a non helical loop (Figure 12.7(b)). The helical regions permit dimerization between two polypeptides. Sometimes HLH motif is adjacent to a stretch of basic amino acids, so that when dimerization occurs, these amino acids can bind to negatively charged DNA. HLH containing transcription factors play a key role in the differentiation of certain types of tissues including skeletal muscle.

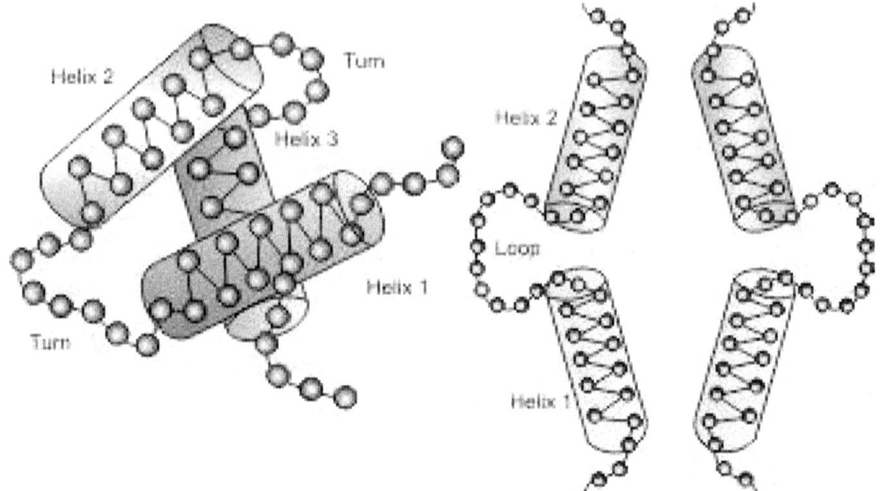

Figure 12.7   (a) Helix-turn-Helix motif (b) Helix-loop-Helix motif

Zinc Finger

Zinc fingers consist of about 30 amino acid residues, four of which (four cys or two cys and two his) coordinate a single Zn^{2+} ion (Figure 12.8). Zinc does not itself interact with DNA. Coordination with zinc stabilizes this small structural motif. Several hydrophobic side chains in the core of the structure also lend stability. Zinc fingers occur in many eukaryotic DNA-binding proteins. If only one zinc finger is involved in binding with the DNA, the interaction will be weak and when multiple zinc fingers are involved, it is strong.

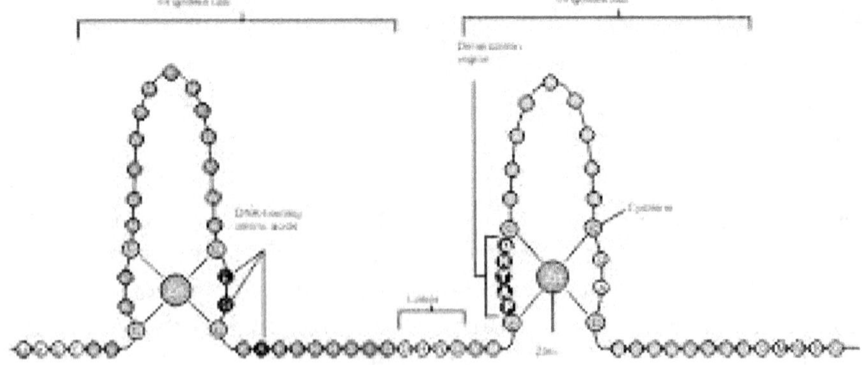

Figure 12.8   Zinc finger

Zinc fingers bind in the major groove of the DNA helix having contact with the bases in G-rich strands. In *Xenopus* there are 37 zinc fingers involved in binding to DNA binding proteins. Zinc fingers also function as RNA binding motifs. There are few examples of the zinc finger motif in prokaryotic proteins. The first zinc finger protein to be discovered was a transcription factor TFIIIA required for the transcription of 5S rRNA gene by RNA polymerase III. This factor requires nine zinc fingers. Each TFIIIA zinc finger contacts about 5 bp of DNA.

The steroid hormones control a wide variety of cellular processes in higher eukaryotes. They first form complexes with specific hormone receptors (HR). This induces a structural alteration in the receptor favouring binding of the complex to a hormone receptor element (HRE) on the DNA. The highly conserved DNA-binding domain of the steroid hormone receptors contains two zinc finger motifs.

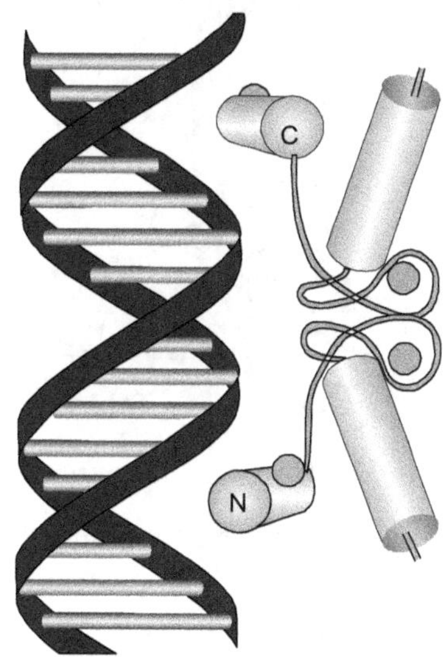

Figure 12.9   Protein–DNA complex formed between a hormone receptor dimer and its HRE

The schematic model for the protein–DNA complex formed between a hormone receptor dimer and its HRE (Figure 12.9) shows how a typical receptor dimer is organized so that the DNA binding domains make identical contacts with adjacent large grooves on the DNA.

Leucine Zipper

A third type of DNA-binding domain was first described for the mammalian enhancer-binding protein C-EBP. These proteins have a highly conserved stretch of about 30 amino acids with a net basic change immediately followed by a region containing four leucine residues at intervals of seven amino acids. This segment is referred to as leucine zipper that is required for dimerization and for DNA binding. Dimerization of proteins in this group is stabilized by hydrophobic interactions between closely apposed α-helical leucine repeat regions of the two proteins. The two helical cylinders are believed to be oriented in parallel in a coiled-coil fashion.

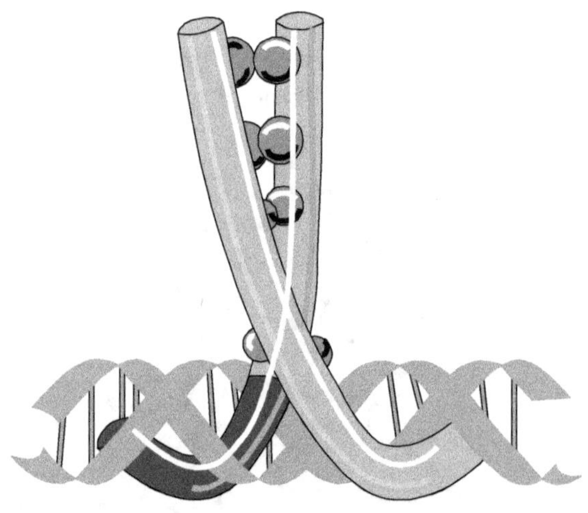

Figure 12.10 Leucine zipper dimer

In Figure 12.10, the knobs represent leucine residues in each protein monomer and the part of the protein monomers that interacts with the major groove of the DNA is shaded.

The main function of the leucine motifs is to bring two proteins together so that they can form homodimers or heterodimers that bind to dissimilar half-sites on the DNA.

The HMG Box

This is a motif present in some proteins called high mobility group (HMG) box. The motif consists of three α-helices arranged in an L shape with two apparent DNA binding sites located on the outside of shape L. Unlike most transcription factors, which bind to DNA and activate transcription by interacting with proteins present at the promoter, the transcription factors that possess HMG boxes activate transcription by distorting the DNA into a conformation that enhances transcription. The best studied HMG protein is called UBF which activates transcription of rRNA genes by RNA polymerase I. UBF binds to the DNA as a dimer. The two subunits contain a total of 10 HMG boxes. They interact with successive sites along the DNA and distort the DNA helix in a way that causes it to loop around the protein (Figure 12.11).

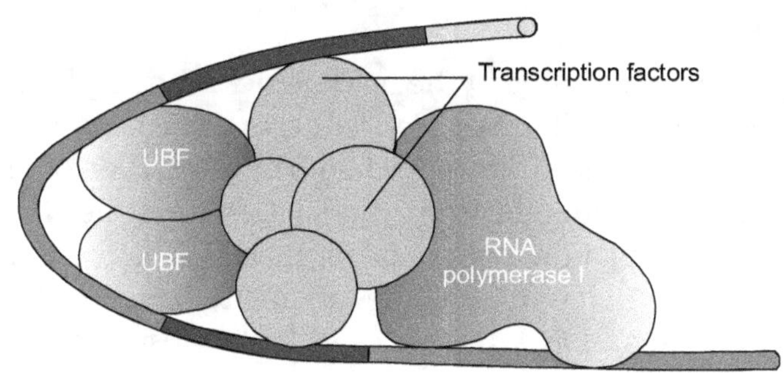

Figure **12.11**   H MG box motif

The looping of the DNA induced by UBF brings two DNA regulatory sequences that are normally separated by 120 base pairs into close proximity where they can be bound cooperatively by a single transcription factor. Thus HMG protein acts by ordering the DNA into a particular spatial arrangement that enhances its access to other regulatory factors.

REPRESSION OF TRANSCRIPTION IN EUKARYOTES

In prokaryotes transcription of structural genes is controlled by repressors which bind to the DNA and block transcription of a nearby gene. Eukaryotes also possess negative regulatory mechanisms. There are two distinct ways in which a gene can be silenced. Specific DNA regulatory sequences can interact with repressor proteins or the DNA can be modified in ways that make it less suitable as a template. A number of negative regulatory proteins, like NC1, NC2, Dr 1 bind to the promoter and block the assembly of the pre-initiation complex required to initiate transcription. Other repressors bind to upstream

DNA sequences and inhibit transcription in some other way. As in prokaryotes, the pattern of transcription of particular genes at a particular time in the life of a eukaryotic cell depends on the balance between positive-acting and negative-acting regulatory factors.

Sometimes, the genes that encode transcription factors are themselves controlled by other transcription factors, whose synthesis is controlled by still other factors. Thus there is a cascade of transcriptional regulation as already described in the Britten–Davidson's model.

DNA Methylation

In many eukaryotes, some of the nucleotides in the DNA in each cell are methylated. Usually methylation occurs in cytosines at the 5th position and 5-methylcytosine is formed. The proportion of methylated cytosines varies considerably from one species to another. In *Drosophila* there is no methylation, whereas about 2–7% of the cytosines in mammals are methylated. Cytosines are methylated most frequently in a CG doublet in the DNA. In most cases, both cytosines are methylated (Figure 12.12).

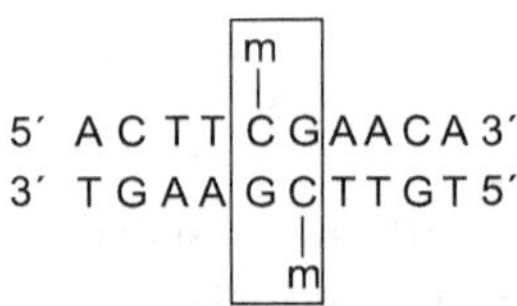

Figure 12.12   Fully methylated CG doublet (m = methyl group)

If only one cytosine in the CG doublet is methylated, the doublet is said to be hemimethylated. Many transcription promoters contain CG doublet-rich regions within them. Absence of methylation of these CG doublet-rich regions is highly correlated with transcriptional activity of the gene. In transcriptionally active genes, the CG doublets are usually not methylated.

When a methylated CG doublet is replicated, the cytosines added in the newly synthesized strand are not methylated resulting in a hemimethylated doublet. The unmethylated cytosines in a hemimethylated doublet are methylated shortly following DNA replication. But if cytosines in the CG doublet are not methylated in the parental DNA, those in the newly synthesized strands are also not methylated following replication. An enzyme, methyl transferase, recognizes only hemimethylated CG doublet and creates fully methylated doublets. If neither of the cytosines in the CG doublet is methylated, the enzyme bypasses the site, leaving the site unmethylated. Thus, methylated and unmethylated sites tend to perpetuate themselves through DNA replication.

This situation is important in cell differentiation. If the regulatory regions of a gene are highly methylated, the gene is not transcribed. The genes that are transcriptionally active in a differentiated cell are undermethylated and remain in the same status when DNA replicates and the cell divides. Therefore, the daughter cells usually have the same active and inactive genes as their parent cell. This may not be a permanent condition. An inactive gene can be demethylated to transform into an active gene or an active gene may be methylated to inactivate it.

Two models have been proposed for the role of methylation in gene inactivation. In the first, methylation acts directly by preventing transcription factors from binding to methylated DNA in the promoter region. In the second model, there is an indirect effect of methylation. Methylated genes can be transcribed, but proteins that are specific for methylated DNA may inhibit transcription.

In mammalian females, one X chromosome is in inactive form, which is totally inactive in gene expression and has a higher level of methylation than the other active X chromosome.

Methylation patterns are tissue-specific and are heritable for all cells of that tissue. The nucleotide 5'-azacytidine is a base analogue of cytidine. If it is incorporated in the DNA in the place of cytidine, it cannot be methylated. This causes the undermethylation of the sites where it is incorporated. Thus, incorporation of 5'-azacytidine causes changes in the pattern of gene expression. This can stimulate expression of alleles on inactivated X chromosomes. In clinical trials for the treatment of sickle cell anaemia, 5'-azacytidine is used. Sickle-cell anaemia is an autosomal recessive condition, where a mutant hemoglobin protein with a defect in the β-subunit causes changes in the shape of red blood cells that in turn causes a cascade of clinical symptoms. During the embryonic stage ε- and γ-globin genes are expressed. They become transcriptionally inactive after birth, when β-globin synthesis begins. When the affected individuals are treated with 5'-azacytidine there is a reduction in the amount of methylation of ε- and γ-globin genes causing a reinitiation of the re-expression of these embryonic and fetal genes. This causes a reduction in the amount of sickling in the red blood cells.

Regulation at the Level of Post-Transcriptional Modification

In eukaryotes, there is abundant opportunity for gene regulation during transcript processing. Approximately 75% of pre-mRNAs are degraded within the nucleus and never become mRNAs. This provides an opportunity for selective degradation as a means of gene regulation.

Alternative splicing of pre-mRNA   In some cases, there may be more than one alternative for pre-mRNA processing into mRNA. Alternative processing permits the synthesis of different versions of a protein. For example, there is variation in proteins between smooth and striated muscle tissues which is due to alternative splicing of the introns of the same pre-mRNA. The α-tropomyosin pre-mRNA in rats includes 13 exons separated by 12 introns. But the final mRNA does not contain all the 13 exons. The predominant form of mRNA in striated muscle cells contains 10 exons, whereas the form found in smooth muscle cells contains 9 exons, 7 of which are common between the two mRNAs. These two mRNAs encode proteins that are similar in some respects but different in others, even though both proteins are encoded by the same gene.

Alternative splicing not only has the potential to produce many different forms of a protein from a single gene, but can also affect the outcome of the development processes. An example is sex determination in *Drosophila*, where splicing involving a single exon determines whether the embryo will develop as a male or a female.

Sex in *Drosophila* is determined by the sex ratio of X chromosomes to sets of autosomes (X : A). If this ratio is 0.5 (1X : 2A), males are produced even if Y chromosome is not present. When the ratio is 1.0 (2X : 2A), females are produced. Intermediate ratios (2X : 3A) produce intersexes. The chromosomal ratios are interpreted by a small number of genes that initiate a cascade of developmental events resulting in the production of male or female somatic cells and the corresponding phenotypes. The major genes involved in this are sex-lethal (sxl), transformer (*tra*) and doublesex (*dsx*).

In females, the sex ratio activates transcription of *sxl* gene and Sxl protein is produced. This protein is not produced in males. In females Sxl protein binds to and controls the splicing of pre-mRNA of the gene *tra*. This pre-mRNA has a stop codon in exon 2. Sxl protein binds to the *tra* pre-mRNA and brings about the splicing and removes exon 2. In males which lack the Sxl protein, exon 2 is not spliced and the stop codon is present in the mature mRNA. Therefore in males, translation of tra mRNA is prematurely terminated by the stop codon, resulting in an inactive gene product. In females, a functional gene product is produced. The result is a female-specific tra protein that leads to the female phenotype. The third gene *dsx* is a critical control point in the development of sexual phenotype, for it produces a functional mRNA and protein in both females and males. However, the pre-mRNA is processed in a sex-specific manner to yield different transcripts. In females, the *tra* protein binds to the *dsx* pre-mRNA and directs its processing in an alternative, female-specific manner. In males, default splicing of the *dsx* pre-mRNA results in a male-specific mRNA and the protein product. The female dsx protein causes female sexual differentiation and the male dsx protein brings about male sexual development. In the absence of dsx function, flies develop into intersexes.

Control of stability of mRNA molecules   All mRNA molecules have a characteristic lifespan. They are degraded sometime after their synthesis, in the cytoplasm. The lifetime of mRNA molecules vary widely. Some are degraded within minutes after synthesis, whereas others last hours or even months and years, as in the case of mRNAs stored in oocytes. An alteration in the stability of mRNA that is involved in translation is significant in controlling eukaryotic gene expression. Longer-lived mRNAs can produce more protein than short-lived ones. Altering the half-life of an mRNA molecule already engaged in translation is a way of regulating gene expression.

Almost all eukaryotic mRNAs have a poly(A) tail added to their 3' ends before they come out from the nucleus (Figure 12.13a). Histone mRNAs are exceptions to this. Poly(A) tail protects the messengers from rapid destruction. A poly(A) binding protein binds to the tail and it protects the mRNA from destruction by a $3' \rightarrow 5'$ exonuclease. This also protects the other end of the mRNA. The poly(A) tail loops around and brings the poly(A) binding protein into contact with the 5'-methyl guanosine cap and prevents the 5' end also from attack by a decapping enzyme.

When the cap is removed, the 5′ end of the mRNA is exposed to digestion by exonuclease. mRNA destruction is preceded by deadenylation. When the poly(A) tail gets reduced in size, poly(A) binding protein cannot attach and destruction of mRNA starts (Figure 12.13b).

In the histone mRNA that lacks the poly(A) tail, a stem-and-loop structure is formed in the 3′ end. Histone synthesis is required only in the S phase of eukaryotic cell cycle. In G2 phase, the level of histone mRNAs rapidly falls due to reduction in the rate of transcription. The destabilization of the histone mRNA requires the presence of free histone monomers. At the end of the S phase, when no more histones are required, histone molecules get accumulated, with cessation of histone gene transcription.

The synthesis of transferrin receptor protein is another case where mRNA stability regulates the synthesis of the protein. The receptor mediates iron transport into the cell. In the 3′-untranslated region (UTR) of the mRNA is a group of five stem loops called an iron-responsive element (IRE) (Figure 12.13c).

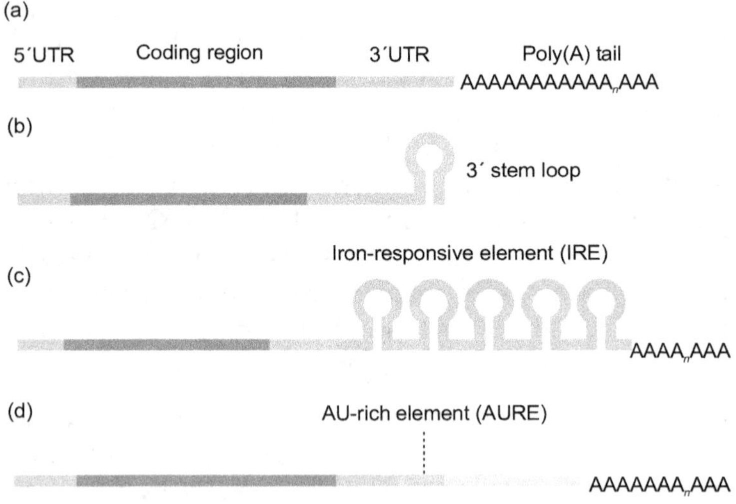

Figure 12.13   Structural stability determinants of mRNAs

In the absence of iron, an IRE binding protein attaches to the IRE and stabilizes the mRNA, thus increasing receptor synthesis with an increase in the import of iron into the cell. When iron is present in large amount, it complexes with the protein which no longer binds to the IRE. This exposes the region to attack by an endonuclease and this results in reduction of the import of iron into the cell.

Some short-lived mRNAs contain "instability sequences" which account for rapid destruction of the mRNA. These sequences contain adenine/uracil-rich element (AURE) found in the 3'-untranslated regions of the molecules about nine bases in length (Figure 12.13d). They cause mRNA destabilization by virtue of AURE-binding proteins. An example is the mRNA from *c-fos* gene which codes for a leucine zipper class of transcriptional factor and contains an AURE in the 3'-untranslated region. The corresponding *v-fos* gene is strongly oncogenic and the mRNA derived from it lacks the destabilizing AURE found in the non-oncogenic c-fos mRNA and has a longer half-life in the cell. This results in prolonged gene-activating signal to the cell. This can be a factor for cancer development.

REGULATION OF GENE EXPRESSION BY INTRACELLULAR AND INTERCELLULAR SIGNALS

Among the known regulators of transcription in higher eukaryotes, the hormones—small molecules or polypeptides that are carried from hormone-producing cells to target cells—have been studied in detail. The steroid hormones, like cortisol, in particular, act by turning on the transcription of specific sets of genes. The hormone penetrates a target cell through diffusion, because steroids are hydrophobic molecules that pass freely through the cell membrane into the cytoplasm. There it encounters a receptor molecule that is complexed with another protein called Hsp82 which functions to mask the receptor. Once cortisol binds to the receptor, it releases the receptor from Hsp82 (Figure 12.14). The hormone–receptor complex migrates to the nucleus where it binds to its DNA target sequences called hormone response elements (HREs) to activate transcription. Non target cells do not contain the receptors and are therefore unaffected by the hormone.

A well-studied example of induction of transcription by a hormone is stimulation of the synthesis of ovalbumin in the chicken oviduct by the steroid sex hormone estrogen. When hens are injected with estrogen, oviduct tissue responds by synthesizing ovalbumin mRNA, which continues as long as estrogen is administered. Since the other tissues lack the hormone receptor, only oviduct synthesizes the ovalbumin mRNA.

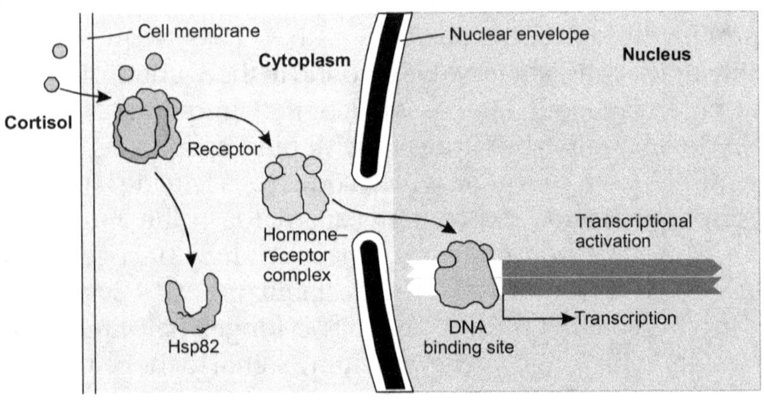

Figure 12.14 Hormonal regulation of gene expression

The hormone response elements of the DNA are similar in length and arrangement, but different in sequence for the various steroid hormones. Each receptor has a consensus HRE sequence to which the hormone–receptor complex binds well, each consensus sequence consisting of two six-nucleotide sequences, either continguous or separated by three nucleotides, in tandem or in a palindromic arrangement. The hormone receptors have a highly conserved DNA-binding domain with two zinc fingers.

The hormone–receptor complex binds to the DNA as a dimer, with the zinc finger domains of each monomer recognizing one of the six-nucleotide sequences. The ability of a given hormone to act through the hormone–receptor complex to alter the expression of a specific gene depends on the exact sequence of the HRE, its position relative to the gene and the number of HREs associated with the gene.

Unlike the DNA-binding domain, there is a ligand-binding domain of the receptor protein, always at the carboxyl terminus which is specific to the particular receptor. In the ligand-binding region, the glucocorticoid receptor is only 30% similar to the estrogen receptor and 17% similar to the thyroid hormone receptor. The size of the ligand-binding region varies dramatically; it has only 25 amino acid residues in the vitamin D receptor, whereas in the mineralocorticoid receptor, it has 603 residues. Some people are unable to respond to cortisol, testosterone, vitamin D or thyroxine during treatment for diseases because of mutations in these regions.

REGULATION OF GENE EXPRESSION AT TRANSLATIONAL LEVEL

Translational regulation may play an important role in regulating certain long eukaryotic genes which may be in millions of base pairs size. Some genes may be regulated at both transcriptional and translational stages. In some anucleate cells like the reticulocytes, transcriptional control is unavailable. Translational control of stored mRNAs become important. There are several major mechanisms of regulation at translational level:

1. The initiation factors involved in protein synthesis are subject to phosphorylation by a number of protein kinases. The phosphorylated forms are less active and inhibit translation. Both eIF-1 and eIF-2 are phosphorylated for controlling translation initiation in many cells. For example, two protein kinases have been

identified specific for the α subunit of eIF-2 in the reticulocytes. One of these kinases, termed the heme-regulated inhibitor (HR1), coordinates the rate of hemoglobin synthesis with the availability of hemin, the precursor of the heme group in hemoglobin. Hemin binds to and inhibits the activity of HRI, thereby enhancing the rate of globin synthesis.

The second eIF-2-specific kinase is activated by double-stranded RNA and hence is referred to as dsRNA-activated inhibition (DAI). DAI may play a role in defending cells against viral infections.

Though HRI and DAI are different proteins, they phosphorylate the same amino acid residue in a subunit of eIF-2. Phosphorylated eIF-2 is capable of catalysing the binding of met-tRNA to the ribosome, but unlike the unphosphorylated eIF-2, it cannot get dissociated from the ribosome.

2. Some proteins bind directly to mRNA and act as translational repressors. Most of them bind at specific sites in the 3'-untranslated region (3'-UTR). These proteins interact with other translation initiation factors bound to the mRNA or with the 40S ribosomal subunit and prevent initiation of translation (Figure 12.15).

3. Specific binding proteins (4E-BPs) disrupt the interaction between eIF-4E and eIF-4G, limiting translation. When cell growth resumes in response to growth factors or other stimuli, the binding proteins are inactivated by protein-kinase-dependent phosphorylation.

4. Translation may be regulated by regulating the lifetime of the mRNA. The silk gland of the silkworm *Bombyx mori* synthesizes a single type of protein, silk fibroin. Since the worm takes several days to construct its cocoon, it is the amounts and not the rate of fibroin synthesis that must be great. This is made possible with a fibroin mRNA molecule that is very long-lived. The oviduct in chicken makes ovalbumin in large amounts by prolonged mRNA lifetime even though it has a single copy of the ovalbumin gene. This is known as translational amplification which is comparable with the gene amplification where there is multiplication in the number of copies of a gene as rRNA-producing genes.

5. Another example of translational regulation is that of masked mRNA. Unfertilized sea urchin eggs store large amount of mRNA for many months in the form of ribonucleoprotein particles.

This mRNA is inactive. Within minutes after fertilization, mRNA molecules are involved in translation.

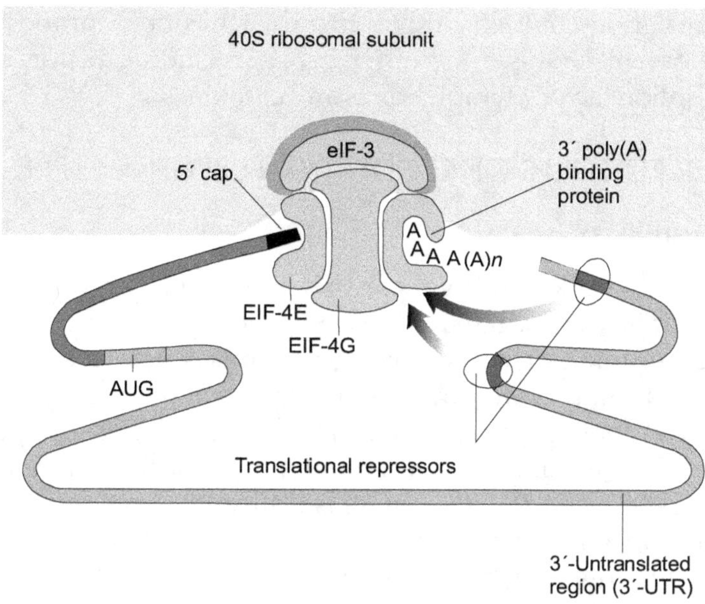

Figure 12.15   Translational regulation of eukaryotic mRNA

6. Autoregulation is another mechanism of translational regulation. One of the best studied examples of this type of regulation is the synthesis of α- and β-tubulins, the subunit components of eukaryotic microtubules. When there is low concentration of α- and β-subunits in the cell, tubulin synthesis is stimulated, whereas at higher concentrations, synthesis is inhibited.

REGULATION AT POST-TRANSLATIONAL LEVEL

In some cases regulation occurs after translation. This can be explained with the example of tubulin mRNA. The first four amino acids (met-arg-glu-ile, MREI) of the tubulin gene product (β-tubulin) constitute a recognition element to which regulatory factors bind. The concentration of α- and β- tubulin subunits in the cytoplasm may serve this regulatory function. This protein–protein interaction involves the action of an RNase which degrades tubulin mRNA shutting down tubulin biosynthesis. Translation must proceed to at least 41 codons because the first 30 to 40 amino acids translated occupy a tunnel within the large ribosomal subunit making the MREI recognition sequence accessible for binding.

Translation-coupled mRNA turnover has been proposed as a regulatory mechanism for other genes, including histones, some transcription factors, lymphokines and cytokines.

OTHER FEATURES OF REGULATION OF GENE EXPRESSION

Gene Families

Many related eukaryotic genes can be functionally grouped as a set of genes called a gene family. Gene families may be classified as simple multigene families, complex multigene families and developmentally controlled complex multigene families. In simple multigene family one or a few genes are repeated in a tandem array. For example, there is a set of genes for 5S rRNA separated by spacers and they form a gene cluster. Each 5S rRNA gene is transcribed as a separate RNA molecule. In *Xenopus*, these clusters themselves exist in several copies, each copy containing hundreds to thousands of 5S rRNA genes. The complex multigene families consist of a cluster of several related genes, each transcribed independently and separated by a spacer. For example the histone gene family of the sea urchin consists of five genes for the five histone proteins separated by spacers. This five-gene unit is tandemly repeated about 1000 times. Each of the five genes is separately transcribed as a single intron-free monocistronic RNA and each gene is transcribed in the same direction from the same DNA strand. In contrast to this, in *Drosophila*, the five genes are transcribed in both directions.

The globin gene is a good example for the developmentally controlled complex gene family. In hemoglobin, the globin chain is of varied types during different stages of development. During the embryonic stage ς_2, an α- like chain appears first which is gradually replaced by ς_1 form and ε chain is formed which is equivalent to β- chains. In the fetal and adult stage α type is present. The two β-like types γ^G and γ^A are roughly equimolar in the fetal stage. There are separate genes for α, β, α-like and β-like chains to be expressed differently depending on the stage of development.

Gene Alteration

Changes in the genes is another feature accounting for regulation of eukaryotic gene expression. Three major types of gene alteration observed are gene loss, amplification and gene rearrangement.

Gene loss   One way to eliminate the activity of certain genes as a cell differentiates is to remove the genes from the cells. In some protozoa, insects and crustaceans, once the required proteins are synthesized, the chromosomes are lost and only the cells involved in the production of germ cells maintain the complete chromosome. Many protozoa have two types of nuclei—micronucleus and macronucleus. Usually, the germ cells contain only a micronucleus, but as the cell differentiates, the micronucleus undergoes a variety of changes and the cell now has both a micronucleus and a macronucleus. The micronucleus contains inactive DNA and serves as a storehouse of genetic information for the production of future germ cells. The DNA of the macronucleus serves as template for transcription that undergoes fragmentation and degradation.

Gene amplification   It is a way by which a cell can produce vast quantities of a specific gene product. The best studied example is the rRNA gene in insects, amphibians, fish, etc. that multiplies about 4000-fold. In *Drosophila*, the genes that produce chorion proteins are amplified in ovarian follicle cells just before maturation of an egg.

Gene rearrangement   A gene could be turned on by moving it from a location far from its promoter to a site nearer to its promoter. The relocation may be regulated or may be responsive to an external or internal signal. The best known example is the synthesis of specific immunoglobulins in response to an antigen by rearrangement of DNA segments. Diversity in L chain and H chain synthesis involves the gene splicing events. Out of the different V genes, J genes, C genes and the D genes, a particular combination of VJC and VDJC obtained by rearrangements gives specific light and heavy chains to constitute a particular type of immunoglobulin. Mating type in yeast is again decided by gene rearrangement.

Spicy Questions and Answers

1. **What do you mean by a master switch gene? How is it related to homeodomain and homeobox?**

   Master switch gene is a eukaryotic gene that controls the expression of many other genes like the operons in prokaryotes. It is responsible for the transcription of a specific transcription factor gene that in turn is responsible for the expression of many other genes. For this to happen, the master switch gene should interact with a DNA homeobox which on transcription and translation gives a homeodomain that is again a part of a transcription factor. The homeotic genes are involved in the development of organisms.

2. **In eukaryotes, why cannot transcription factors alone initiate transcription?**

   Transcription factors are proteins which are necessary for initiation of transcription. But they are not sufficient to start transcription. They contain at least two functional domains—one to bind to the p site and/ or the enhancers and the other interacts with RNA polymerase or other transcription factors.

3. **Geetha, a young immunologist, told her friend, Rita, a student of molecular biology that there are 10^8 different antibodies produced by human B cells. To her astonishment, Rita strongly disagreed, saying that there are not so many genes in the human chromosome to code for so many antibodies. Who is correct? Give reasons for your answer.**

   Geetha is correct. Human body can produce a huge number of types of antibodies in order to combat against the huge number of types of antigens entering into the body and this is possible by gene rearrangement-specific VJC joining in case of light chain and VDJC joining in case of heavy chain synthesis.

4. **Why is the Britten and Davidson model more acceptable than the operon model for explaining regulation of gene expression in cells of higher animals?**

   In eukaryotes, the genes encoding for proteins are not present as clusters as the structural genes in the prokaryotes. The complex pattern of gene expression during development in higher animals require complex integrated controls that can govern expression. Genomes of higher organisms contain single copy DNA sequences that are interspersed with middle repetitive sequences. The Britten and Davidson model is consistent with these observations.

5. **How do histones regulate gene activity?**

    Histones are thought to be nonspecific repressors or inhibitors of transcription. This could be proved from the rate of RNA synthesis from chromatin which is found to be inversely related to the amount of histone present for a constant amount of DNA. Also histone modifications like phosphorylation, acetylation and so on are found to play a role in eukaryotic gene regulation.

6. **If all cells in a given organism carry the same genes, how can gene expressions that are localized in time and space be explained?**

    The genome remains the same in all the cells in an organism. But the molecular composition of the cytoplasm of the cells is constantly changing in response to both the genome and the environment. Many of the metabolic processes occurring in the cell take place in the cytoplasm. Thus changes in the cytoplasm may induce or repress the gene expression.

7. **Enhancers and classical promoters both act as positive regulators of gene expression. By what criteria can these two classes of regulatory sequences be distinguished?**

    Classical promoters can regulate the transcription of the genes only when they are located just 5′ to the gene and have the proper orientation. But enhancers are orientation independent, position independent and can function over relatively large distances.

8. **Based on the available evidences, does methylation of DNA sequences in a gene tend to enhance or depress the level of gene expression?**

    Methylation is found to decrease the level of expression of those genes where its effect has been studied.

Review Questions

1. Outline the different levels of regulation of eukaryotic gene expression.

2. What is meant by chromatin remodelling?

3. What is the role of transcription factors in the regulation of eukaryotic gene expression ?

4. How do enhancers control initiation of transcription?

5. Write a note on Britten–Davidson's model for the regulation of eukaryotic gene expression.

6. Describe the features of structural motifs in many regulatory proteins that play a major role in DNA binding.

7. Discuss the strategies of regulation of eukaryotic gene expression at the level of mRNA processing.

8. Comment on hormonal regulation of gene expression.

9. Explain regulation of eukaryotic gene expression at translational and post-translational levels.

10. What are gene families?

11. Explain with examples of gene alterations in higher organisms.

13

ANALYTICAL TECHNIQUES USED IN THE STUDY OF NUCLEIC ACIDS

Several physical and chemical techniques exist for the separation and analysis of macromolecular components of the nucleus and cytoplasm in the cells and these have been instrumental in understanding molecular genetics. The background and theoretical basis of some of these techniques will help in the study of nucleic acid chemistry and various genetic analyses and concepts.

EXTRACTION OF DNA AND RNA FROM CELLS AND TISSUES

Nucleic acids are mostly associated with proteins. When the cells are broken and opened, the nucleic acids must be deproteinized. This is done by shaking the nucleic acid–protein mixture with a phenol solution and/or a chloroform–isoamyl alcohol mixture so that the protein is precipitated and can be removed by centrifugation (Figure 13.1). Also detergents, guanidium chloride or high salt concentrations may help in the dissociation of proteins from nucleic acids. Proteins can also be degraded enzymatically by treating the mixture with proteases. In all these cases, the nucleic acids can be isolated by precipitation with ethanol. If this mixture of nucleic acids has both RNA and DNA, DNA can be eliminated by treatment with deoxyribonuclease, or RNA can be eliminated by treatment with ribonuclease enzyme. If both the nucleic acids have to be separated and isolated, the mixture can be subjected to ultracentrifugation.

During experimental operation, the nucleic acids must be protected from degradation by nucleases that may be present in the solution or in human hands. Nucleases may be inhibited by using chelating agents such as EDTA, which sequester the divalent metal ions that nucleases require for activity. All glassware must be autoclaved to heat-denature the nucleases and the experimenter should wear plastic gloves. Nucleic acids are generally easier to handle than proteins because they lack a complex tertiary structure that makes them relatively tolerant to extreme experimental conditions.

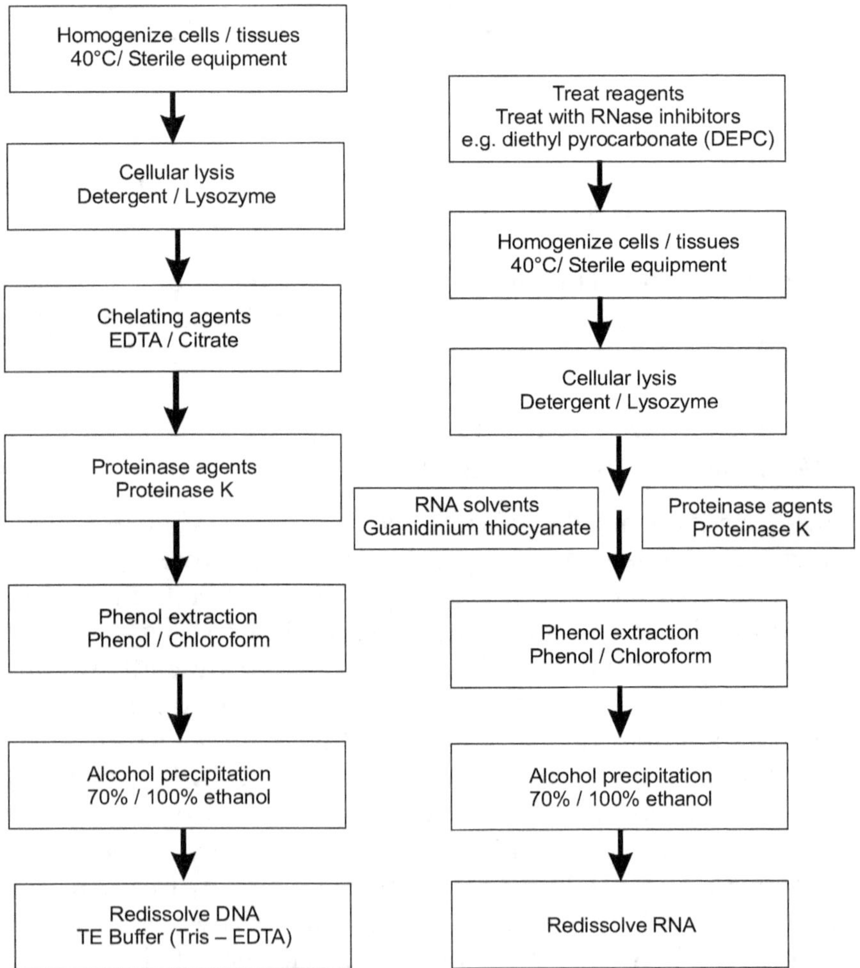

Figure 13.1 General steps involved in extracting DNA and RNA from cells or tissues

> ## BOX 13.1   DNA FROM SALIVA
>
> The traditional DNA collection methods such as blood extraction can be replaced by the most effective, non-invasive specimen collection using saliva. DNA in saliva comes from buccal epithelial cells and white blood cells found in the mouth. DNA from saliva is of high quality and high quantity. Researchers have optimized methodologies to preserve and stabilize saliva sample for long term storage at room temperature without DNA degradation. DNA yield from 250 µl of saliva is similar to the yield from 1.0 ml of whole blood. Forensic studies have shown that DNA from saliva on stamps or cigarettes can be used in PCR analyses to identify the suspects. Purification is based on spin column chromatography and the purified saliva DNA is compatible with a number of downstream application including PCR, Southern blot analysis, sequencing and microarray analysis.

CHROMATOGRAPHY

Many of the chromatography techniques that are used to separate proteins are also applicable to nucleic acids. **Paper chromatography** and **thin layer chromatography** are useful in fractionating oligonucleotides. These techniques are now replaced by **high-pressure liquid chromatography,** particularly **reverse-phase chromatography**. Larger nucleic acids are often separated by **ion-exchange chromatography** and **gel filtration.**

Hydroxyapatite, a form of calcium phosphate, is useful in the chromatographic purification and fractionation of DNA. Double-stranded DNA binds to hydroxyapatite more tightly than single-stranded DNA and most other molecules. Therefore, DNA can be rapidly isolated by passing a cell lysate through a hydroxyapatite column, washing the column with a phosphate buffer of concentration low enough to release only the RNA and proteins and then eluting the DNA with a concentrated phosphate solution. Single-stranded DNA elutes from hydroxyapatite at a lower phosphate concentration than does double-stranded DNA (Figure 13.2). This phenomenon forms the basis of a technique known as **thermal chromatography**, for separating DNA according to its base composition. When eluted with the phosphate buffer, only single-stranded DNA is released as the temperature of the column is gradually increased. That is, as the DNA melts and is converted to the single-stranded form, it is eluted from the column. The T_m value, i.e., the melting temperature of a duplex DNA, varies with its G + C content. Therefore, thermal chromatography permits the fractionation of double-stranded DNA according to its base composition.

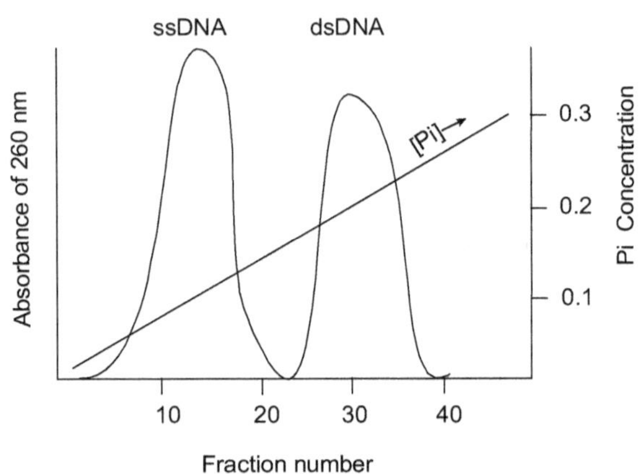

Figure 13.2 Chromatographic separation of ss and dsDNAs on hydroxyapatite by elution with a solution of increasing phosphate concentration

Affinity chromatography is useful in isolating specific nucleic acids. For example, most eukaryotic messenger RNAs have a poly(A) tail at their 3′ ends. They can be isolated on agarose or cellulose to which poly(U) is covalently attached. The poly(A) sequences specifically bind to the complementary poly(U) in high salt and at low temperatures and can later be released by altering these conditions. Also, if the partial sequence of an mRNA is known, derived from corresponding protein's amino acid sequence, the complementary DNA strand may be synthesized and used to isolate that particular mRNA (Figure 13.3).

Hydrophobic interaction chromatography (HIC) delivers 95% supercoiled pDNA. By using matrices derived with mildly hydrophobic ligands (e.g. polypropylene glycol) it is possible to separate supercoiled and relaxed plasmid DNA from the impurities on the basis of differences in surface hydrophobicity. HIC takes advantage of the higher hydrophobicity of single-stranded nucleic acids that show a high exposure of the hydrophobic aromatic bases when compared with double-stranded nucleic acids. In double-stranded plasmid molecules, the hydrophobic bases are packed and shielded inside the helix and thus hydrophobic interaction with the HIC support is minimal.

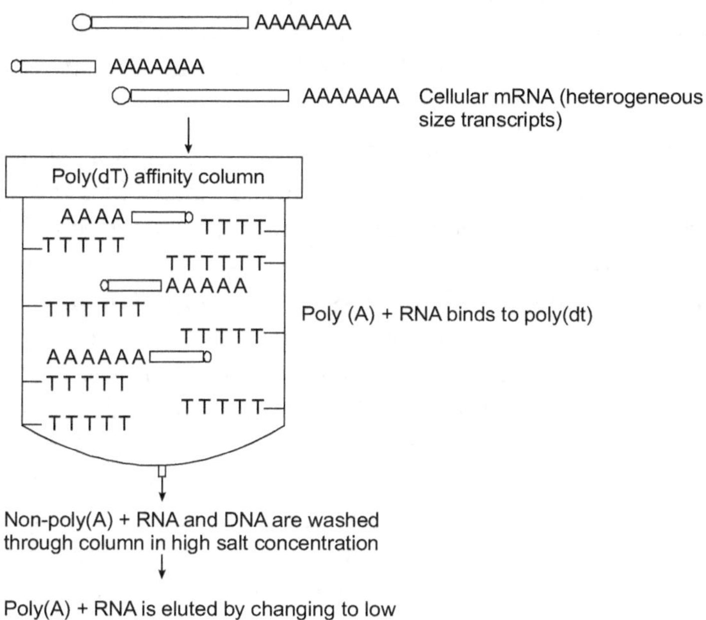

Figure 13.3   Affinity chromatography of poly(A)+ RNA

ELECTROPHORESIS

Nucleic acids are separated by **polyacrylamide gel electrophoresis** because their electrophoretic mobilities in such gels vary inversely with their molecular masses. But DNAs with more than a few thousand base pairs cannot penetrate polyacrylamide gel which is overcome by the use of **agarose gel**. Even large DNAs in a varied size range are fractionated using low agarose content.

In the electrophoretic separation of RNA, a smaller pored acrylamide gel is used to enhance resolution. To resolve transfer RNAs (4S) from 5S ribosomal RNA, a 2.5–5% acrylamide gradient gel with an overnight run is preferred. Prior denaturation and disruption of hydrogen bonds of some secondary structures within RNA molecules can be promoted by using denaturing agents like formaldehyde, glyoxal and methyl mercuric hydroxide that are compatible with both RNA and agarose. The sample is first heat-denatured followed by the addition of these agents which form adducts with the amino groups of guanine and uracil, thereby preventing hydrogen bond reformation during electrophoresis. Denatured RNA stains weakly with ethidium bromide. Acridine orange is commonly used to visualize RNA on denaturing gels. If RNA sample is radiolabelled, bands are identified by autoradiography.

Capillary electrophoresis (CE) involves electrophoresis of samples in very narrow bore tubes (Figure 13.4). An advantage of using capillaries is that they reduce problems resulting from heating effects. Because of the small diameter of the tube, there is a large surface-to-volume ratio which gives enhanced heat dissipation. High-purity synthetic oligodeoxyribonucleotides are necessary for a range of applications including use as hybridization probes in diagnostic and gene cloning experiments, use as primers for DNA sequencing and for polymerase chain reaction, use in site-directed mutagenesis and use as antisense therapeutics. Capillary electrophoresis can provide a rapid method for analysing the purity of such samples. For example, analysis of an 18-mer antisense oligonucleotide containing contaminant fragments (8-mer to 17-mer) can be achieved in just five minutes. Point mutations in DNA, such as those occurring in a range of human diseases can be identified by CE. CE can be used to quantify DNA. For example, CE analysis of PCR products from HIV patients allowed the identification of 200,000–500,000 viral particles per cubic centimetre of serum.

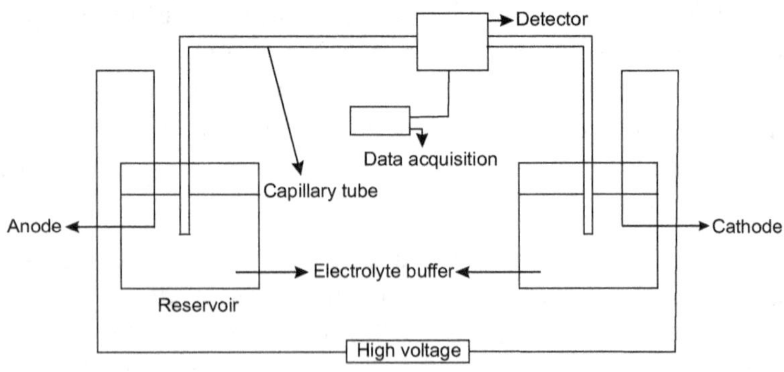

Figure 13.4   Capillary electrophoresis apparatus

Beckman Coulter CE technology is a modified version that helps to analyse a molecule as it behaves in free solution. This is possible with the high sensitivity of laser-induced fluorescence (LIF) and direct on-line quantitation. With the addition of polymers, the capillary tube can be further transformed into a sieving environment, allowing nucleic acids to be separated by size.

Manipulating and analysing DNA are fundamentals in the field of molecular biology. Separating complex mixtures of DNA into different sized fragments by electrophoresis was a well established technique by the early 1970s. In 1984, Schwartz and Cantor described **pulsed field gel electrophoresis (PFGE),** which resolved extremely large DNA for the first time, raising the upper size limit of DNA separation in agarose from 30–50 kb to well over 10 Mb (10,000 kb). The electrophoretic apparatus used in PFGE has two or more pairs of electrodes arrayed around the periphery of an agarose slab gel. The method basically involves electrophoresis in agarose where two electric fields are applied alternately at different angles for defined time periods. Activation of the first electric field causes the coiled molecules to be stretched in the horizontal plane and to start to move through the gel (Figure 13.5).

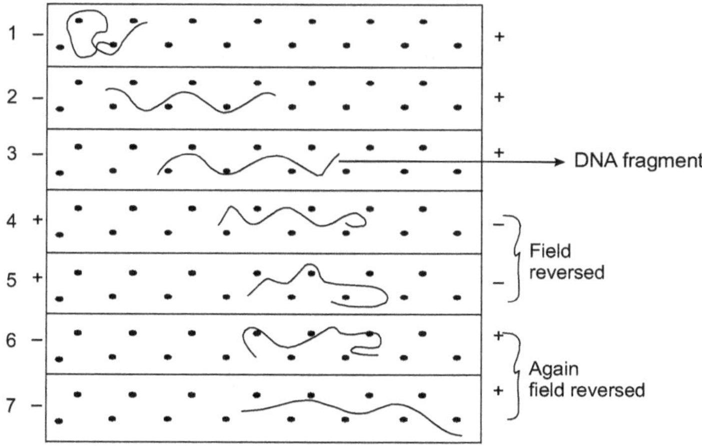

Figure 13.5   Pulsed field gel electrophoresis

Interruption of this field and application of the second field, force the molecule to move in the new direction. Since there is a length-dependent relaxation behaviour when a long-chain molecule undergoes conformational change in an electric field, the smaller a molecule, the quicker it realigns itself with the new field and is able to move continuously through the gel. Larger molecules take longer to realign. In this way, with continual reversing of the field, smaller molecules draw ahead of larger molecules and separate according to size. In recent years a number of different developments on the same basic theme have come into practice like orthogonal field alternating gel electrophoresis (OFAGE), field inversion gel electrophoresis (FIGE), transverse alternating field gel electrophoresis (TAFE), contour-clamped homogeneous electric field electrophoresis (CHEF) and rotating field electrophoresis (RFE).

Linear DNA fragments above a certain size (above 20 kb) are not separated on a conventional agar gel. Their separation is achieved by changing the direction of electric field by PFGE (Figure 13.6). Initially the DNA molecule is too large to migrate through the gel. Under the influence of an electric field, it slowly orients itself. Again on reversing the direction of the field it gets tangled up. Before it gets re-oriented once again the field direction is changed. The DNA molecule again tries to resume its linear form. The mobility of the molecule is independent of its size. The main determinant is how quickly it can re-orient itself which is a function of its length.

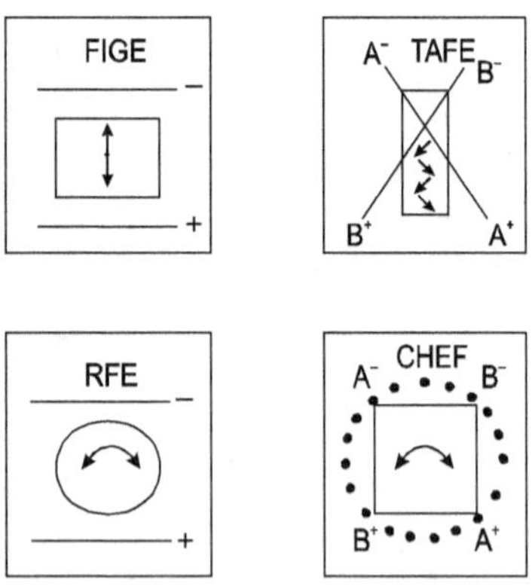

Figure 13.6   Types of PFGE

The applications of PFGE include cloning large plant DNA using yeast artificial chromosomes (YACs) and PI cloning vectors, identifying restriction fragment length polymorphisms (RFLPs) and construction of physical maps, detecting *in vivo* chromosome breakage and degradation and determining the number and size of chromosomes—

"electrophoretic karyotype" from yeasts, fungi and parasites such as *Leishmania*, *Plasmodium* and *Trypanosoma*.

During continuous field electrophoresis in agarose gels, DNA above 30–50 kb migrates with the same mobility regardless of size. This is seen in a gel as a single large diffuse band. If the DNA is forced to change direction during electrophoresis, different sized fragments within this diffuse band begin to separate from each other. With each reorientation of the electric field relative to the gel, smaller sized DNA will begin to move in the new direction more quickly than the larger DNA.

Field inversion gel electrophoresis (FIGE) includes periodical inversion of the polarity of the electrodes during electrophoresis. This subjects DNA to a 180° re-orientation and DNA moves backward for a short while. Only an electric field switching module is needed. A standard vertical or horizontal gel box that has temperature control can be used to run the gel. This method is used for the separation of high molecular weight DNA.

Zero integrated field electrophoresis (ZIFE) also resembles FIGE, but is very slow. ZIFE is capable of resolving larger DNA and giving a larger useful portion of the gel.

In **transverse alternating field electrophoresis (TAFE)**, the electrodes are positioned on opposite sides of a vertical gel. The DNA moves first towards one electrode and then towards the other, forming a zigzag pattern, and the movement is faster. The vector of this oscillation is a straight line from the loading well to the base of the gel. Variation in voltage and pulse time allows separation of DNAs ranging in size from 2 kb to > 6,000 kb.

Contour-clamped homogeneous electric field electrophoresis (CHEF) is needed for the resolution of DNA fragments > 6,000 kb.

Single cell gel electrophoresis (SCGE) /comet assay is a sensitive microscopic gel electrophoresis method capable of measuring DNA strand breaks in individual cells. An **alkaline comet assay** was then introduced. A number of variations of this assay greatly increase the flexibility and utility of this technique for detecting various manifestations of DNA damage (e.g. single- and double-strand breaks, oxidative induced base damage, and DNA–DNA/DNA–protein crosslinking) and DNA repair in virtually any eukaryotic cell. It has been used widely in genetic toxicology, radiation biology and medical and environmental research for the detection of single-strand breaks, double-strand breaks and alkali-labile sites in DNA and can be employed also to visualize DNA degradation due to necrosis or apoptosis. The assay essentially involves embedding of eukaryotic cells in low-melting agarose gel on a slide, followed by cell lysis and electrophoresis. Fragmented DNA is visualized after appropriate staining as a tail projecting from the original cell that resembles the tail of a comet. Visual display of the detected comet shape and calculation of comet head radius, tail length and integral intensity of fluorescence can provide information about the nature and degree of DNA fragmentation.

BOX 13.2 NOVEL NUCLEIC ACID ASSAYS AND THEIR APPLICATIONS

Apoptosis is associated with the fragmentation of chromosomal DNA into multiples of 180 bp nucleosomal unit known as DNA laddering. In DNA laddering assay, small fragments of oligonucleosomal DNA is extracted selectively from the cells whereas the higher molecular weight DNA stays associated with the nuclei. The isolated DNA is separated by electrophoresis and visualized using ethidium bromide. DNA ladder is considered a hallmark of cells undergoing apoptosis.

Branched-chain DNA assay is employed to measure RNA in formalin-fixed paraffin-embedded (FF-PE) tissues. This assay does not require RNA isolation and enzymatic preamplification. The assay helps in the correct identification of the over expression of known cancer genes. It also appears to be well suited for the clinical analysis of FF-PE tissues with diagnostic or prognostic gene expression biomarker panels for use in patient treatment and management.

Direct detection of base sequence in duplex nucleic acid has long been unfulfilled. The heteropolymeric triplex based genomic assay helps in the detection of pathogens or single nucleotide polymorphisms in human genomic samples.

Conformation-sensitive gel electrophoresis (CSGE) is a method that requires no special equipment or preparation of the DNA. It detects conformational differences between heteroduplexes and homoduplexes, which are detected in differences in electrophoretic migration that are accentuated by the use of a mild denaturing agent.

In **2-D DNA gel electrophoresis (2-DDGE)** high-molecular-weight DNA fragments restricted with a rare-cutter restriction enzyme are separated. After staining and photography, the lane is cut from the gel and incubated in a solution containing another restriction enzyme. Subsequently, the gel-strip is placed on top of a new gel for electrophoretic separation in the second dimension.

Denaturing gradient gel electrophoresis (DGGE) resolves partially denatured double-stranded DNA in precisely defined conditions of temperature and denaturant concentration. Different alleles may denature to varied extent under such conditions and migrate differently on DGGE acrylamide gels.

Denaturing detergent gradient gel electrophoresis (DDGE) is based on differential electrophoretic migration of single- versus double-stranded DNA, which can easily be detected by commonly used ethidium bromide staining methods.

Temperature-gradient gel electrophoresis (TGGE) is based on differential electrophoretic migration of single- versus double-stranded DNA, which can easily be detected by commonly used ethidium bromide staining methods.

ISOLATION OF INDIVIDUAL HUMAN CHROMOSOMES

High Speed Sorting Method Based on Fluorescent Staining

DNA is treated with several fluorescent dyes and individual chromosomes are recognized by their relative fluorescent intensities. The dyes Hoechst 33258 and chromomycin A3 in combination respond to light of different wavelengths and also bind to DNA differently. Hoechst binds to A-T-rich regions of DNA whereas chromomycin binds to G-C-rich regions. Chromomycin fluoresces in the presence of a laser tuned to 458 nm and Hoechst fluoresces in the presence of a UV laser. By plotting the fluorescence intensity of chromosomes with these two dyes on the X and Y axes, they can be identified from the flow karyotype. Flow cytometry techniques allow the isolation of the identified chromosomes.

Chromosomes are isolated from cells that have been arrested in metaphase by treatment with colcemid which inhibits spindle formation. These chromosomes are purified in buffer and treated with the two dyes. As the chromosome containing buffer passes through the laser beams, they are identified.

High Speed Sorting Method Based on Flow Cytometry

The chromosomes in the buffer are passed with force to form minute droplets of about 215,000 per second by passing through a vibrator. Specific droplets carrying the identified chromosomes are then charged either positively or negatively and passed between deflection plates. Positively charged and negatively charged droplets are thus isolated. At a rate of 200 chromosomes per second, it is possible to isolate 0.1 g of DNA in less than an hour (Figure 13.7).

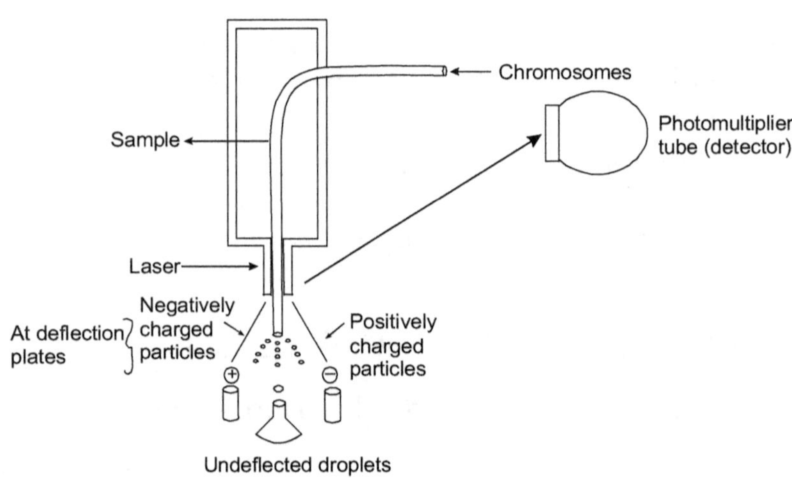

Figure 13.7 Flow cytometry device used to separate chromosomes at high speed

By fluorescent staining, all chromosomes are resolved except chromosome 9–12. Chromosome 21 also is difficult to be separated which forms a debris and clumps during sorting (Figure 13.8). Such problems can be overcome by using hybrid cell lines of hamsters containing only one human chromosome.

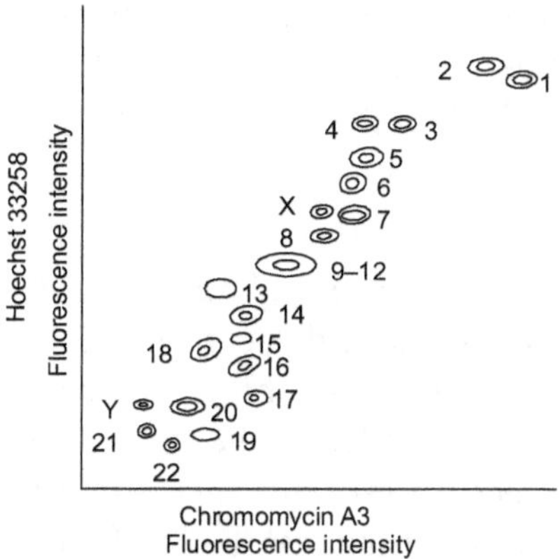

Figure 13.8 Flow karyotype of human chromosomes at high resolution under low speed sorting

BIOSENSORS

A biosensor is an analytical device that makes use of biological material to detect chemical species directly without any requirement for complicated processes. High specificity and sensitivity of biomolecules and biological systems have made this device possible.

A biosensor consists of a biological sensing element (antibody, enzyme, cells) which is in contact with a physical or chemical transducer (an electrode or optical detector). A desired analyte can be measured by selective transduction of a parameter of the biological sensing element–analyte reaction into a quantifiable electrical signal (Figure 13.9).

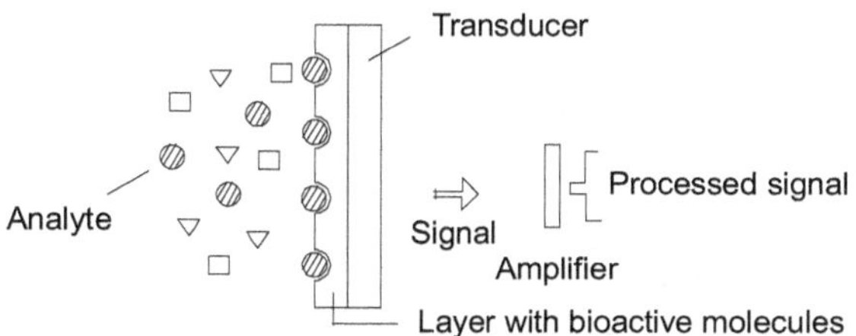

Figure 13.9 Biosensor

If the analyte is a nucleic acid, the bioreceptor can be a DNA or RNA, nucleic acid binding proteins or intercalating dyes. Efficient transducer systems are used in biosensor technology. Surface plasmon

(SP) and waveguide devices is the system for DNA/RNA and the resonant signal is the mode of measurement. SP resonance is used to screen for single-nucleotide polymorphisms and DNA hybridization.

DNA BIOCHIPS

Miniaturization offers a number of advantages. Thousands of DNA hybridization reactions can be done in parallel on a single glass slide replacing Southern and northern blots. An integrated laboratory can be provided on a simple CD-ROM-like disk, a lab-on-a-chip. Reaction volumes can be reduced and surface-to-volume ratios can be increased. Experiments can be performed faster and at reduced cost. Nanoplates with volumes between 100 and 1000 nl per well are used. DNA tests based on thin-film electrodes were the first developed DNA sensors. Thin-film sensor with oligonucleotides or cDNA were attached to a metal electrode. Glucose oxidase was attached to the oligonucleotide which produced hydrogen peroxide in the presence of the substrate, glucose. The hydrogen peroxide formed was then quantified by the sensor.

A microarray was then designed as a glass plate with 96 wells. The plate is also made of silicon, quartz or plastic. Classical membranes like nitrocellulose and nylon offer strong binding. For detection, an enzyme-linked immunosorbent assay (ELISA) and a scanning CCD detector were used.

Biochips represent a special class of biosensors. They bear various transducer elements and are based on integrated circuit microchips. It resembles the computer chip. In general, a biochip is a device that bears a high-density array of probes used for a variety of assays. Any device comprising a two-dimensional array and having biological materials on a solid substrate is referred to as a biochip. Biochips offer the advantage of combining miniaturization and the possibility of low-cost mass production.

DNA chips rely on DNA hybridization. One of the interacting partners is immobilized on the chip surface. Subsequent probing of the chip surface with a solution containing a biomolecule with an affinity for the immobilized partner results in a recognition and binding event.

Usually, DNA is spotted on activated glass surface coated with poly L-lysine or aminosilane. The efficiency of this step depends on the homogeneity of the DNA spots and the chemical properties of the solution in which the DNA is dissolved. The spotty buffer is mostly saline sodium citrate buffer. Another step of DNA chip manufacture is the processing of the glass surface after spotting. The remaining unreacted amino residues of poly L-lysine polymer or aminosilane have to be deactivated or blocked. For this, the surface is probed with succinic anhydride in aqueous borate-buffered 1-methyl 2-pyrrolidinone, when the amines are converted to carboxylic moieties. To prevent the interaction of the immobilized DNA with the blocking solution, a nonpolar solvent is used.

Direct spotting of cDNA involves immobilization of pre-synthesized oligonucleotides. This is used in specific diagnostic tests and drug discovery. *In situ* oligonucleotide synthesis makes use of a parallel on-chip synthesis of oligonucleotide probes by photochemistries.

DNA microarrays are used for the analysis of both single nucleotide polymorphism (SNP) and mutations and also for the detection of microorganisms. DNA chips are used for gene-expression studies. Unlike protein chips, DNA chips can be operated even under denaturing conditions, and can be subjected to harsh treatments such as heating, freezing and drying. Immobilization at the chip surface can be done in any direction using DNA. Probes can be produced easily and fast and are easily labelled, mostly by fluorophores.

CHROMOSOME BANDING TECHNIQUES

Until the 1970s, the mitotic chromosomes when viewed under the light microscope were distinguished only by their relative size and the positions of their centromeres. New cytological procedures were developed for the differential staining along the longitudinal axis of mitotic chromosomes and referred to as **chromosome banding techniques**. The technique was first devised by Mary Lou Pardue and Joe Gall. When chromosome preparations were heat-denatured and then treated with Giemsa stain, a unique staining pattern was got. The centromeric regions of mitotic chromosomes preferentially took up the stain, which were heterochromatic regions. The staining pattern is referred to as **C-banding.** When fluorescent dyes were used, still greater staining differentiation of metaphase chromosomes was achieved. These fluorescent dyes are found to bind to nucleoprotein complexes and produce unique banding patterns. When the chromosomes are treated with the fluorochrome quinacrine mustard and viewed under a fluorescent microscope, each of the 23 human chromosome pairs were distinguished. The bands produced by this method are called **Q-bands**. Another method producing **G-bands**, involves the digestion of the mitotic chromosomes with the proteolytic enzyme trypsin, followed by Giemsa staining. Another technique results in the reverse G-band staining pattern, called an **R-band** pattern.

A variety of staining reactions under different conditions reflect the heterogeneity and complexity of chromosome compositions.

TECHNIQUES BASED ON THE SEDIMENTATION BEHAVIOUR OF NUCLEIC ACIDS

DNA and RNA molecules are extensively analysed by centrifugation techniques (Figure 13.10). In **velocity sedimentation,** nucleic acids are separated according to their nucleotide length. The sample containing the mixture of nucleic acid molecules is carefully layered over a solution containing an increasing concentration of sucrose which provides a density or viscosity gradient column from top to bottom. When high centrifugal force is applied, the molecules move through the gradient at a rate determined by their sedimentation coefficient. By using marker molecules of known sedimentation coefficient, the S value of unknown components can be determined. In **equilibrium or isopycnic sedimentation,** nucleic acids are separated based on their buoyant density. The technique uses a caesium chloride solution because caesium is a very heavy atom of molecular mass 132 and a density of 1.873 g/ml. Its chloride salt is highly soluble in water. Therefore CsCl concentrations greater than 6 M will have a very high density. DNA has a density of about 1.70 g/ml and thus CsCl solution is prepared at this density. Upon centrifugation at very high speeds, caesium atoms are sufficiently heavy to move in response to the high gravitational forces and form a density gradient in the centrifuge tube, of density 1.6 g/ml at the top to 1.8 g/ml at the bottom. The DNA molecules equilibrate at a density that equals the density of the surrounding CsCl and forms a sharp band in the tube. Using this technique, in 1958 Matthew Meselson and Franklin Stahl used heavy isotope of nitrogen ^{15}N to label the parental DNA. The new DNA synthesized was made to take up ^{14}N precursors in order to distinguish from the parental strands. The bands of DNA obtained at different regions of the density gradient column confirmed the semi-conservative replication of DNA.

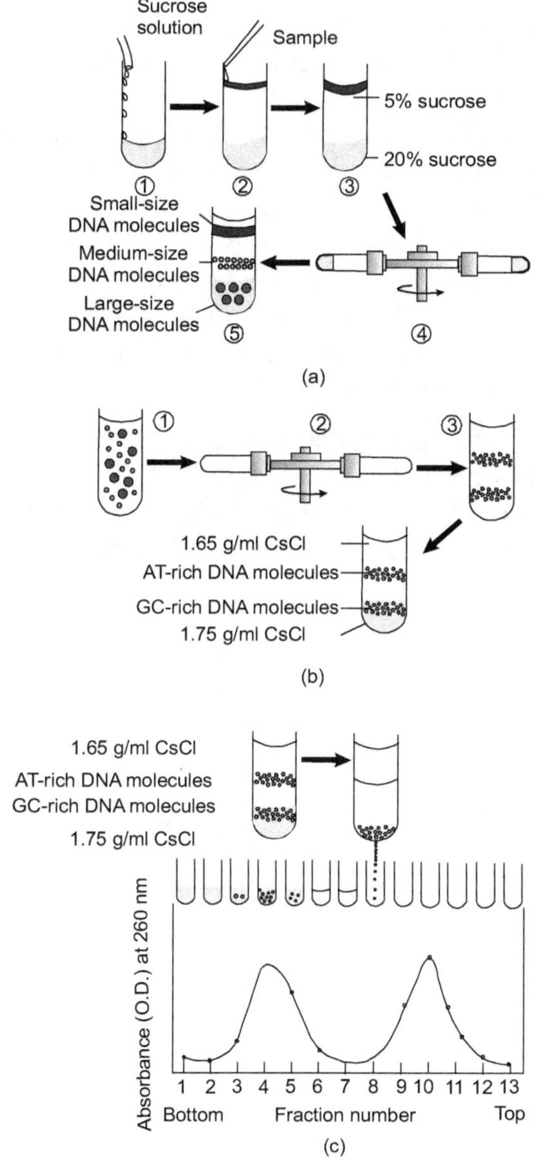

Figure 13.10   Techniques of nucleic acid sedimentation

X-RAY DIFFRACTION ANALYSIS OF DNA

An X-ray pattern of DNA is obtained by holding a stretched fibre with many DNA molecules in a vertical direction and exposing it to a collimated monochromatic beam of X-rays. A small fraction of these rays get diffracted while most of the beam travels through the specimen with no change in direction. A photographic film is held on the other side of the specimen. When a hole in made in the centre of the film, the incident undiffracted beam passes through it (Figure 13.11). As per Bragg's law, diffraction patterns are got when $2d\sin\theta = n\lambda$, where d is the distance between identical repeating structural elements, θ is the angle between the incident beam and the regularly spaced diffracting planes, λ is the wavelength of the X-ray used and n is the order of diffraction. When $\sin\theta \approx \theta$ and $d \approx 1/\theta$, with a spot far out on the photographic film, it indicates the presence of a repeating element of small dimension and vice versa.

When Watson and Crick studied the X-ray diffraction pattern of DNA, they observed strong 3.4 Å and 34 Å spacings with central crosslink pattern. This suggested a helical structure for DNA. They interpreted this as due to the hydrogen-bonded antiparallel double helix structure.

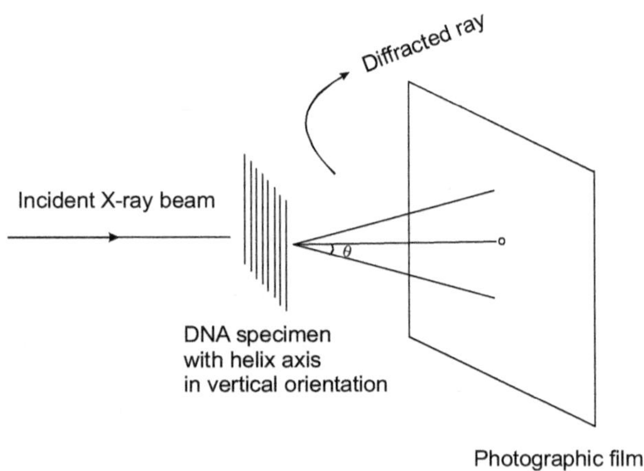

Figure 13.11   Set-up for obtaining DNA diffraction pattern

USE OF RADIOISOTOPES AS INDISPENSABLE TOOLS FOR THE DETECTION OF BIOLOGICAL MOLECULES AND BIOLOGICAL EVENTS

Labelling the precursors of macromolecules has simplified many standard biochemical assays and enhanced the researchers to follow biochemical events in whole cells and in cell extracts. The radioisotopes commonly used in biological research are tritium (^{3}H), carbon (^{14}C), phosphorus (^{32}P), sulphur (^{35}S) and iodine (^{125}I, ^{131}I) with half-life of 12.35 years, 5730 years, 14.3 days, 87.5 days, 60 days and 8.07 days respectively. The presence of a radioisotope does not change the chemical properties of a molecule. Radioisotopes emit easily detectable particles and the fate of radiolabelled molecules can be traced in cells and cellular extracts.

Some compounds, if radiolabelled, are not suitable for studies with whole cells because they do not enter the cells. For example, if ATP is labelled with ^{32}P, although it contributes to DNA and RNA synthesis in a cell free-system, it cannot do so in whole cells because it cannot get into the cell. But labelled orthophosphate (^{32}PO$_4^{3-}$) in the medium can enter into bacterial and animal cells and then be incorporated into phosphorylated proteins, nucleotides and later into DNA and RNA.

Many biological compounds like amino acids, nucleosides and a number of metabolic intermediates labeled with radioisotopes are commercially available. They vary in their specific activity, which is the amount of radioactivity per unit of material, measured in disintegrations per minute per millimole. Shorter the half-life of a radioisotope, higher is its specific activity. The specific activity of a labelled compound must be high enough so that sufficient radioactivity is incorporated into the molecule. For example, cellular protein preparations are labelled with ^{35}S containing methionine and cysteine, which have high specific activity. Commercial preparation of ^{3}H-labelled nucleic acid precursors have higher specific activities than the corresponding ^{14}C-labelled preparations. ^{32}P-labelled nucleotides are routinely used to label the nucleic acids in cell-free systems.

AUTORADIOGRAPHY

By the autoradiography technique, a radiolabel can be localized and recorded and may be tagged in a specimen. It involves production of an image in a photographic emulsion. The emulsion consists of silver halide crystals suspended in gelatin. The radioactive substance emits radiations namely β-particles or γ-rays. When these radiations pass through the emulsion, the silver ions are converted to silver atoms. This causes formation of a latent image which is converted to a visible image.

The silver atoms cause the entire silver halide crystal to be reduced to metallic silver.

There are two types of autoradiography—direct and indirect. In direct autoradiography, the sample is placed in contact with the film and the radioactive emissions produce black areas on the developed autoradiograph. By this type, weak to medium strength β-emitting radionuclides like ^{3}H, ^{14}C, and ^{35}S can be detected. High-energy β-particles like those from ^{32}P or γ-rays emitted from isotopes like ^{125}I cannot be detected by this type.

In indirect autoradiography, the emitted energy is converted to light by means of a scintillator by techniques like fluorography. In fluorography, the sample is impregnated with a liquid scintillator. The radioactive emissions transfer their energy to the scintillator molecules. These molecules emit photons which expose the photographic emulsion.

A single hit by a β-particle or γ-ray can produce hundreds of silver atoms. But a single hit by a photon of light produces only a single silver atom. But a single silver atom will be unstable and will revert to a silver ion very rapidly. The probability of a second photon being captured before the first silver atom has reverted is greater for larger amounts of radioactivity than for small amounts. Therefore, small amount of radioactivity is under-represented by using fluorography. This problem can be solved by using a pre-exposed film (to flash of light) and exposing the autoradiograph at –70°C.

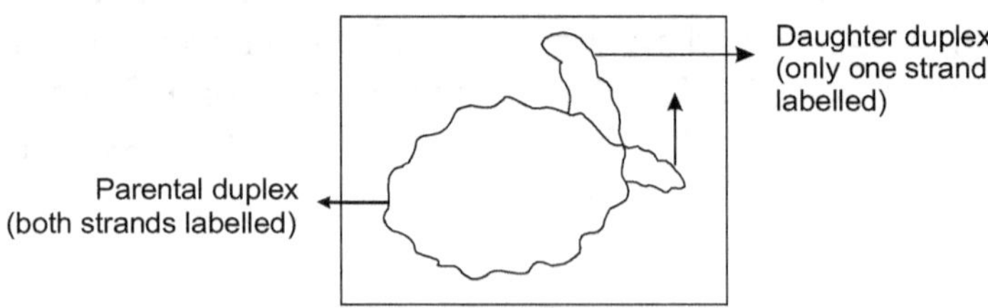

Figure 13.12 Autoradiogram of replicating *E. coli* chromosome

The first proof that *E. coli* DNA replicates as a circle came from an autoradiographic experiment. Cells were grown in a medium containing radioactive thymine (^{3}H-thymine) so that all DNA synthesized would be radioactive. Without fragmentation, the DNA was isolated and placed on photographic film. Each radioactive decay caused a tiny black spot on the film. After many months, there were enough spots to visualize the DNA with a microscope. The replicating DNA was found to take a θ (theta) structure. Therefore this mode of replication was referred to as θ replication. In some places, only one strand showed radiolabelling indicating that they are the new strands synthesized (Figure 13.12).

MOLECULAR HYBRIDIZATION

The property of renaturation of complementary single strands of nucleic acid is the basic principle of molecular hybridization which is a powerful analytical technique in molecular genetics. The renaturing single strands need not have to come from the denatured double-stranded molecule. If DNA strands are isolated from different organisms and if there is some degree of base complementarity between them, there will be formation of molecular hybrids when renatured. Also when single-stranded DNA and RNA are slowly cooled, hybridization may occur. The RNA molecules which have been transcribed from one of the double strands of DNA will form hybrid with the DNA template strand. A DNA–RNA hybrid will result. Hybridization can occur in solution or when DNA is bound either to a gel or to a special type of filter, facilitating recovery of the newly formed hybrids DNA:RNA hybridization can be performed using cytological preparations by a technique called *in situ* **hybridization**. DNA is fixed to a slide, denatured and hybridized to radioactive RNA. The hybrid that is formed is detected using autoradiography (Figure 13.13).

In situ hybridization (ISH) is used to visualize defined nucleic acid sequences in cellular preparations by hybridization of complementary probe sequences. Probe sequences can be labelled with isotopes, but nonisotopic ISH is used increasingly as it is considerably faster, usually has greater signal resolution and provides many options to simultaneously visualize different targets by combining various detection methods. The most popular protocols use fluorescence detection. These protocols have many applications, from basic gene mapping and diagnosis of chromosomal aberrations to detailed studies of cellular structure and function, such as the painting of chromosomes in three-dimensionally preserved nuclei. The hybridized probes are detected and visualized using fluorochrome-conjugated reagents (Figure 13.13a).

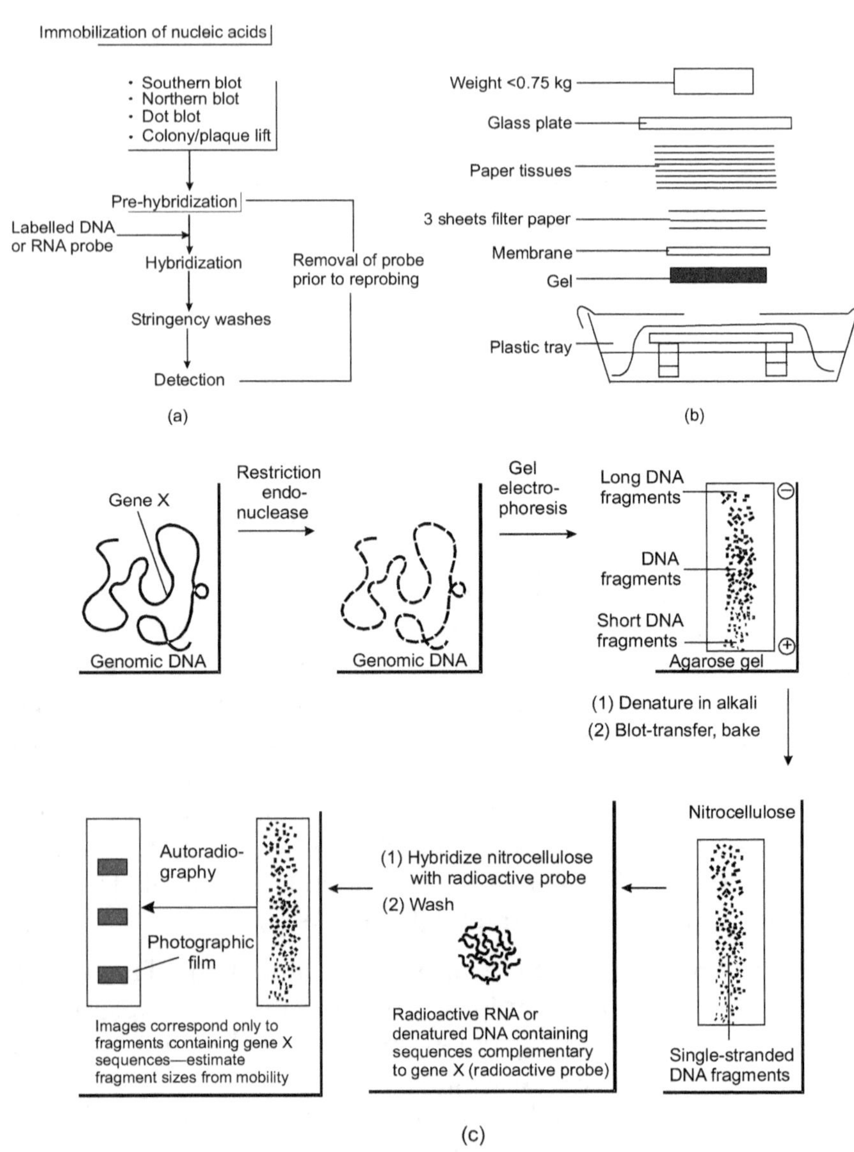

Figure 13.13 Blotting techniques

Southern blot is a technique for transferring DNA fragments separated by agarose gel electrophoresis to a nitrocellulose filter, on which specific fragments can then be detected by their hybridization to probes, which are labelled radioactively or using nonradioactive methods.

Restriction endonucleases are used to cut high molecular weight DNA strands into smaller fragments. The DNA fragments are then electrophoresed on an agarose gel to separate them by size. If some of the DNA fragments are larger than 15 kb, then prior to blotting, the gel may be treated with an acid, such as dilute HCl, which depurinates the DNA fragments, breaking the DNA into smaller pieces, thus allowing more efficient transfer from the gel to membrane. If alkaline transfer methods are used, the DNA gel is placed into an alkaline solution containing sodium hydroxide to denature the double-stranded DNA. Denaturation in an alkaline environment provides for improved binding of the negatively charged DNA to a positively charged membrane, separates it into single DNA strands for later hybridization to the probe and destroys any residual RNA that may still be present in the DNA.

A sheet of nitrocellulose or nylon membrane is placed on top (or below, depending on the direction of the transfer) of the gel. Pressure is applied evenly to the gel (either using suction, or by placing a stack of paper towels and a weight on top of the membrane and gel), to ensure good and even contact between gel and membrane (Figure 13.13b). Buffer transfer by capillary action from a region of high water potential to a region of low water potential (usually filter paper and paper tissues) is then used to move the DNA from the gel on to the membrane. Ion-exchange interactions bind the DNA to the membrane due to the negative charge of the DNA and positive charge of the membrane. The membrane is then baked, i.e., exposed to high temperature (60 to 100°C) (in the case of nitrocellulose) or to ultraviolet radiation (nylon) to permanently and covalently crosslink the DNA to the membrane. The membrane is then exposed to a hybridization probe—a single DNA fragment with a specific sequence whose presence in the target DNA is to be determined. The probe DNA is labelled so that it can be detected, usually by incorporating radioactivity or tagging the molecule with a fluorescent or chromogenic dye. In some cases, the hybridization probe may be made from RNA, rather than DNA. To ensure the specificity of the binding of the probe to the sample DNA, most common hybridization methods use salmon testes (sperm) DNA for blocking of the membrane surface and target DNA, deionized formamide and detergents such as SDS to reduce non-specific binding of the probe. After hybridization, excess probe is washed from the membrane and the pattern of hybridization is visualized on X-ray film by autoradiography in the case of a radioactive or fluorescent probe, or by development of colour on the membrane if a chromogenic detection method is used (Figure 13.13c).

Northern blot is a technique analogous to Southern blot technique, but performed on fragments of RNA instead of DNA; a nylon membrane is often substituted for the nitrocellulose filter. Northern blotting and hybridization are used to study gene expression by detecting RNA species of interest and to identify alternate RNA splicing patterns. The protocol describes the transfer of RNA from agarose gels or denaturing polyacrylamide gels depending on the size of the RNA to be detected to nylon membranes and the fixation of the RNA to the membrane. A notable difference in the procedure in case of agarose gels, (as compared with the Southern blot) is the addition of formaldehyde which acts as a denaturant. For smaller fragments, denaturing polyacrylamide urea gels are employed. As in the Southern blot, the hybridization probe may be made from DNA or RNA.

POLYMERASE CHAIN REACTION (PCR)

PCR is a technique used to amplify (i.e., replicate) a piece of DNA by *in vitro* enzymatic replication. As PCR progresses, the DNA generated is itself used as template for replication. This sets in motion a chain reaction in which the DNA template is exponentially amplified. With PCR, it is possible to amplify a single or few copies of a piece of DNA across several orders of magnitude, generating millions or more copies of the DNA piece. PCR can be performed without restrictions on the form of DNA and it can be extensively modified to perform a wide array of genetic manipulations.

Almost all PCR applications employ a heat-stable DNA polymerase, such as *Taq* polymerase, an enzyme derived from the bacterium *Thermus aquaticus*. This DNA polymerase enzymatically assembles a new DNA strand from DNA building blocks, the nucleotides, using single-stranded DNA as template and DNA oligonucleotides (also called DNA primers) required for initiation of DNA synthesis. Majority of PCR methods use thermal cycling, i.e., alternately heating and cooling the PCR sample to a defined series of temperature steps. These different temperature steps are necessary to bring about physical separation of the strands in a DNA double helix (DNA melting), and permit DNA synthesis by the DNA polymerase to selectively amplify the target DNA. PCR is now a common and often indispensable technique used in medical and biological research laboratories for a variety of applications. These include DNA cloning for sequencing, DNA-based phylogeny, or functional analysis of genes, the diagnosis of hereditary diseases, the identification of genetic fingerprints (used in forensics and paternity testing) and the detection and diagnosis of infectious diseases.

PCR is used to amplify specific regions of a DNA strand (the DNA target). This can be a single gene, a part of a gene, or a non-coding sequence. Most PCR methods typically amplify DNA fragments of up to 10 kilo base pairs, although some techniques allow for amplification of fragments up to 40 kb in size.

Components of PCR

A basic PCR set-up requires several components and reagents.

- DNA template that contains the DNA region (target) to be amplified.

- One or more primers, which are complementary to the DNA regions at the 5' and 3' ends of the DNA region.

- A DNA polymerase such as *Taq* polymerase or another DNA polymerase with a temperature optimum at around 70°C.

- Deoxyribonucleoside triphosphates (dNTPs), the building blocks from which the DNA polymerases synthesize a new DNA strand.

- Buffer solution, providing a suitable chemical environment for optimum activity and stability of the DNA polymerase.

- Divalent cations, magnesium or manganese ions; generally Mg^{2+} is used, but Mn^{2+} can be utilized for PCR-mediated DNA mutagenesis, as higher Mn^{2+} concentration increases the error rate during DNA synthesis.

- Monovalent cation potassium ions.

PCR Cycle

(1) Denaturing at 94–96°C (2) Annealing at ~65°C (3) Elongation at 72°C (Figure 13.14).

⊿ *Initialization step*   This step consists of heating the reaction mixture to a temperature of 94–96°C (or 98°C if extremely thermostable polymerases are used), which is held for 1–9 minutes.

⊿ *Denaturation step*   This step is the first regular cycling event and consists of heating the reaction to 94–98°C for 20–30 seconds. It causes melting of DNA template and primers by disrupting the hydrogen bonds between complementary bases of the DNA strands, yielding single strands of DNA.

⊿ *Annealing step*   The reaction temperature is lowered to 50–65°C for 20–40 seconds allowing annealing of the primers to the single-stranded DNA template. Typically the annealing temperature is about 3–5°c below the T_m of the primers used. Stable DNA–DNA hydrogen bonds are formed only when the primer sequence very closely matches the template sequence. The polymerase binds to the primer–template hybrid and begins DNA synthesis.

⊿ *Extension/elongation step*   The temperature at this step depends on the DNA polymerase used; *Taq* polymerase has its optimum activity temperature at 75–80°C and commonly a temperature of 72°C is used with this enzyme. At this step the DNA polymerase synthesizes a new DNA strand complementary to the DNA template strand by adding dNTPs that are complementary to the template in 5′ to 3′ direction, condensing the 5′-phosphate group of the dNTPs with the 3′-hydroxyl group at the end of the nascent (extending) DNA strand. The extension time depends both on the DNA polymerase used and on the length of the DNA fragment to be amplified. As a rule-of-thumb, at its optimum temperature, the DNA polymerase will polymerize a thousand bases in one minute.

⊿ *Final elongation*   This single step is occasionally performed at a temperature of 70–74°C for 5–15 minutes after the last PCR cycle to ensure that any remaining single-stranded DNA is fully extended.

⊿ *Final hold*   This step at 4–15°C for an indefinite time may be employed for short-term storage of the reaction.

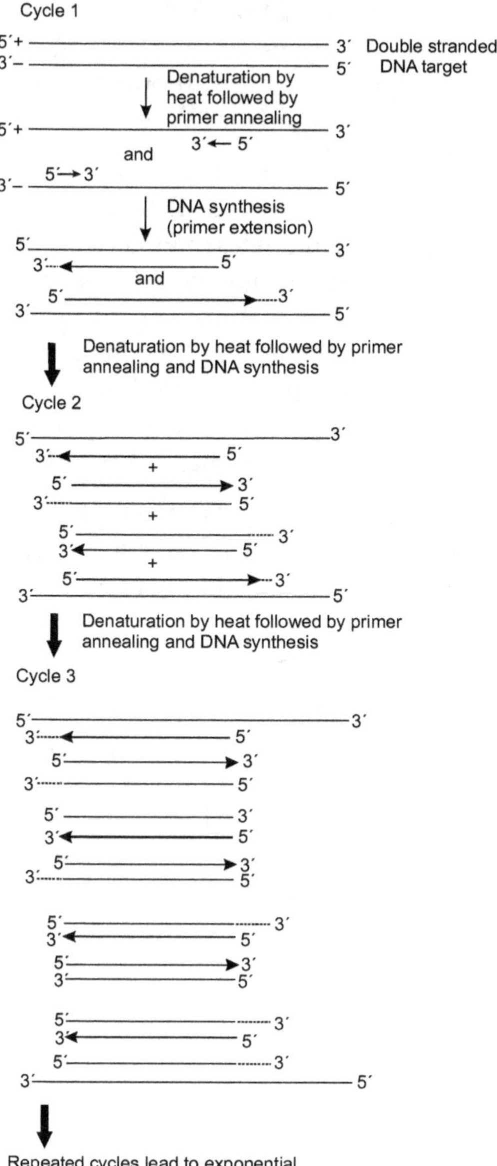

Figure 13.14   Schematic drawing of the PCR cycle

PCR allows isolation of DNA fragments from genomic DNA by selective amplification of a specific region of DNA. This use of PCR augments many methods, such as generating hybridization probes for

Southern or northern hybridization and DNA cloning, which require larger amounts of DNA, representing a specific DNA region. PCR supplies these techniques with high amounts of pure DNA, enabling analysis of DNA samples even from very small amounts of starting material.

Other applications of PCR include DNA sequencing to determine unknown PCR-amplified sequences in which one of the amplification primers may be used in Sanger sequencing, isolation of a DNA sequence to expedite recombinant DNA technologies involving the insertion of a DNA sequence into a plasmid or the genetic material of another organism. PCR may also be used for genetic fingerprinting, a forensic technique used to identify a person or organism by comparing experimental DNAs through different PCR-based methods.

Some PCR "fingerprint" methods have high discriminative power and can be used to identify genetic relationships between individuals, such as parent–child or between siblings and are used in paternity testing. This technique may also be used to determine evolutionary relationships among organisms.

Because PCR amplifies the regions of DNA that it targets, PCR can be used to analyse extremely small amounts of sample. This is often critical for forensic analysis, when only a trace amount of DNA is available as evidence. PCR may also be used in the analysis of ancient DNA that is thousands of years old. PCR-based techniques have been successfully used on animals, such as a forty-thousand-year-old mammoth and also on human DNA, in applications ranging from the analysis of Egyptian mummies to the identification of a Russian Tsar.

Viral DNA can likewise be detected by PCR. The primers used need to be specific to the targeted sequences in the DNA of a virus and PCR can be used for diagnostic analyses or DNA sequencing of the viral genome. The high sensitivity of PCR permits virus detection soon after infection and even before the onset of disease. Such early detection may give physicians a significant lead in treatment. The amount of virus in a patient can also be quantified by PCR-based DNA quantitation techniques.

Quantitative PCR methods allow the estimation of the amount of a given sequence present in a sample. It is a technique often applied to quantitatively determine levels of gene expression. Real-time PCR is an established tool for DNA quantification that measures the accumulation of DNA product after each round of PCR amplification.

Types of PCR

Allele-specific PCR   This diagnostic or cloning technique is used to identify or utilize single nucleotide polymorphisms (SNPs) (single base differences in DNA). It requires prior knowledge of a DNA sequence, including differences between alleles, and uses primers whose 3'ends encompass the SNP. PCR amplification under stringent conditions is much less efficient in the presence of a mismatch between template and primer. Therefore successful amplification with an SNP-specific primer signals presence of the specific SNP in a sequence.

Assembly PCR   Assembly PCR is the artificial synthesis of long DNA sequences by performing PCR on a pool of long oligonucleotides with short overlapping segments. The oligonucleotides alternate between sense and antisense directions and the overlapping segments determine the order of the PCR fragments thereby selectively producing the final long DNA product.

Asymmetric PCR   Asymmetric PCR is used to preferentially amplify one strand of the original DNA more than the other. It finds use in some types of sequencing and hybridization probing where only one of the two complementary stands is required. PCR is carried out as usual, but with a great excess of the primers for the chosen strand. Due to the slow (arithmetic) amplification later in the reaction after the limiting primer has been used up, extra cycles of PCR are required. A recent modification on this process, known as **Linear-After-The-Exponential**-PCR (LATE-PCR), uses a limiting primer with a higher melting temperature (T_m) than the excess primer to maintain reaction efficiency as the limiting primer concentration decreases mid-reaction.

Colony PCR   Bacterial colonies can be rapidly screened by PCR for correct DNA vector constructs. Selected bacterial colonies are picked with a sterile toothpick and dabbed into the PCR master mix or sterile water. The PCR is started with an extended time at 95°C when standard polymerase is used or with a shortened denaturation step at 100°C and special chimeric DNA polymerase.

Helicase-dependent amplification   This technique is similar to traditional PCR, but uses a constant temperature rather than cycling through denaturation and annealing/extension cycles. DNA helicase, the enzyme that unwinds DNA, is used in place of thermal denaturation.

Hot-start PCR   This is a technique that reduces non-specific amplification during the initial set-up stages of the PCR. The technique may be performed manually by heating the reaction components to the melting temperature (e.g. 95°C) before adding the polymerase. Specialized enzyme systems have been developed that inhibit the polymerase's activity at ambient temperature, either by the binding of an antibody or by the presence of covalently bound inhibitors that dissociate only after a high-temperature activation step. Hot-start/cold-finish PCR is achieved with new hybrid polymerases that are inactive at ambient temperature and are instantly activated at elongation temperature.

Intersequence-specific (ISSR) PCR   A PCR method for DNA fingerprinting that amplifies regions between some simple sequence repeats to produce a unique fingerprint of amplified fragment lengths.

Inverse PCR   A method used to allow PCR when only one internal sequence is known. This is especially useful in identifying flanking sequences to various genomic inserts. This involves a series of DNA digestion and self ligation, resulting in known sequences at either end of the unknown sequence.

Ligation-mediated PCR   This method uses small DNA linkers ligated to the DNA of interest and multiple primers annealing to the DNA linkers. It has been used for DNA sequencing, genome walking and DNA footprinting.

Methylation-specific PCR (MSP)   The MSP method is used to detect methylation of CpG islands in genomic DNA. DNA is first treated with sodium bisulphite, which converts unmethylated cytosine bases to uracil, that is recognized by PCR primers as thymine. Two PCR reactions are then carried out on the modified DNA, using primer sets identical except at any CpG islands within the primer sequences. At these points, one primer set recognizes DNA with cytosines to amplify methylated DNA and one set recognizes DNA with uracil or thymine to amplify unmethylated DNA. MSP using Q-PCR can also be performed to obtain quantitative rather than qualitative information about methylation.

Multiplex ligation-dependent probe amplification (MLPA) PCR   This permits multiple targets to be amplified with only a single primer pair, thus avoiding the resolution limitations of multiplex PCR.

Multiplex-PCR   The use of multiple, unique primer sets within a single PCR reaction produces amplicons of varying sizes specific to different DNA sequences. By targeting multiple genes at once, additional information may be gained from a single test run that otherwise would require several times the reagents and more time to perform. Annealing temperatures for each of the primer sets must be optimized to work correctly within a single reaction, and amplicon sizes, i.e., their base pair length, should be different enough to form distinct bands when visualized by gel electrophoresis.

Nested PCR   This type increases the specificity of DNA amplification, by reducing background due to non-specific amplification of DNA. Two sets of primers are being used in two successive PCR reactions. In the first reaction, one pair of primer is used to generate DNA products, which besides the intended target, may still consist of non-specifically amplified DNA fragments. The product(s) are then used in a second PCR reaction with a set of primers whose binding sites are completely or partially different from and located 3' of each of the primers used in the first reaction. Nested PCR is often more successful in specifically amplifying long DNA fragments than conventional PCR, but it requires more detailed knowledge of the target sequences.

Overlap-extension PCR   This is a genetic engineering technique allowing the construction of a DNA sequence with an alteration inserted beyond the limit of the longest practical primer length.

Quantitative PCR (Q-PCR)   This is used to measure the quantity of a PCR product (preferably real-time). It is the method of choice to quantitatively measure starting amounts of DNA, cDNA or RNA. Q-PCR is commonly used to determine the presence of a DNA sequence in a sample and the number of copies in the sample. The method with currently the highest level of accuracy is **quantitative real-time PCR**. It is often known as RT-PCR (**Real Time** PCR)/RQ-PCR/QRT-PCR/ RTQ-PCR. RT-PCR commonly refers to reverse transcription PCR, which is often used in conjunction with Q-PCR. QRT-PCR methods use fluorescent dyes, such as Sybr Green, or fluorophore-containing DNA probes to measure the amount of amplified product in real time.

RT-PCR   **Reverse Transcription** PCR is a method used to amplify, isolate or identify a known sequence from a cellular or tissue RNA. The PCR is preceded by a reaction using reverse transcriptase to convert RNA to cDNA. RT-PCR is widely used in expression profiling, to determine the expression of a gene or to identify the sequence of an RNA transcript, including transcription start and termination sites and, if the genomic DNA sequence of a gene is known, to map the location of exons and introns in the gene. The 5′ end of a gene (corresponding to the transcription start site) is typically identified by an RT-PCR method, named **RACE-PCR**, short for Rapid Amplification of cDNA ends PCR.

TAIL-PCR   **Thermal asymmetric interlaced** PCR is used to isolate unknown sequence flanking a known sequence. Within the known sequence TAIL-PCR uses a nested pair of primers with differing annealing temperatures; a degenerate primer is used to amplify in the other direction from the unknown sequence.

Touchdown PCR   This is a variant of PCR that aims to reduce nonspecific background by gradually lowering the annealing temperature as PCR cycling progresses. The annealing temperature at the initial cycles is usually a few degrees (3–5°C) above the T_m of the primers used, while at the later cycles, it is a few degrees (3–5°C) below the primer T_m. The higher temperatures give greater specificity for primer binding and the lower temperatures permit more efficient amplification from the specific products formed during the initial cycles.

Taq polymerase, a DNA polymerase purified from the thermophilic bacterium *Thermus aquaticus* that naturally occurs in hot (50 to 80°C/ 120 to 175°F) environments, paved the way for dramatic improvements of the PCR methods. The DNA polymerase isolated from *T. aquaticus* is stable at high temperatures remaining active even after DNA denaturation, thus obviating the need to add new DNA polymerase after each cycle. This allowed an automated thermocycler-based process for DNA amplification.

SEQUENCING OF DNA

In the 1940s, techniques were available to determine the base composition of DNA. It was only in the 1960s that methods were developed to sequence the nucleotides in nucleic acids. Initially, the amino acid sequence of proteins was determined from which the nucleotide sequence was determined either in the RNA or DNA. In 1965, the sequence of a tRNA molecule with 74 nucleotides was determined.

Two methods were developed to determine the sequence of nucleotides in DNA—one, a chemical method developed by Allan Maxam and Walter Gilbert in 1977 in which DNA is cleaved at specific bases and another method developed by Fred Sanger and his colleagues in which DNA is synthesized but is made to terminate at a given base. In both the methods, the DNA to be sequenced is subjected to four individual reactions, one for each base. The products of the four reactions are a series of DNA fragments that differ in length by only one nucleotide. By electrophoresis, these reaction products are separated in four adjacent lanes on a gel. Each band on the gel corresponds to a base and its position from which the sequence of the DNA segment can be read.

Maxam–Gilbert Chemical Method of DNA Sequencing

In this method, the complementary strands of the DNA fragments to be sequenced are separated and recovered. The strand to be sequenced is labelled at its 5' end with radioactive ^{32}P using the enzyme polynucleotide kinase. By this, the specific DNA fragment can be identified after gel electrophoresis.

Aliquots of DNA are subjected to four different chemical treatments which cleave the strand at a specific nucleotide. The reactions are carried out for a limited time in order to cleave a given DNA at small number of target nucleotides. Finally, a number of fragments are got and all will be labelled at the 5'end. These fragments differ in length depending on the cleavage point.

Four different reactions cut DNA at guanine (G > A), adenine (A > G), cytosine alone (C) or cytosine and thymine (C+ T). Purines are cleaved by using dimethylsulphate, which methylates guanine more efficiently than adenine and on heating, the cleavage occurs at the methylated site. Therefore, a DNA fragment is produced that is cleaved at a G residue (G > A). The reaction can be reversed by cleaving the strand in acid which produces fragments cleaved at an A residue (A > G). Pyrimidines are cleaved by using hydrazine which cleaves both cytosine and thymine. In high salt solution, only cytosine reacts. Cleavage at the site of the hydrazine reaction is done by treatment with piperidine. Therefore two types of fragments are got on treatment with hydrazine, one representing cytosine and the other representing C + T.

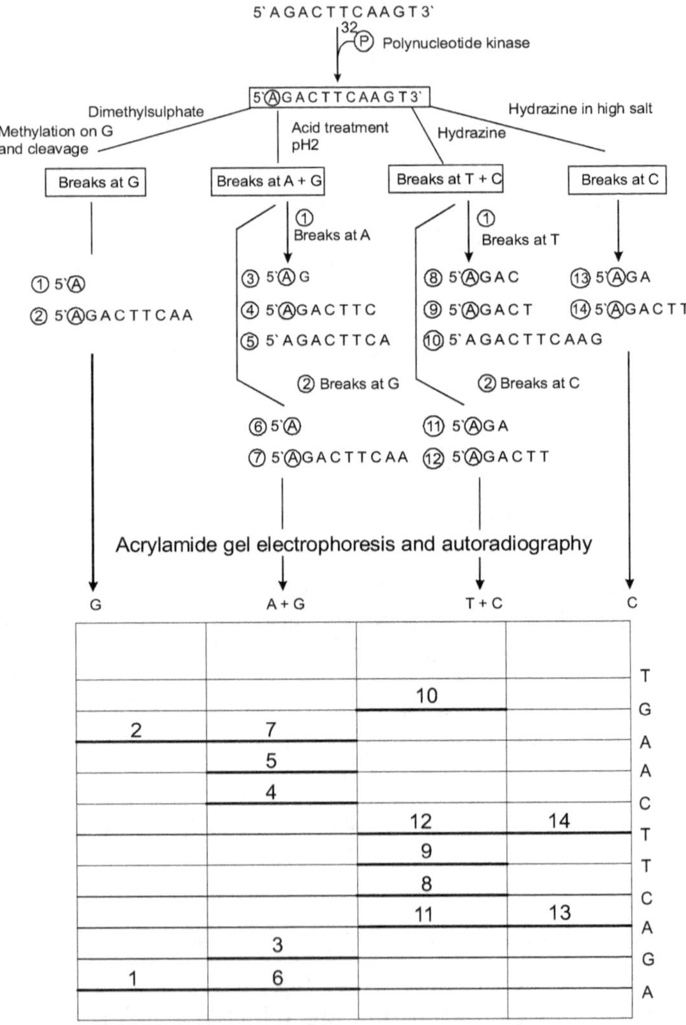

Figure 13.15　Maxam and Gilbert method of DNA sequencing

The mixture from each reaction containing the fragments are subjected to gel electrophoresis in adjacent lanes on the gel. In an electric field, smallest fragments migrate the farthest.

Let us consider a DNA strand which is radiolabelled with ^{32}P in 5'end. The production of the different fragments are given in figure. The fragments obtained on treatment with dimethylsulphate and hydrazine are subjected to electrophoresis and autoradiography. The above cleavage protocols are base-specific. Therefore, the rule is, if we get a fragment with n nucleotides and if that fragment was generated by a treatment active against a particular base, then that base is present in the position $n + 1$ in the DNA strand from 5' end. The four types of fragments are cleaved at sites G, A+G, T+C and C. In gel electrophoresis, a series of labeled fragments of varying length are formed. For each fragment, the actual length depends on the site at which the base is destroyed from the labeled end. The mixture containing the four types of fragments are subjected to electrophoresis side by side in four lanes and then the position of the bands studied by autoradiography (Figure 13.15).

Sanger's Method of DNA Sequencing

Sanger devised this method in 1975. It is also known as the plus-minus method. This is an enzymatic technique. Specific terminators of DNA chain elongation, 2'3'-dideoxynucleoside triphosphates, were synthesized. These ddNTP molecules can be incorporated normally into a growing DNA chain through their 5'-triphosphate group. But they cannot form phosphodiester bond with the next incoming dNTP. When a small amount of a specific dideoxy NTP (for example ddATP) is added along with the four normal substrates dATP, dCTP, dGTP and dTTP in the reaction mixture for DNA synthesis along with DNA polymerase, the products will be a series of fragments specifically terminated at the dideoxy residue. The DNA strand whose sequence has to be determined is used as the template and a primer is added to start synthesis.

A very low concentration of each of the ddNTP is added to each of the four separate reaction mixtures, each of which also contains the four normal dNTPs. Because of the low concentration, ddATP is not always incorporated at all locations opposite to T in the template. dATP which is at a higher level can also be incorporated. If we consider an example of a template sequence, 3'T A G C C G T A T G C A T G G C T A 5', ddATP can compete with dATP to be incorporated in positions either at 5, 6, 11, or 17 (Figure 13.16). In case they are added in one of these positions, DNA synthesis cannot proceed further. Therefore in the reaction mixture containing ddATP, a number of fragments of newly synthesized DNA strands of varying lengths are obtained. If the short primer is labelled with ^{32}P at the 5' end, these fragments when added to a separate lane in gel electrophoresis, the smallest fragments move the farthest and they are separated into discrete bands of varying length. The positions of the bands are indicative of the position of A in the new strand. The experiment is repeated separately with each of the other three ddDTPs.

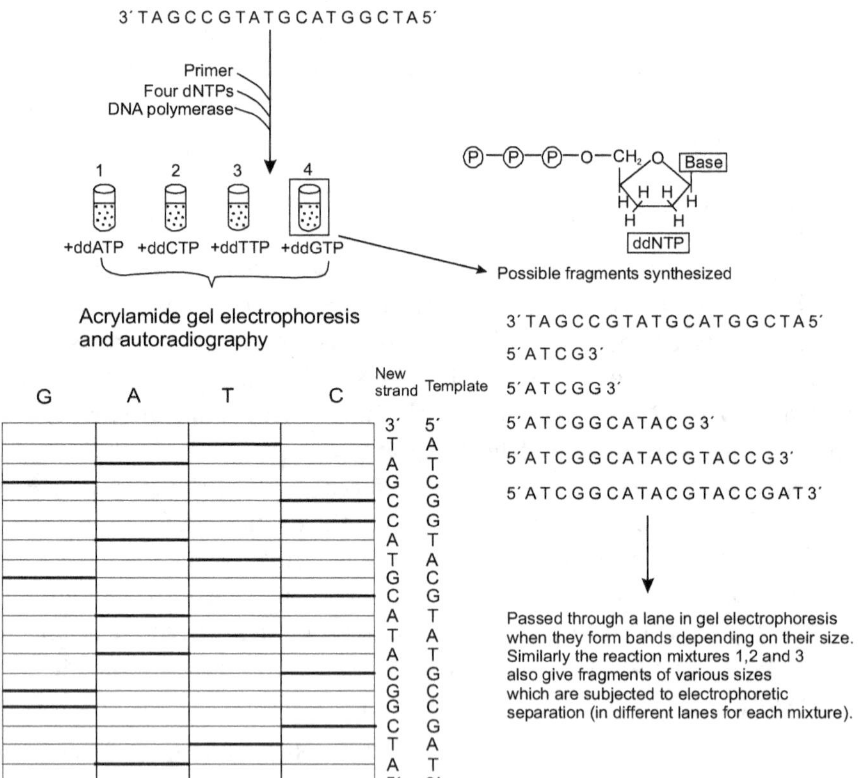

Figure 13.16   Sanger's method of sequencing of DNA

Fragments obtained in those cases are also passed through separated adjacent lanes and the positions of their bands are identified, which is possible by autoradiography. From the ladder pattern of the bands obtained, the nucleotide positions are identified (Figure 13.16).

RFLP IDENTIFICATION

RFLP is used for the detection of differences in genomic DNA. Restriction endonucleases cleave DNA at specific nucleotide recognition sequences. If the sequence changes even by just 1 bp, the enzyme cannot cut at that site. Variations in DNA sequence occur among individuals within a population at a frequency of about 1 in every 1000 bp. These small differences in the DNA sequence can be used to construct a map of any genome and are useful in identifying an individual. DNA fragments of varied length can be easily identified using agarose gel electrophoresis. Shorter DNA fragments migrate faster than longer DNA fragments. If a DNA fragment loses or gains a restriction endonuclease cleavage site, the fragment size obtained after digestion is likely to vary. In addition to changes in nucleotide sequences at the restriction site, changes in distances between the cut sites may reflect deletions, additions or rearrangements of DNA between the cut sites and result in different restriction fragment patterns after separation by agarose gel electrophoresis. Restriction fragment length polymorphism refers to the differences in the distance between adjacent restriction endonuclease cut sites of individuals in a population. These RFLP changes in DNA sequence are inherited in a codominant fashion, unlike other morphological markers. Therefore, if one copy of DNA from a diploid organism has an RFLP and the other does not have, both fragment patterns can be seen in the electrophoretic gel. Like RFLPs, VNTRs and microsatellite DNA can be identified in a population and used in mapping experiments for the purpose of identification.

When DNA is digested with a specific restriction enzyme, fragments of different sizes result, each possessing different lengths of VNTRs. These VNTRs are at different locations in the genome and represent genetic landmarks. After digestion, a Southern blot of the separated DNA fragments containing the VNTRs is hybridized to a probe containing sequences complementary to VNTR sequence. In this way, usually 10–50 bands are identified which is distinctive for each person, except identical twins. Thus they act as "fingerprints" of our DNA and help in the identification of criminals.

In Figure 13.17, the DNA fingerprinting of three individuals A, B and C are shown with varied pattern of bands (a) due to variations in the restriction enzyme cuts on their respective DNA molecules (b) H1 refers to *Hinf* I cut regions on the DNA. In this example, the RFLPs detected are due to variation in the number repeat units (VNTR polymorphism) between the restriction sites, rather than changes in the location of the restriction sites.

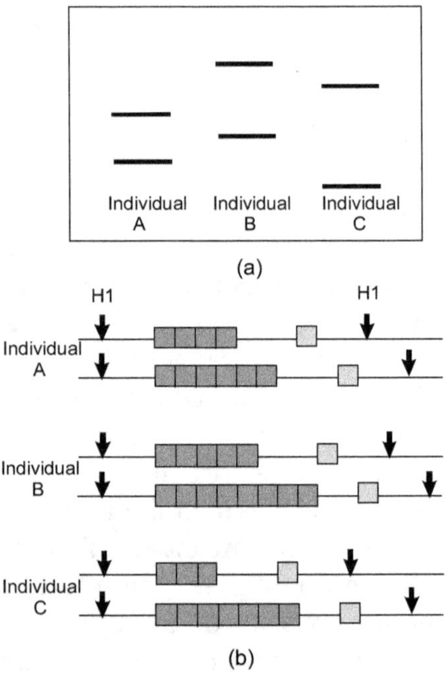

Figure 13.17   DNA fingerprinting

CHROMOSOME PAINTING

This is a method for visualizing each of the human chromosomes in distinct bright colours. It simplifies distinction between chromosomes of similar size and shape. This technique involves use of probes specific for sites scattered along the length of each chromosome. The probes are labelled with one of two dyes that fluoresce at different wavelengths. Probes specific for each chromosome are labelled with a predetermined fraction of each of the two dyes. The probes are hybridized to the chromosomes and the excess is removed. The sample is placed in a fluorescent microscope. A detector determines the fraction of each dye present. The information is passed on to a computer. A special program assigns a false colour image to each type of chromosome. A combination of chromosome painting and fluorescent *in situ* hybridization (FISH), called **multicolour FISH** is used to detect translocations in chromosomes.

CHROMOSOME WALKING

It is a sequencing method of choice for sequencing DNA fragments between 1.3 and 7 kilobases. Such fragments are too long to be sequenced in a single sequence read using the cyclic dideoxy terminator method. The DNA of interest may be a plasmid insert, a PCR product or a fragment representing a gap when sequencing a genome. The traditional chain termination method may not allow long DNA strands to be sequenced. Chromosome walking works by dividing the long sequence into several consecutive short ones. The term "primer walking" is used where the main aim is to sequence the genome. The term "chromosome walking" is used instead when we know the sequence but do not have a clone of a gene. For example, the gene for a disease may be located near an RFLP on DNA. The fragment is first sequenced as if it were a shorter fragment—sequencing will be performed from each end using either universal primers or primers designated by the customer. This should identify approximately the first 1000 bases.

The basic technique is as follows:

1. A primer that matches the beginning of the DNA to sequence is used to synthesize a short DNA strand adjacent to the unknown sequence, starting with the primer.

2. The new short DNA strand is sequenced using the chain termination method. The end of the sequenced strand is used as a primer for the next part of the long DNA sequence.

That way, the short part of the long DNA that is sequenced keeps "walking" along the sequence. The method can be used to sequence entire chromosomes (thus, chromosome walking). A different method with the same purpose which becomes more popular for large-scale sequencing (e.g. the Human Genome Project) is shotgun sequencing.

CHROMOSOME JUMPING

Chromosome jumping is used as a tool in the physical mapping of genomes. It is related to several other tools used for the same purpose, including chromosome walking.

Chromosome jumping is used to bypass regions such as those containing repetitive DNA, that cannot be easily mapped by chromosome walking, and is useful in moving along a chromosome rapidly in search of a particular gene.

In chromosome jumping, the DNA of interest is identified, cut into fragments with restriction enzymes and circularized (the beginning and end of each fragment are joined together to form a circular loop). From a known sequence, a primer is designed to sequence across the circularized junction. This primer is used to jump 100–300 kb intervals: a sequence 100 kb away would have come near the known sequence on circularization. Thus, sequences not reachable by chromosome walking can be sequenced. Chromosome walking can be used from the new jump position (in either direction) to look for gene-like sequences, or additional jumps can be used to progress further along the chromosome.

ELECTRON MICROSCOPY

Two basically different types of electron microscopes have been developed—transmission electron microscopes (TEMs) which form images from electrons that are transmitted through a specimen and scanning electron microscopes (SEMs) which utilize electrons that have bounced off the surface of the specimen. Preparation of the specimen for observation is very essential in electron microscopy.

Nucleic acids which are larger than protein molecules are embedded in a protein film. If nucleic acids are present in a protein solution, they will spread out as the protein film expands. This enables the entire nucleic acid molecule to be seen without the tangling and intramolecular aggregation that is observed if spreading does not occur. Proteins form a thick coating on nucleic acid molecules. To make nucleic acids visible, the protein-coated molecules are additionally coated with a thick layer of a heavy metal. This is done by evaporating platinum or uranium atoms over them, a process called shadowing or chemically depositing uranium atoms on the protein coating, a process called staining. These heavy atoms are good absorbers of electrons and the background support absorbs relatively few electrons.

Spicy Questions and Answers

1. If ^{15}N labelled *E. coli* were grown for three generations (eight fold increase in population) in a ^{14}N medium, which of the following distribution of DNA among the three major density classes would you expect to find after DNA extraction, shearing to average molecular weight of 10^6 and density determination by equilibrium sedimentation?

Pure ^{15}N	^{15}N ^{14}N hybrid	Pure ^{14}N
(a) 1/8	1/8	6/8
(b) 1/8	0	1/8
(c) 0	1/8	7/8
(d) 0	2/8	6/8

   (d)

2. By spectrophotometric assay, a pure DNA solution gave an absorbance of 0.35 at 260 nm when measured in a 1 cm quartz cuvette. What is the concentration of the DNA in µg/ml?

   A solution of dsDNA at a concentration of 50 µg/ml in a one cm cuvette will give an absorbance of 1.0. Therefore the concentration of a DNA solution showing an absorbance of 0.35 will be 50/1 × 0.35 = 17.5 µg/ml

3. What physical characteristics of a biomolecule influence its rate of movement in an electrophoresis matrix?

   The size and the charge of the molecule.

4. What is the advantage of slab gel over column gel for PAGE?

   The gel matrix in slab gels is more uniform than the column gels, which are made individually.

5. Can we use polyacrylamide gels for the analysis of plasmid DNA with more than 2000 base pairs?

   No, polyacrylamide gels can be used for nucleic acids up to 2000 base pairs only. The large DNA molecules cannot penetrate through polyacrylamide gel. Even gels with low percentage cross linking are not useful.

6. **Explain why ethidium bromide is used to detect nucleic acids on electrophoretic supports.**

Ethidium bromide is a dye that displays enhanced fluorescence when intercalated between stacked nucleic acid bases. Fluorescence is much greater for double-stranded nucleic acid. The dye can be incorporated in the gel or buffer. The gel can be illuminated with UV light during electrophoresis to view the extent of separation.

7. **The density of DNA in 7 M CsCl containing 0.15 M $MgCl_2$ is less than the density of the same DNA in 7 M CsCl. Why?**

Magnesium ions bind to DNA and reduce the density of DNA.

8. **What is the importance of viscoelastometry in genetic analysis?**

The technique is used to measure the viscosity of molecules in solution. Also, it is used to estimate the size of large DNA molecules in aqueous solutions. By this technique, the presence of giant eukaryotic chromosomes were identified.

9. **What is the advantage of polymerase chain reaction technique over other methods in the analysis of nucleic acid structure and function?**

PCR technique has high sensitivity than other methods in the analysis of nucleic acid. A very small amount of DNA sample is sufficient for the technique. PCR protocols permit detection of the presence of rare gene transcripts that could not be detected by less sensitive procedures such as northern blot analysis or *in-situ* hybridization studies.

10. **What common experimental procedure is carried out in Southern, northern and western blot analysis? What is the difference between them?**

In all the procedures, macromolecules, DNA, RNA or proteins are separated by gel electrophoresis to a solid support—nitrocellulose or nylon membrane—for further analysis. The major difference is the macromolecule involved—DNA for Southern blot, RNA for northern blot and proteins for western blot.

11. **It is best if two primers in a PCR reaction mixture have similar G+C contents. Why?**

The specificity of a PCR depends on precise recognition of the complementary regions on the target DNA strands. On increasing the temperature of annealing, regions of high complementarity alone can bind. In case the primers have different G+C content, they will have different melting temperatures and may have different stringency requirements.

12. **Differentiate between sedimentation velocity and sedimentation equilibrium centrifugation. How are these techniques useful in nucleic acid chemistry?**

The rate of sedimentation of molecules in a centrifugal field is called sedimentation velocity. Though it is expressed in terms of density of the solvent ρ, partial specific volume v and frictional force f, these are difficult to measure. Therefore sedimentation coefficient, S, the ratio of sedimentation velocity to centrifugal force is used.

Sedimentation equilibrium denotes the sedimentation of particles until they reach an equilibrium position at the point where the centrifugal force is equal to the frictional component opposing their movement.

13. **If you centrifuge in a density gradient a sample of DNA that contained both closed circular DNA and supercoiled DNA, would you expect to find two bands in the sedimentation pattern?**

Yes. The two forms of sediment have different densities. Supercoiled DNA gives a more compact band.

Review Questions

1. Outline the steps involved in the extraction of DNA and RNA from cells and tissues.

2. List and explain the different chromatographic techniques available for the study of nucleic acids.

3. What is meant by electrophoretic mobility? What is the net charge on nucleic acids and how do they behave in an electric field? Compare and contrast the different electrophoretic methods.

4. Suggest a method for the isolation of individual human chromosomes.

5. Write notes on biosensors and DNA biochips.

6. Comment on chromosome banding techniques.

7. Highlight on the sedimentation behaviour of nucleic acids in a centrifugal field.

8. Differentiate between analytical and preparative centrifugation with special reference to the study of nucleic acids.

9. Differentiate between sedimentation velocity method and sedimentation equilibrium method with reference to nucleic acid analysis.

10. What do you mean by X-ray diffraction? How is the diffraction analysis helpful in the structure analysis of DNA?

11. Explain the applications of the radio isotopes in the field of molecular biology.

12. Write a note on autoradiography.

13. Explain the basic principles of molecular hybridization. Compare and contrast the different blotting techniques.

14. Write an essay on the principle, technique, types and application of polymerase chain reaction in the study of nucleic acids.

15. Discuss Sanger's method of DNA sequencing.

16. Discuss Maxam–Gilbert method of DNA sequencing.

17. Describe the principle of RFLP analysis and mention its applications.

18. Write short notes on

    i.   Chromosome walking

    ii. Chromosome jumping

    iii. Chromosome painting

GLOSSARY

Affinity labelling A technique in which a reactive group is attached to a natural ligand of the system of interest, such as an antibiotic that binds to a specific site on the ribosome. The reactive group, if photolabile, reacts upon UV illumination by photoaffinity labelling. The technique has helped in the identification of the functional components of the ribosomes.

Alternative splicing Inclusion or exclusion of certain exons in different tissues of an organism or at different developmental stages. For example, alpha-tropomyosin is a protein in different kinds of contractile systems in various cell types. The exons present in the gene for this protein are included or excluded in different ways in different cells.

Anti-insulator Sequence that allows an enhancer to overcome the effect of an insulator.

Antisense RNA The transcription product of the DNA sense strand— the strand that does not encode a protein. This RNA can pair with an mRNA species and inhibit translation of that mRNA. Therefore it is shown to control the expression of several prokaryotic and eukaryotic genes.

Antitermination A mechanism of transcriptional control in which termination is prevented at a specific terminator site, allowing RNA polymerase to read into the genes beyond it. Sometimes RNA polymerase pauses at ρ-dependent termination sites. A protein called NusA is involved in causing antitermination in phage lambda. Antitermination occurs early in the phage transcription, when two rho-dependent termination sites are inactivated. RNA polymerase moves past these sites and transcribes genes essential to phage development. For this inactivation, a viral protein (product of gene N) interacts with NusA (N utilization substance).

Antitermination proteins Proteins that allow RNA polymerase to transcribe through certain terminator sites.

Autogenous control Action of a gene product that either inhibits or activates expression of the gene coding for it.

Bam **islands** A series of short, repeated sequences found in the non-transcribed spacer of *Xenopus* rDNA genes. The name reflects their isolation by use of the *Bam* I restriction enzyme.

Basal factor   A transcription factor required by RNA polymerase II to form the initiation complex at all promoters. These factors are identified as TFIIX where X is a letter.

Basal transcription apparatus   The complex of transcription factor that assembles at the promoter before RNA polymerase binds.

Buoyant density of DNA   Density that depends on the base composition of DNA $\rho = 1.660 + 0.098\ X_{G+C}$. Equilibrium density gradient ultra-centrifugation separates DNA molecules based on their buoyant density on caesium chloride density gradient columns.

bZIP proteins   Proteins that have basic DNA-binding regions adjacent to leucine zipper dimerization motif.

Centrosome   Organelle located near the nucleus of animal cells that is the primary microtubule-organizing centre (MTOC) and contains a pair of centrioles. It divides during mitosis forming the spindle poles.

Cleavage and polyadenylation specificity factor (CPSF)   A protein that recognizes the AUUAAA sequence in eukaryotic mRNA and activates the enzyme poly(A) polymerase to form the 3´ poly (A) tail.

Condensin   Protein components of a complex that bind to chromosomes to cause condensation for mitosis or meiosis. They are members of the SMC family of proteins.

Copy number   Number of copies per cell of a gene or other DNA sequence. Most eukaryotic cells are diploid, meaning that the copy number of most chromosomal genes is two, while most prokaryotic cells have a haploid genotype and a copy number of one for most genes. Extrachromosomal genes can have much higher copy numbers, whether they are carried on organelle DNA in eukaryotic cells or on an extrachromosomal DNA element such as a bacterial plasmid.

Cryptic satellite   A satellite DNA sequence not identified as such by a separate peak on a density gradient. It remains present in main-band DNA.

C-value   Total amount of DNA in the genome per haploid set of chromosomes.

C-value paradox   Factor that describes the lack of relationship between the DNA content (C-value) of an organism and its coding potential.

Dalton   Unit of molecular mass approximately equal to the mass of a hydrogen atom (1.66×10^{-24} g).

Degradosome   A complex of bacterial enzymes, including RNase and helicase activities, involved in mRNA degradation.

Dinucleotide fold   An active site pocket is formed by this common protein structural motif, which frequently acts as a nucleotide binding region. The protein is aminoacyl-tRNA synthetase. This motif also binds ATP. It provides a binding site for glutamine also.

Divergence The per cent difference in nucleotide sequence between two related DNA sequences or in amino acid sequences between two proteins.

DNA-PK DNA-dependent protein kinase which has a binding site for DNA free ends and another for dsDNA just inside these ends which has to be joined with the other DNA fragment.

DNA unwinding element (DUE) An 80-bp AT-rich sequence that can easily unwind. ORE is located adjacent to this sequence. DUE is the origin of replication in yeast.

Down mutation A mutation in a promoter that decreases the rate of transcription.

Editosome A high-molecular mass particle present in the enzymatic machinery used in RNA editing. The gRNA's internal segment is used as a template to correct the transcript. gRNA's 3' oligo(U) tail acts as a source of U's to be added to the edited mRNA.

Equilibrium labelling A tracer experiment used to measure rates of biological processes from the rates of labelling by low-molecular-weight precursors. Radio-labelled thymidine is used to follow rates of DNA synthesis in cell cultures or in intact organisms.

Evolutionary clock The rate at which mutations accumulate in a given gene.

Excisionase An enzyme, λ xis gene product, required for prophage excision, which is in concert with integrase.

Extein Sequences that remain in the mature protein that is produced by processing a precursor via protein splicing.

Fluctuation test Basis of the Luria–Delbrück experiment to determine the origin of bacterial resistance. Variations in the generation of mutant colonies showed up as fluctuations in the number of resistant colonies growing in different petri plates.

Footprinting A technique used in the identification of protein-binding sites on DNA. These DNA sites are protected from attack by DNase I (pancreatic deoxyribonuclease). Recent improvement in the technique involves cleavage with chemical agents such as methidiumpropyl EDTA–Fe^{2+} (MPE–Fe^{2+}). This compound intercalates between DNA bases like ethidium bromide and catalyses oxidation leading to cleavage at a nearby site. Unlike DNase I, MPE–Fe^{2+} does not have any sequence selectivity and therefore gives cleaner footprints.

GAP GTPase activating protein which stimulates the GTP-binding protein like EF-Tu to hydrolyse its bound GTP during protein biosynthesis.

GDPNP Guanosine-5'-(β,γ–imido) triphosphate, a GTP analog which is found to hydrolyse slowly that forms complex with EF-Tu in the thermophilic bacterium *Thermus aquaticus*.

Gene machines Automated DNA synthesizers where the monomers used are phosphoramidites, highly reactive molecules with trivalent phosphorus. Potentially reactive groups like NH_2 group of bases are blocked using compounds with dimethoxytrityl (DMTr), benzoyl or isobutyryl groups.

Genome Totality of genetic information in an organism that include the base sequences of all of the DNA molecules in a cell. These sequences consist of individual genes each on a chromosome or on a small piece of extrachromosomal DNA.

GMP-PCP An analog of GTP that cannot be hydrolysed. It is used to test which stage in a reaction requires hydrolysis of GTP.

GRF Guanosine nucleotide releasing factor, otherwise referred to as guanine nucleotide release protein (GNRP) or guanine nucleotide exchange factor (GEF), induces the GTP-binding protein to exchange its bound GDP for GTP. In the case of EF-Tu, ribosome is its GAP and EF-Ts is its GRF.

GTPase superfamily Group of GTP-binding proteins that cycle between an inactive state with bond GDP and an active state with bound GTP. These proteins—including G proteins, Ras proteins and certain polypeptide elongation factors— function as intracellular switch proteins.

Guide RNA A small RNA molecule involved in RNA editing, discovered in mitochondrial mRNAs of some unicellular eukaryotes. It inserts the uridine residues into mRNA by a kind of reverse splicing mechanism at certain points.

Hammerhead ribozymes Simplest known ribozymes embedded in the RNAs of certain plant viruses with some of their secondary structures resembling a carpenter's hammer.

Head piece DNA-binding domain of the lac repressor.

Heat shock response element (HSE) Sequence in a promoter or enhancer that is used to activate a gene by an activator induced by heat shock.

Helix-turn-helix (HTH) motif Supersecondary structure exhibited by specialized DNA-binding proteins like CAP, is a common DNA recognition element in prokaryotes. DNA-binding proteins containing an HTH motif, associate with their target base pairs mainly via the side chains extending from the second helix of their HTH motif, so called recognition helix.

Heteromers Multimeric proteins composed of non-identical subunits.

Homeobox Conserved DNA sequence that encodes a DNA binding domain (homeodomain) in a class of transcription factors encoded by certain homeotic genes.

Homeodomain   A conserved DNA-binding motif found in many developmentally important transcription factors.

Homeosis   Transformation of one body part into another arising from mutation in or misexpression of certain developmentally critical genes.

Homeotic gene   A gene in which mutations cause cells in one region of the body to act as though they were located in another, giving rise to conversion of one cell, tissue, or body region into another.

Hox complex   Clusters of homologous selector genes, which help to determine the body plan in animals.

Indirect read-out   Phenomenon in which a protein senses the base sequence of DNA through the DNA's backbone conformation and flexibility.

Insulators   DNA elements in association with one or more proteins that prevent an enhancer from acting on a promoter on the other side of an insulator in another transcription domain.

Enhancesome   An assembled LCR-transcription factor complex.

Exon shuffling   Exchange of exons among different genes, producing mosaic proteins with two or more distinct functions.

Isoaccepting tRNA   Different tRNAs that are specific for the same amino acid.

Isoalleles   Genes containing mutations with small effects that can be recognized only by special techniques. Because of the degeneracy of the genetic code, some base-pair changes do not change the protein products encoded by the genes.

Isoform   One of several forms of the same protein whose amino acid sequences differ slightly but whose general activity is identical.

Knockin gene   Technique in which the coding sequences of one gene are replaced by those of another.

Knockout gene   Technique for selectively inactivating a gene by replacing it with a mutant allele in a normal organism.

Ku   A heterodimer of 70- and 86-kDa subunits, a protein involved in double-strand break repair of DNA. It binds to free DNA ends and has latent ATP-dependent helicase activity.

Locus control regions (LCR)   Complex DNA elements involved in controlling expression of clusters of genes in association with bound proteins. The best defined LCR regulates expression of the globin gene family over a large region of DNA.

Luxury genes   Genes coding for specialized functions and synthesized usually in large amounts in particular cell types.

Matrix attachment regions (MARs) A region of DNA that attaches to the nuclear matrix. It is also known as a scaffold attachment region (SAR).

Maturase A protein coded by a group I or group II intron that is needed to assist the RNA to form the active conformation required for self splicing.

Megabase One million base pairs of DNA.

Micro RNAs Very short RNAs that may regulate gene expression.

Mitogen Any extracellular substance, like a growth factor, that promotes cell proliferation.

Mixed reconstitution An approach to analyse the function of each subunit of an enzyme like RNA polymerase. Subunits are isolated from two or more different RNA polymerases and recombined. For example, when β from rifampicin-resistant form of the enzyme is recombined with α, β and σ from wild type, the reconstituted enzyme becomes rifampicin-resistant, confirming the role of β-subunit in rifampicin resistance/inhibition.

Multicatalytic proteinase complex (MPC) or proteosome Giant complexes of different kinds of proteolytic enzymes involved in protein destruction in the cytoplasm.

Nicking–closing enzyme Topoisomerase I which catalyses the relaxation of negative supercoils in DNA by increasing its linking number in increments of one turn.

Nomadic sequences Intermediate repetitive DNA sequences capable of moving from one location to another in a chromosome. They are the transposable elements identified in yeast and *Drosophila* and considered to play a role in the regulation of gene expression during development and in the evolution of eukaryotic genomes. They are known to be responsible for a large number of mutations in yeast and *Drosophila*.

Null mutation A mutation that completely eliminates the function of a gene.

Origin replication complex (ORC) A set of proteins in eukaryotes which bind to ORE analogous to the dnaA protein of *E.coli*.

Origin replication element (ORE) A somewhat degenerate 11-bp sequence present in the autonomously replicating sequences (ARS) in eukaryotes.

PEST proteins Proteins with segments rich in Pro(P), Glu(E), Ser(S) and Thr(T) residues which are degraded rapidly.

Pre-protein A protein to be imported into an organelle or secreted from bacteria until its signal sequence has been removed.

Promoter–Proximal element Any regulatory sequence in eukaryotic DNA that is located within ⊔ 200 base pairs of the transcription start site. Transcription of many genes is controlled by multiple promoter–proximal elements.

Protamines Basic proteins present in some sperms, playing a major structural role in the chromatin similar to histones.

Pulse-chase experiment A labelling technique in which a label is administered for a short time and then rapidly reduced to prevent further incorporation. The reduction is done by adding unlabelled precursor at a molar excess of about 1000-fold. Then the metabolic fate of the material labelled is studied. mRNA was originally detected by pulse labelling. It was difficult to detect mRNA because of low abundance and metabolic instability. When label incorporated into mRNA by a pulse of ^{32}P orthophosphate was chased out, the label was found to be distributed uniformly in all cellular RNA species.

Quick-stop mutant A type of temperature-sensitive mutant (dna) in *E. coli* that immediately stops DNA replication when the temperature is increased to 42°C.

Ras protein A monomeric GTP-binding protein that functions in intracellular signalling pathways and is activated by ligand binding to receptor tyrosine kinases and other cell-surface receptors.

Recoding Events that occur when the meaning of a codon or series of codons is changed from that predicted by the genetic code. It may involve altered interactions between aminoacyl-tRNA and mRNA that are influenced by the ribosome.

Relaxase An enzyme that cuts one strand of DNA and binds to the free 5′ end.

Replacement sites Sites in a gene at which mutations alter the amino acid that is coded.

Replicase RNA-dependent RNA polymerase which replicates the plus (+) strand RNA of virus to form a new minus (–) strand. The new (–) strand serves as template for synthesis of (+) strand which are packaged into progeny virions or virus particles. Sometimes the virion RNA may itself be a (–) strand, referred to as negative-strand viruses. RNA replicases lack proofreading activity.

Riboswitch A catalytic RNA whose activity responds to a small ligand.

RNA-11 An RNA molecule, also called A-RNA, having 11-base pair double-helical conformation like A-DNA. The complementary sequences in many RNAs, like tRNA and rRNA form such double-helical stems.

RNA ligase An enzyme that functions in tRNA splicing to make a phosphodiester bond between the two exon sequences that are generated by cleavage of the intron.

RNA silencing Ability of a dsRNA to suppress expression of the corresponding gene systemically in a plant.

RNA World Hypothesis that nucleic acids but not proteins can direct their own synthesis, that cells contain batteries of protein-based enzymes for manipulating DNA but few for processing RNA and that many coenzymes are ribonucleotides establishing that RNAs were the original biological catalysts in pre-cellular times and that the chemically more versatile proteins were relative latecomers in macromolecular evolution.

Rotational positioning The location of the histone octamer relative to turns of the double helix, which determines which face of DNA is exposed on the nucleosome surface.

r-protein A ribosomal protein.

R segments Sequences that are repeated at the ends of a retroviral RNA. They are called R-U5 and U3-R.

rut An acronym for rho utilization site, the sequence of RNA that is recognized by the rho termination factor.

S1 nuclease mapping A technique used for mapping transcriptional start points. Fungal S1 nuclease is used which specifically and quantitatively cleaves single-stranded DNA and RNA.

Scarce mRNA A large number of individual mRNA species each present in very few copies per cell. This accounts for most of the sequence complexity in RNA.

S domain A sequence of 7S RNA of the signal recognition particle (SRP) that is not related to *Alu* RNA.

Selfish DNA Sequences that do not contribute to the genotype of the organism but have self-perpetuation within the genomes as their sole function.

Serum response element (SRE) A sequence in a promoter or enhancer that is activated by transcription factor(s) induced by treatment with serum. This activates genes that stimulate cell growth.

Sign inversion A model that describes the mechanism of DNA gyrase. DNA gyrase binds a positive supercoil, breaks both strands in one duplex, passes the other duplex through and re-seals the strands.

Silencer A short sequence of DNA that can inactivate expression of a gene in its vicinity.

Silencer sequence A sequence in eukaryotic DNA that promotes formation of condensed chromatin structures in a localized region. This blocks access of protein required for transcription of genes within several hundred base pairs of the silencer sequences.

Silencing The repression of gene expression in a localized region, usually as the result of a structural change in chromatin.

Silent site A site in a coding region where mutation does not change the sequence of the protein.

Simple-sequence DNA Short, tandemly repeated sequences that are found in centromeres and telomeres as well as other chromosomal locations and are not transcribed.

SL RNA A small RNA that donates an exon in the trans splicing reaction of trypanosomes and nematodes.

SR protein A protein that has a variable length of an arg-ser-rich region and is involved in splicing.

Stringent response Synthesis of rRNAs and tRNAs in bacteria is inhibited when protein synthesis is blocked by amino acid starvation. During amino acid starvation, a regulatory nucleotide, guanosine 3',5'-tetraphosphate (ppGpp) accumulates and inhibits rRNA and tRNA synthesis.

Structural maintenance of chromosomes (SMC) A group of proteins that include the cohesins, which hold sister chromatids together and the condensins, which are involved in chromosome condensation.

Substitutional editing A new class of RNA editing, which resembles post-transcriptional modification of tRNA. Both apo B-48 and apo B-100 are expressed from the same gene. The mRNAs encoding the two proteins differ by a single $C \rightarrow U$ change. The apo B-100 mRNA glutamine 2153(CAA) codon is replaced by a stop codon (UAA) in apo B-48 mRNA.

Sugar ring pucker The sugar ring becoming slightly non-planar so as to reorient the ring substituents. Originally ribose ring has a certain amount of flexibility which affects the conformation of the sugar-phosphate backbone. The ribofuranose ring is expected to be nearly flat. But the ring substituents are eclipsed when the ring is planar. To relieve the resultant crowding, the ring puckers.

SWI/SNF A chromatin remodelling complex. It uses hydrolysis of ATP to change the organization of nucleosomes.

Synteny A relationship between chromosomal regions of different species where homologous genes occur in the same order.

Synthetic genetic array analysis (SGA) An automated technique in budding yeast whereby a mutant is crossed to an array of approximately 5000 deletion mutants to describe if the mutation interacts to cause a synthetic lethal phenotype.

TAFs Subunits of TFIID that assist TBP in binding to DNA. They also provide points of contact for other components of the transcription apparatus.

Terminase An enzyme that cleaves multimers of a viral genome and then uses hydrolysis of ATP to provide the energy to translocate the DNA into an empty viral capsid starting with the cleaved end.

Third-base degeneracy   Describes the lesser effect on the meaning of the codon due to the nucleotide present in the third codon position.

Transcriptome   Complete set of RNAs present in a cell, tissue or organism. Its complexity is mostly due to mRNAs, but it also includes non-coding RNAs.

Trailer A   A non-translated sequence at the 3′ end of an mRNA following the termination codon.

Up-promoter mutations and down-promoter mutations   In prokaryotes there are two conserved sequences called –35 region and –10 region in the promoters. Natural promoters do not have sequences identical to the consensus sequence. The more closely these regions in a promoter resemble the consensus sequence, the more efficient is that promoter in initiating transcription. Up-promoter mutations increase promoter strength because of mutation in either –35 or – 10 region bringing the sequence close to consensus sequence. Down-promoter mutations, which decrease promoter strength, change the sequence away from the consensus.

Variable arm of tRNA   The pseudouridine arm which has 3 to 21 nucleotides and has a stem consisting of up to 7 bp.

Virion   An individual viral particle.

Viscoelastometry   A technique for analysing viscosity parameters of molecules in solution. The size of DNA molecules can be determined by measuring the recoil time of the DNA in solution.

Zooblot   Southern blotting used to test the ability of a DNA probe from one species to hybridize with the DNA from the genomes of a variety of other species.

REFERENCES

Alan, G. Atherley, Jack, R. Girton and John, F. McDonald. *The Science of Genetics*. Saunders College of Publishing, Fort Worth. 1999.

Anthony, J.F. Griffiths, William, M. Gelbart, Jeffrey, H. Miller and Richard, C. Lewontin. *Modern Genetic Analysis*. W.H. Freeman and Company. 2000.

Benjamin Lewin. *Genes VII*. Oxford University Press. 2000.

Brown, T.A. *Genomes*. John Wiley & Sons (Asia) Pte. Ltd. 1999.

Christopher, K. Mathews and van Holde, K.E. *Biochemistry*. The Benjamin/Cummings Publishing Inc. 1996.

Daniel, J. Fairbanks and Ralph Anderson, W. *Genetics-The Continuity of Life*. Brooks/Cole Publishing Company. 1999.

David Freifelder. *Molecular Biology*, 2nd Edition. Narosa Publishing House. 1987.

Donald Voet, Judith, G.Voet and Charlotte, W. Pratt. *Fundamentals of Biochemistry-Life at the Molecular Level*. John Wiley & Sons, Inc. (Asia). 2006.

Eldon John Gardner, Michael, J. Simmons and Peter, D. Snustad. *Principles of Genetics*, 8th Edition. John Wiley & Sons, Inc. 2002.

Geoffrey, M. Cooper. *The Cell-A Molecular Approach*, 2nd Edition. ASM Press, Sinauer Associates, Inc. 2000.

Gerald Karp. *Cell and Molecular Biology: Concepts and Experiments*. John Wiley & Sons, Inc. 1996.

Harvey Lodish, Arnold Berk, Paul Matsudaira, Chris, A. Kaiser, Monty Krieger, Matthew, P. Scott, Lawrence Zipursky, S. and James Darnell. *Molecular Cell Biology*, 5th Edition. W.H. Freeman and Company. 2004.

James, D.Watson, Nancy, H. Hopkins, Jeffrey, W. Roberts, John Argetsinger Steitz and Alan, M. Weiner. *Molecular Biology of the Gene*, 4th Edition. The Benjamin/Cummings Publishing Company, Inc. 1987.

Jeremy, M. Berg, John, L. Tymoczko and Lubert Stryer. *Biochemistry*, 6th Edition. W.H. Freeman and Company. 2006.

Keith Wilson and John Walker. *Principles and Techniques of Biochemistry and Molecular Biology*, 6th Edition. Cambridge University Press. 2007.

Leland Hartwell, Leroy Hood, Michael, L.Goldberg, Ann, E. Reynolds, Lee, M. Silver and Ruth, C. Veres. *Genetics-From Genes to Genomes*. McGraw-Hill Publishers. 2000.

Monroe, W. Strickberger. *Genetics*, 3rd Edition. Prentice-Hall of India Pvt. Ltd. 2006.

Nelson, L. David. and Michael, M.Cox. *Lehninger's Principles of Biochemistry*, 4th Edition. W.H. Freeman Publishers. 2005.

Peter, D. Snustad and Michael, J.Simmons. *Principles of Genetics*, 3rd Edition. John Wiley & Sons, Inc. 2003.

Primrose, S.B. Twyman, R.M. and Old, R.W. *Principles of Gene Manipulation*, 6th Edition. Blackwell Publishing Company. 2001.

Primrose, S.B. and Twyman, R.M. *Principles of Gene Manipulation and Genomics*, 7th Edition. Blackwell Publishing. 2006.

Robert H. Horton, Laurence, A. Moran, Gray Scrimgeour, K., Marc, D. Perry and David, J. Rawn. *Principles of Biochemistry*, 4th Edition. Pearson Education International. 2006.

Robert, H. Tamarin. *Principles of Genetics*. Tata McGraw-Hill Publishing Co. Ltd. 2007.

Robert, J. Brooker. *Genetics: Analysis and Principles*. An Imprint of Addison Wesley Longman, Inc. 1999.

Robert, K. Murray, Daryl, K. Granner and Victor, W. Rodwell. *Harper's Illustrated Biochemistry*, 27th Edition. McGraw-Hill Publishers. 2006.

Roger, L.P. Adams, John, T. Knowler and David, P. Leader. *The Biochemistry of Nucleic Acids*, 11th Edition. Chapman & Hall. 1992.

Turner, P.C., Mc Lennan, A.D., Bates, A.D. and White, M.R.H. *Instant Notes-Molecular Biology*, 2nd Edition. Viva Books Pvt. Ltd. 2002.

William, H. Elliott and Daphne, C. Elliott. *Biochemistry and Molecular Biology*, 2nd Edition, Oxford University Press. 2001.

William, S. Klug and Michael, R. Cummings. *Concepts of Genetics*, 6th Edition. Prentice Hall. 1994.

INDEX

A

ABC excinuclease 147

Acetylation 309, 324, 353, 354, 379

Acridine 142, 215, 258, 280, 386

Acridine dyes 253, 258

Acridine orange 258

Acriflavine 258

Acrocentric 87, 89

Actinomycin D 142, 198

Activator 318, 323, 325, 336, 357, 359, 360

Adaptation hypothesis 234

Adaptive mutation 235, 236

Adaptor molecules 181

A-DNA 40, 41, 42

ADP-ribosylation 309

Affinity chromatography 384, 385

Agarose gel 385, 389, 390, 402, 403, 417

Alkaline comet assay 389

Alkylating agents 252, 257, 280, 281, 283

Allele-specific PCR 409

Allolactose 326, 327, 328, 329, 332

Allosteric protein 326

Altered bases 280

Alternative splicing 368

Amanita phalloides 199

Amatoxins 199

Ames test 268

2-aminopurine 253, 255

6-aminopurine 35

Antibacterial agents 302

Antibiotics 197, 199, 279, 302, 304, 305, 322

Anticodon 181, 209, 210, 211, 212, 217, 220, 228, 246, 282

Anticodon arm 181

Anticodon deaminase 210

Antineoplastic agent 198

Antiparallel 28, 213, 281, 398

AP endonucleases 145

AP1 357

Aphidicolin 142

Aporepressor 324, 339, 341

AP site 145, 258

Apurinic 257, 280

Apurinic or apyrimidinic site 145, 257

Arabinose operon 334, 335, 350

*ara*C 334

*ara*E 336

*ara*F 336

*ara*G 336

*ara*I 335

arao$_2$ 335

ara operon 334, 335, 336

Arylhydrazinopyrimidines 142

A site 290, 296, 297, 298, 299, 300, 301, 302, 303, 313, 323

Assembly PCR 409

Asymmetric PCR 409

Atelocentric chromosome 89

ATF 357

Attenuation 279, 341, 342, 343, 351

D

Q

R